THE ESSENTIAL
AUDREY EAGLE

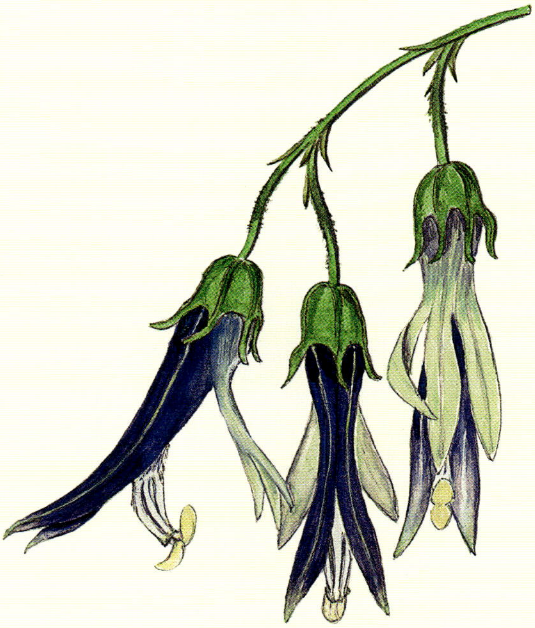

THE ESSENTIAL
AUDREY EAGLE

BOTANICAL ART
OF NEW ZEALAND

TE PAPA PRESS

First published in New Zealand in 2013 by
Te Papa Press, PO Box 467, Wellington, New Zealand

All images in the introduction are held by the Museum of New Zealand Te Papa Tongarewa, except for the following: page 3 Natural History Museum (London), page 11 Alexander Turnbull Library (Wellington), page 13 Auckland War Memorial Museum.
Plant details accompanying the text are from the following plates: 25, 44, 53, 60, 137, 147.
Imaging of Audrey Eagle's plates by New Zealand Micrographics Services Ltd, imaging of introduction images by Jeremy Glyde, Te Papa.

A catalogue record for this book is available from the National Library of New Zealand.
ISBN 978-1-877385-90-2

Designed by Margaret Cochran
Typeset by Geoff Norman
Printed by Everbest Printing Co, China

Front cover image: Ipomoea cairica *(panahi, pōwhiwhi)*
Back cover images, clockwise from top left: Sophora tetraptera *(kōwhai),* Fuchsia excorticata *(kōtukutuku, kōnini),* Metrosideros fulgens *(akatawhiwhi, scarlet rātā vine),* Phormium cookianum *(wharariki, mountain flax, coastal flax); inside flap:* Knightia excelsa *(rewarewa, New Zealand honeysuckle)*

CONTENTS

Preface

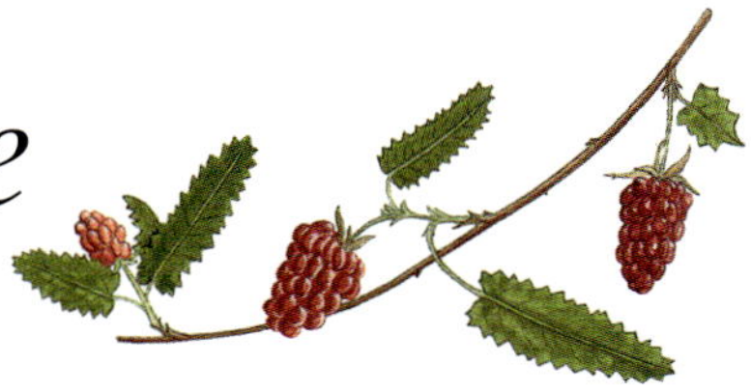

AUDREY EAGLE

I've been delighted by the way New Zealanders have championed *Eagle's Complete Trees and Shrubs of New Zealand*. Winning the Montana Medal for Non-Fiction in 2007 (and a Goodman Fielder Wattie Book Award in 1983 for *Trees and Shrubs of New Zealand*) was a great honour, it's humbling to see a life's work embraced in this way. And in May this year I was blessed to be awarded an honorary doctorate in science from the University of Otago; this formal recognition is really very special to me.

The demand for such a book is perhaps not so surprising when you consider that New Zealanders have a deep love for their natural world. This I discovered when I arrived back in the country, with my husband Harold in 1949, determined to follow my intuition that I had work to do here. Some of my engineering colleagues at the State Electricity Department, where I worked as a draughtswoman, were plant enthusiasts in their spare time. They showed me a whole new world of flora, and I became enraptured by the richness and diversity of the dense and beautiful New Zealand bush.

A desire to keep learning about New Zealand's flora has been a constant throughout my life, and the only way to do this was to get out into the wilderness. In the mid-1950s, I became a founding member of the Waikato branch of the Royal Forest and Bird Protection Society and it was then that I really began building my knowledge – I rarely missed an opportunity to explore the forests and ranges nearby on monthly field trips or to go on the excursions of longer duration.

Harold and I also went on numerous camping trips, later undertaken in the luxury of a caravan, around the North and South Islands searching for those last elusive specimens to complete the illustrations of all our native woody plants. A pursuit that began for my own education had become a task with its own agenda – life became hectic at times, when publication for the books neared, and many generous people sent me plants to include that I had been unable to find.

Highlights of my research were trips to Stewart and Great and Little Barrier Islands, and many of the other small offshore islands around New Zealand. These places were exciting to visit both historically and botanically, the remote Auckland Island group especially so because details of some of the plants growing there have rarely been illustrated.

I have been fortunate to combine my love of plants with a more structured role in their preservation. I was involved in the Royal Forest and Bird Protection Society for fifty-eight years, and in the early 1980s I served on the Nature Conservation Council and the Loder Cup Committee. It was an honour to give my colleagues the recognition they deserved for the essential work of conserving our flora.

Conservation goes hand in hand with education. As a member of the Waikato branch of the Royal Forest and Bird Protection Society, I was called upon to give talks on native plants and to promote the work of the society. I also led excursions and took plant-lovers of all ages on many botanical and conservation trips. At home, together with other members, I grew species of native seedlings, and we used these to raise funds for conservation activities. My strongest wish is that my books have inspired New Zealanders to see first hand our exceptional plants in their natural habitat and to have strengthened their attitudes towards protecting our native species.

Now the *Complete* has become the *Essential*, with 163 illustrations chosen especially for readers. The beauty and variety of New Zealand's native trees and shrubs, from the grand kauri to the tiny mountain snowberry, are still plain to see. All major genera are represented, including some of our most-loved plants, such as kōwhai, tōtara, nīkau and pōhutukawa. My life in botany sustains me still, driven by a motivation to make New Zealand plants accessible to as many people as possible. I hope that *The Essential Audrey Eagle* will contribute to doing just that.

June 2013

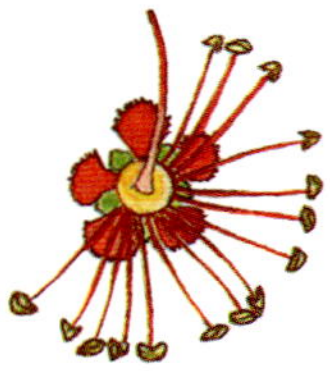

Introduction

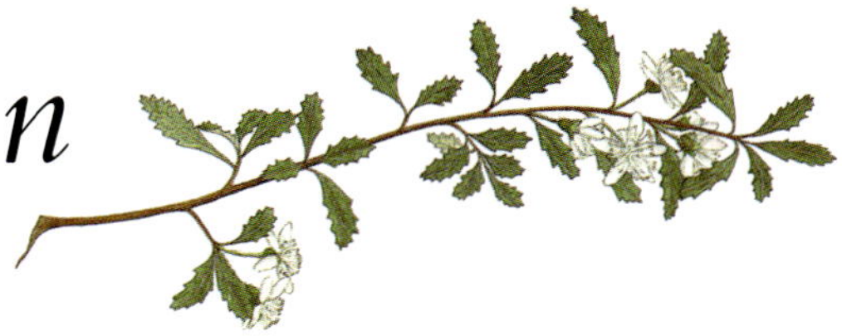

PATRICK BROWNSEY

Within hours of Joseph Banks and Daniel Solander stepping ashore from the *Endeavour* at Poverty Bay on 8 October 1769, the work of illustrating New Zealand's flora was under way. As the first plants were brought on to the ship, the expedition's artist, Sydney Parkinson, started painting them in watercolour before they wilted. However, because of the sheer volume of new material, he resorted to sketching the plants and making annotations that would allow him to complete the paintings in quieter moments.

The voyage, captained by James Cook, to observe the transit of Venus and to search for the great southern continent, was underwritten by the British Government, but Banks, a young and very wealthy aristocrat, paid for his own private party of nine to assist him in his pursuit of scientific discovery. The would-be botanist ensured that the *Endeavour* was better equipped for biological study than any previous expedition, with an impressive array of collecting gear, preserving materials, microscopes, a library and, of course, paper and paints to record the keenly anticipated discoveries. However, Banks could do little about conditions on board, where the only suitable working space was the great cabin. Usually reserved for the ship's commander, this small area was taken over by Banks's party and soon became a hive of botanical and artistic activity. Banks wished

> that our friends in England could by the assistance of some magical spying glass take a peep at our situation; Dr Solander sits at the cabin table describing, myself at my bureau journalising, between us hangs a large bunch of sea weed, upon the table lays the wood and barnacles ...[1]

Parkinson sketched at the table, too, and while working in such a confined environment was certainly problematic, the bigger concern was keeping specimens and artwork safe on a perpetually leaking boat, where insects and rodents ruled.

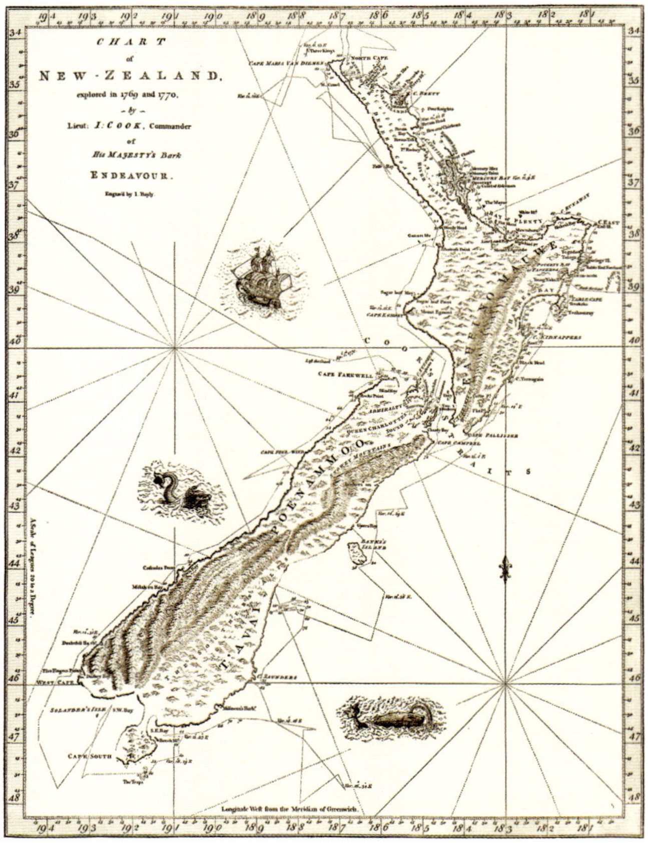

James Cook's map of New Zealand, 1769–70.

The New Zealand flora was unlike anything European botanists had seen before: markedly different juvenile and adult stages on predominantly woody plants; 'divaricating' shrubs (with small leaves, and widely branching stems); and floral characters dominated by inconspicuous white flowers, often separated into male and female forms. But in Solander they had just the man for the challenge. Trained by the Swedish botanist Carl Linnaeus, who had revolutionised plant taxonomy with his 'sexual system' of classification and binomial system of nomenclature, Solander was among a group of his most committed students (known as 'apostles') who were encouraged to travel far and wide to collect and classify the world's plant diversity. During the six months that Cook circumnavigated New Zealand, Banks and Solander made more than 2000 collections of around 350 species, most of them new to science.

Parkinson sketched 204 plants and although he completed only about thirty paintings, they demonstrate that at just twenty-four years old, he was a gifted artist with an eye for botanical detail, even in a flora that was so totally unfamiliar. Sadly, he died at Batavia

(Jakarta) on the voyage home and, because of the difficulty of access, his original work now housed at the Natural History Museum in London has only rarely been seen by New Zealand botanists. His unfinished sketches were completed by a team of artists, and formed the basis of 184 copper-plate engravings intended to illustrate Solander's descriptions in Banks's proposed *Primitiae Florae Novae Zelandiae.* Had this book been published, New Zealand botany would have had a very confident start, but unfortunately it never eventuated. Banks lost interest as he took on other scientific roles and gained a status that no longer required him to prove his botanical standing, but it was the untimely death of Solander in 1782 and an economic recession that finally sealed the book's fate.

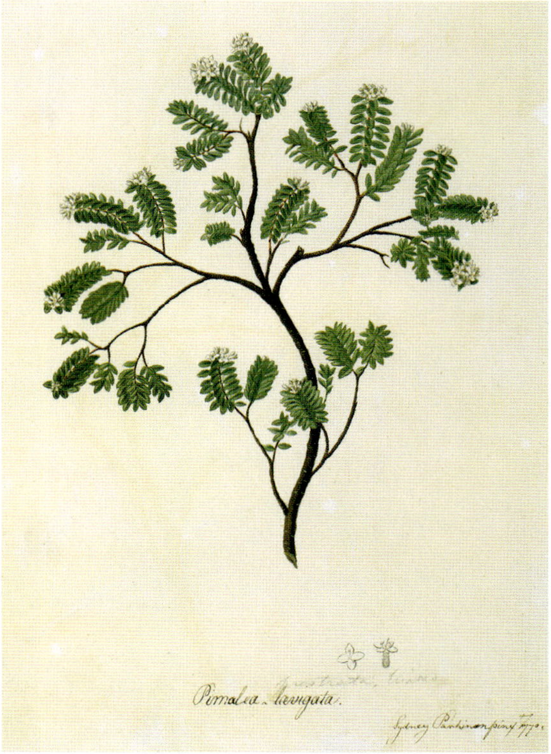

A watercolour of *Pimelea prostrata* (pinātoro) by Sydney Parkinson, 1770, based on a specimen collected during the *Endeavour* expedition.

It was another 200 years before the surviving plates were used in *Banks' Florilegium* – a limited edition published jointly by the Natural History Museum and Alecto Historical Editions. The plates were produced *à la poupée*, a form of printing in which two or more different coloured inks are added to the same plate by means of a wad of fabric similar to a modern-day cotton bud. Contemporary accounts indicate that the chosen colours were based on those of Parkinson's sketches and notes, but the prints, when displayed, are overwhelmingly blue-green and bright green (see overleaf) compared to the bronze and brown-green colours of New Zealand's forest plants, and to the colours in Parkinson's originals. The true quality of Parkinson's work has therefore never been fully appreciated.

The sheer beauty of flowers has attracted artists from all civilisations and ages: Minoans used lilies, palm trees and the saffron crocus in their decorative art, and the Egyptians, lotus and papyrus. Much later, after the Renaissance, the genre of still life elevated

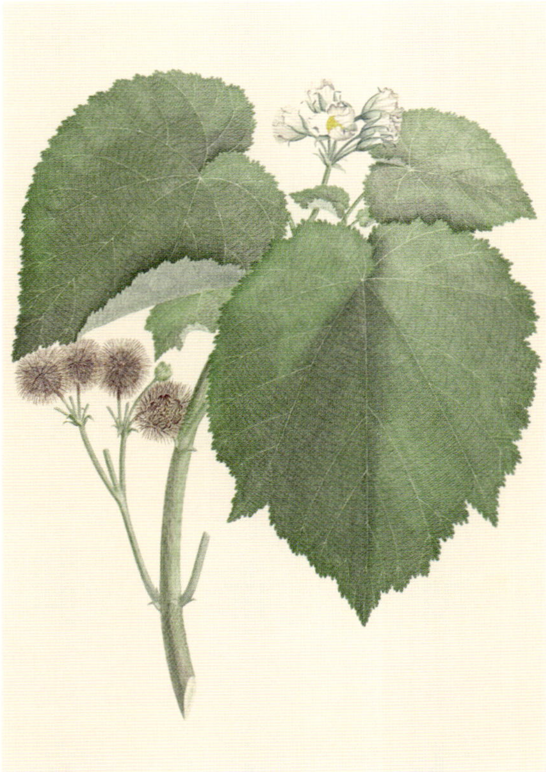
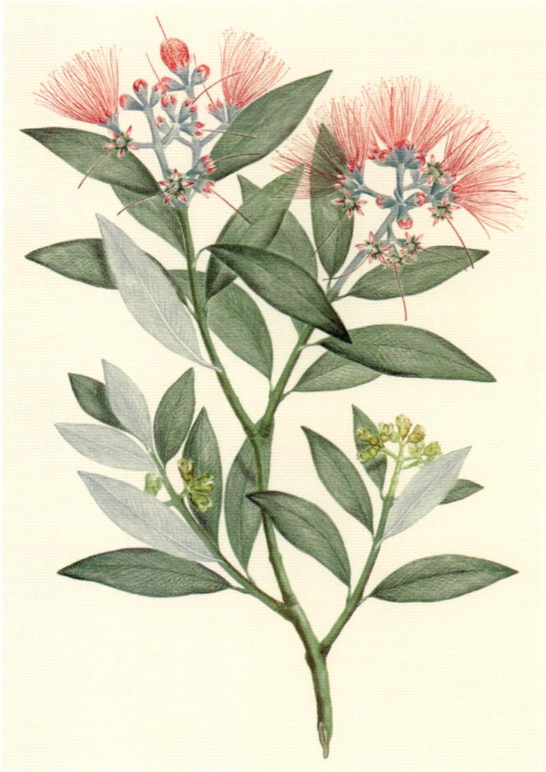

Entelea arborescens (whau) and *Metrosideros excelsa* (pōhutukawa) from *Banks' Florilegium*, 1980–90, showing the predominant blue-green colour of the leaves in these plates. The engravings were made by Gerald Sibelius and Gabriel Smith, respectively, from sketches by Sydney Parkinson, completed by Banks's artists.

flora to an artistic subject in its own right, while, at the same time, flowers and foliage became popular decorative subjects for ceramics, domestic furnishings, textiles and even buildings. The enormous popularity of botanical art in such a range of media is attributable to different factors in different ages, and includes the aesthetic qualities of flowers themselves; the everyday familiarity of plants; a love of, or a desire to feel closer to, nature; increased participation in gardening by the general public, and reinforcement of national identity by association with iconic local plants (for example, the pōhutukawa or kōwhai in New Zealand).

Botanical illustration is a subset of botanical art; it dates back more than 2000 years to ancient Greek civilisation, and was driven by the need to describe and illustrate plants of medicinal value in what came to be known as herbals. Probably the most influential was *De Materia Medica*, written around AD 65 by Pedanios Dioscorides, a physician in the Roman army, detailing the properties of some 500 medicinal plants. The original

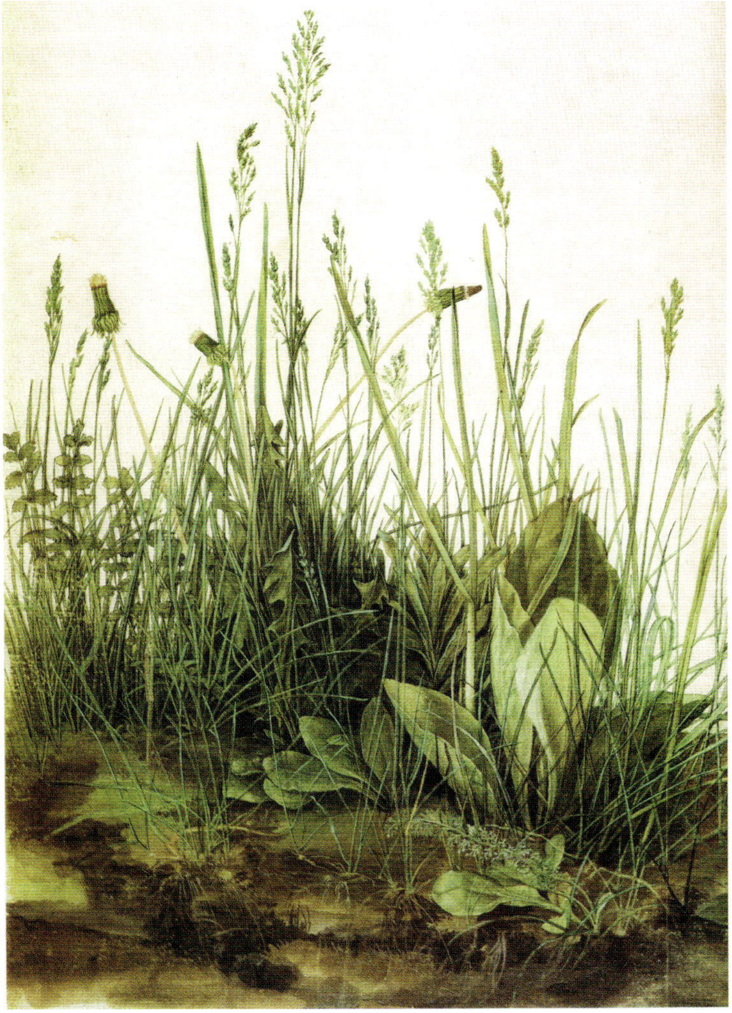

Albrecht Dürer's *Das große Rasenstück* (The Great Piece of Turf), 1503, a painting that demonstrates the renewed interest in observing from nature that came with the Renaissance.

illustrations in this and other early herbals were highly naturalistic, but they were copied and recopied in subsequent herbals for the next 1000 years, becoming stylised and debased to the point where they were almost useless. However, with the Renaissance came a revival of naturalism, perhaps nowhere better exemplified than in Albrecht Dürer's *Das große Rasenstück* (1503), a painting of a clump of turf with grasses, plantains and dandelions drawn in stunningly life-like detail. Initially, the main purpose of flower painting remained to illustrate books on medicinal plants. Only in the seventeenth century did botanical illustration extend to recording rare plants in cultivation and then later, in the Age of Enlightenment, to scientific documentation of new floras, as European explorers journeyed to distant parts of the world.

Botanical illustration went hand in hand with the business of finding, describing, classifying, and later exploiting, the plants of these new territories as they were occupied. Documenting New Zealand's flora was no exception, and a long line of resourceful artists has showcased the beauty and diversity of our plants to the wider world. These artists are a varied assemblage of amateurs and professionals, men and women, visitors and residents, New Zealanders and foreigners, some well known, others deserving to be better known. Regardless of their backgrounds, all have helped to unravel the complexities of a flora that has evolved on a distant island archipelago in the south-west Pacific where 82 percent of native seed plants are endemic – that is, they occur naturally nowhere else in the world.

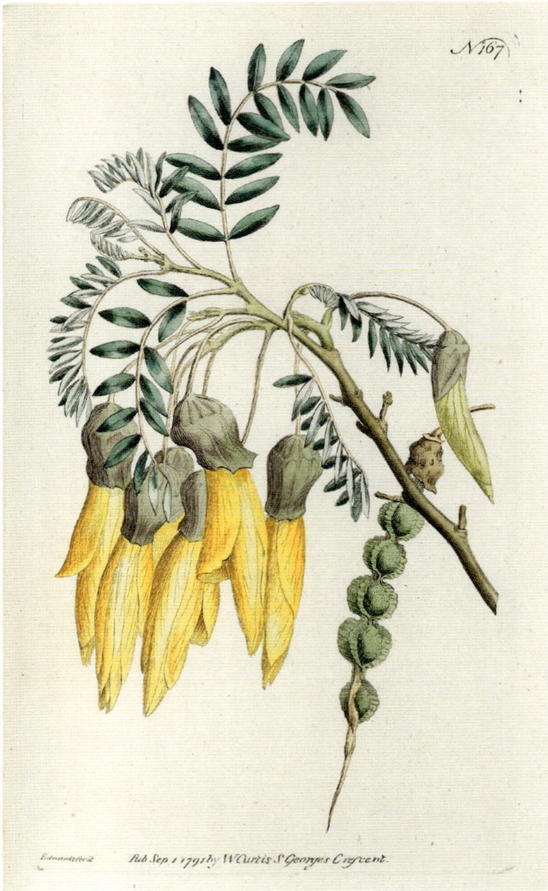

The first published colour illustration of a New Zealand plant, *Sophora tetraptera* (kōwhai), a hand-coloured plate from *Curtis's Botanical Magazine*, 1791, drawn and engraved by Sydenham Edwards.

Despite Banks's failure to publish *Primitiae Florae Novae Zelandiae*, collections gathered during Cook's voyages did slowly introduce Europeans to the botanical novelties of the Antipodes. The earliest published hand-coloured engravings of New Zealand plants appeared in *Curtis's Botanical Magazine*, a journal of remarkable longevity first published in 1787, surviving under various titles to the present day, and illustrating well over 10,000 species. *Sophora tetraptera* (kōwhai) was drawn and engraved in 1791 by Sydenham Edwards, based on a plant raised from seed brought back on the *Endeavour*. In the accompanying description, an exuberant William Curtis wrote:

> A finer sight can scarcely be imagined than a tree of this sort, extending to a great breadth on a wall with a western aspect, in the Apothecaries Garden at Chelsea,

> where it was planted by Mr. Forsyth about the year 1774, and which at this moment (April 28, 1791) is thickly covered with large pendulous branches of yellow, I had almost said golden flowers; for they have a peculiar richness, which is impossible to represent in colouring.[2]

Solanum laciniatum (poroporo) was illustrated by an unknown artist in 1796, and *Wahlenbergia gracilis* drawn by Edwards and engraved by Francis Sansom in 1803. Edwards was a precocious talent, fortunate to be alive at a time when scientific expeditions were being sent to all parts of the world. At a young age he came to the attention of William Curtis, who had him trained in both botany and botanical illustration. He worked at a rapid rate, preparing 1700 plates for the *Botanical Magazine* alone, and many more for other publications. Edwards' skilful illustrations became enormously popular as the public clamoured for information about new plants, even inspiring ceramic decoration by major manufacturers of the day. Although he is only peripherally associated with New Zealand botany, his contribution came at a formative moment, just as this distinctive flora was being revealed to the rest of the world.

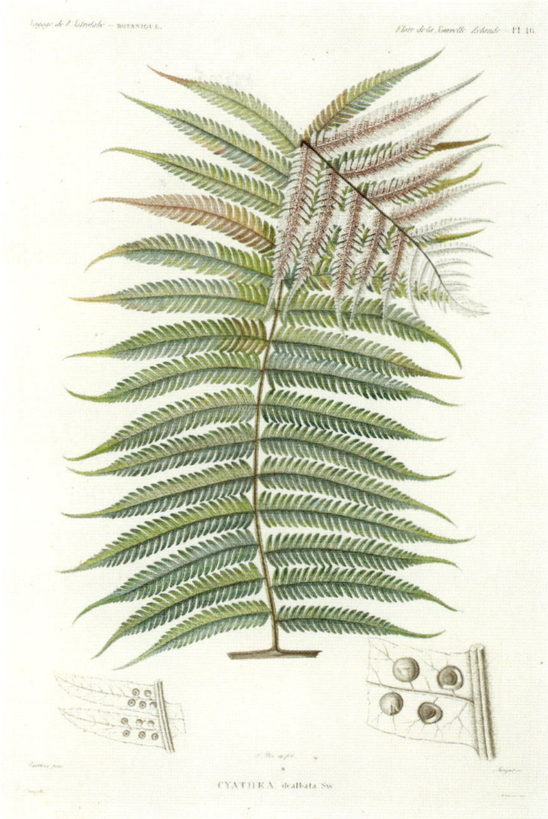

Cyathea dealbata (silver fern, ponga) from JSC Dumont d'Urville's 'Voyage de découvertes de l'*Astrolabe*', 1832–34, drawn by Antoine Vauthier and engraved by Jean Joseph Mougeot.

Following Cook, French voyages of exploration dominated the South Pacific in the 1820s and 1830s. The naturalist and naval officer JSC Dumont d'Urville, was pre-eminent among the French explorers, participating in three expeditions and leading two of them. The published results of his latter two voyages, 1826–29 and 1837–40, respectively, were prodigious – the botany alone running to four volumes and two atlases of illustrations. *Essai d'une Flore de la Nouvelle-Zélande*, published by Achille Richard in 1832, described more than 380 species of plants, and was the first publication to cover not only flowering plants but also ferns, mosses, liverworts, lichens and seaweeds. Some forty-one species were illustrated by Eulalie Delile and Antoine Vauthier, while a team of engravers prepared the plates that were later hand-coloured. Although none of the team had visited New Zealand, many of the plates were exquisitely executed, successfully capturing the essence of the living plant. The original paintings are in the Muséum national d'Histoire naturelle, Paris, but, even today, regrettably little can be gleaned about the artists. The botanical publications of the last voyage are less comprehensive, and although the atlas contains sixty-six plates, only twenty of these are coloured and none are of flowering plants.

Coinciding with the last of Dumont d'Urville's expeditions was another British expedition, the Antarctic voyage of James Clark Ross in the *Erebus* and *Terror* (1839–43). Joseph Hooker, the ship's assistant surgeon and naturalist, was a man destined to have tremendous influence on New Zealand botany, and would later become Director of Kew Gardens in London. The expedition circumnavigated the world through the Southern Ocean, visiting all the major land areas, including the sub-Antarctic islands of New Zealand, and refitting at the Bay of Islands. Like Banks's party on the *Endeavour*, Hooker worked in the great cabin of the *Erebus*, perhaps in slightly greater comfort, because one of the tables under the stern window was exclusively his. Here he examined all the plant collections from his forays ashore, made notes and sketches of them, and then pressed them for future reference. On his return to England, Hooker published *The Botany of the Antarctic Voyage* (1844–60) in six magnificent volumes, observing that 'The traveller from whatever country, on arriving in New Zealand, finds himself surrounded by a vegetation that is almost wholly new to him.'[3]

Flora Antarctica (1844) included about 120 seed plants and ferns from the highly distinctive flora of the Auckland and Campbell Islands, which Hooker thought 'to be more remarkable for their beauty and novelty than the flora of any other country'.[4] *Flora Novae-Zelandiae* (1855) described another 850 seed plants and ferns from the main islands of New Zealand, as well as many mosses, algae, liverworts, lichens and fungi. Hooker later condensed all this into a single volume, the *Handbook of the New Zealand Flora*. Commissioned by the New Zealand Government, it was published in 1864–67 and contained more than 1000 descriptions of native plants. The importance of this accessible, affordable and comprehensive catalogue of New Zealand's plants, not only to the economy of the country, but also to the general public in the nineteenth century, cannot be underestimated. In a gesture that would be inconceivable today, an admiring audience of wealthy and influential citizens commissioned a beautiful writing bureau

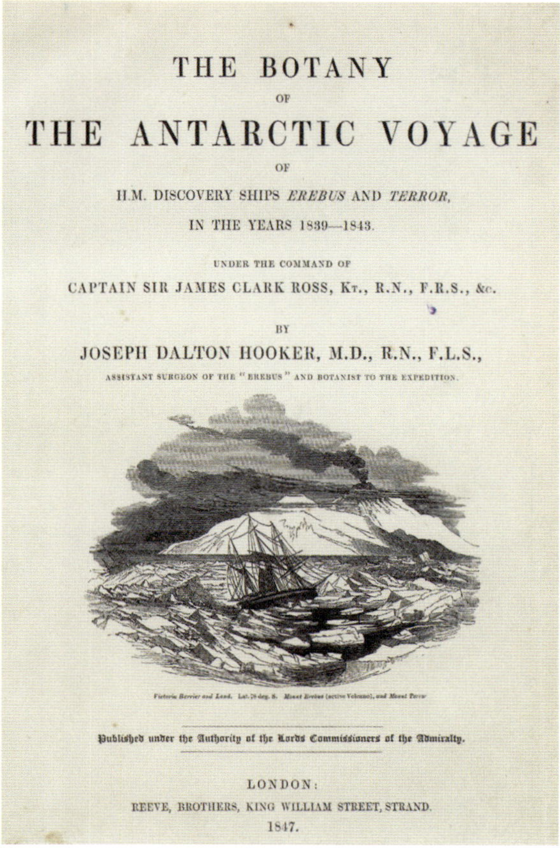
THE BOTANY
OF
THE ANTARCTIC VOYAGE
OF
H.M. DISCOVERY SHIPS *EREBUS* AND *TERROR*,
IN THE YEARS 1839—1843.
UNDER THE COMMAND OF
CAPTAIN SIR JAMES CLARK ROSS, Kt., R.N., F.R.S., &c.
BY
JOSEPH DALTON HOOKER, M.D., R.N., F.L.S.,
ASSISTANT SURGEON OF THE "EREBUS" AND BOTANIST TO THE EXPEDITION.

Published under the Authority of the Lords Commissioners of the Admiralty.

LONDON:
REEVE, BROTHERS, KING WILLIAM STREET, STRAND.
1847.

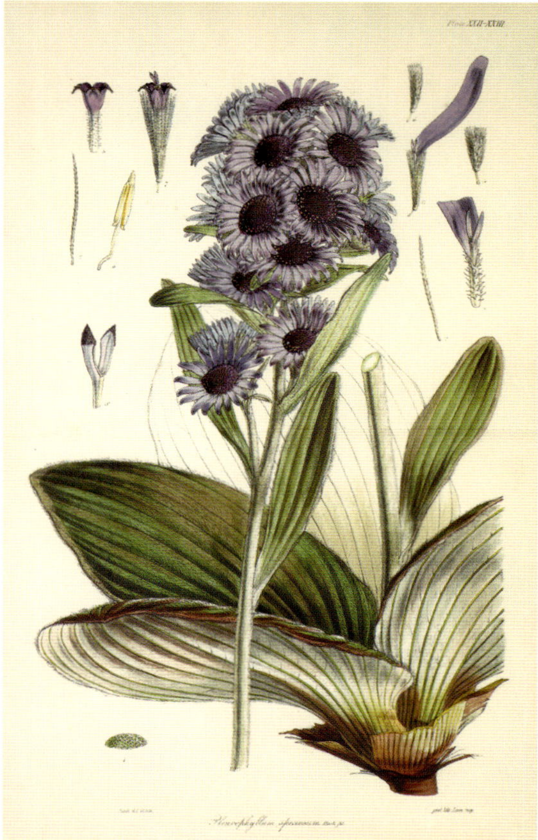

The title page of *The Botany of the Antarctic Voyage*, 1844–60 (left) and *Pleurophyllum speciosum* (Campbell Island daisy), from Joseph Hooker's *Flora Antarctica*, 1844, drawn and engraved by Walter Fitch. Hooker was the naturalist on the *Erebus* voyage, where he sketched the arresting megaherbs of the Auckland and Campbell Islands.

from Anton Seuffert, a Bohemian cabinetmaker based in Auckland, and inscribed it with the words: 'From grateful colonists in New Zealand to the Eminent Botanist, the discoverer of their Flora, Joseph Dalton Hooker MD FRS, 1867.'

Significantly, the original *Flora Antarctica* and *Flora Novae-Zelandiae* were accompanied by 130 hand-coloured plates, illustrating for the first time a substantial component of the flora. They were mostly based on Hooker's preliminary sketches from the expedition, but were later redrawn and lithographed by Walter Fitch. Trained in Scotland as a designer of textile fabrics, Fitch also worked in the herbarium for Hooker's father, William, who was at the time Professor of Botany at the University of Glasgow. William Hooker recognised Fitch's talent as a botanical illustrator and, from 1834, Fitch became the principal artist for *Curtis's Botanical Magazine*, which Hooker had taken over in 1826. By 1877, when he resigned, Fitch had prepared more than 3000 illustrations for the *Botanical Magazine*, including many of New Zealand plants. Altogether he produced in excess of 10,000 illustrations, was widely regarded as one of the most outstanding botanical artists of his day and, in partnership with Joseph Hooker, generated an authoritative basis for identifying New Zealand plants. Hooker himself summed Fitch up as an 'incomparable botanical artist' with an 'unrivalled skill in seizing the natural character of a plant ... I don't think that Fitch *could* make a mistake in his perspective and outline, not even if he tried.'[5]

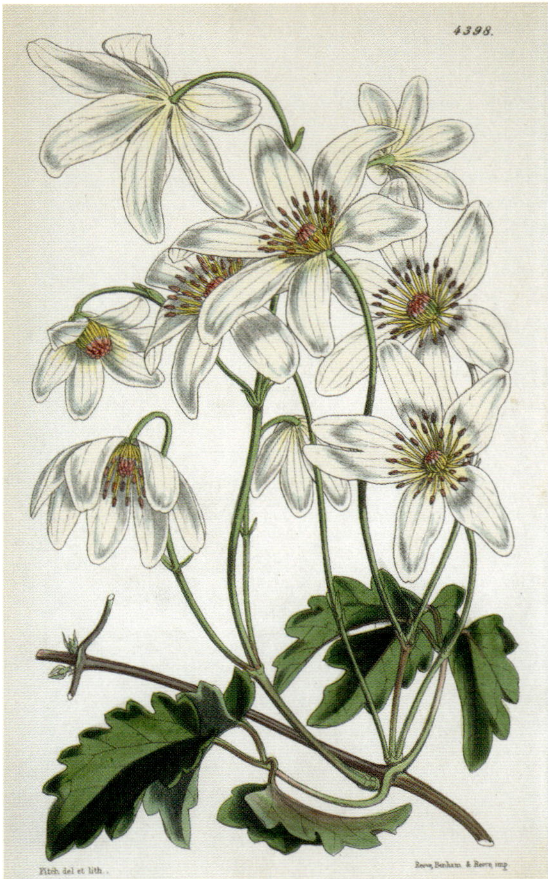

Clematis paniculata (white clematis, puawānanga) drawn and lithographed by Walter Fitch, from *Curtis's Botanical Magazine*, 1848.

Until 1840, illustration of the New Zealand flora had been undertaken only by European-based artists, or by botanists on short visits to New Zealand. Martha King changed that when she took up residence in New Zealand, arriving with her sister and brother on a New Zealand Company ship in 1840, and remaining here for the rest of her life. The family initially lived in a raupō building in Whanganui, and the two sisters opened a primary school in another. Soon afterwards, in 1842, King was commissioned by the Wellington Horticultural and Botanical Society to prepare two sets of drawings of native plants – one for the New Zealand Company and the other for the London Horticultural Society. Exactly why King was chosen for this commission is unknown, but in the small European community in which she lived her artistic abilities must have soon become apparent. Certainly she was interested in gardening and later bequeathed her own garden to the New Plymouth Recreation Grounds Board.

King made two sets of forty watercolours, a few of which were reproduced in early works by New Zealand Company agent Edward Jerningham Wakefield. The New

Zealand Company set was acquired by the Alexander Turnbull Library in 1981. Moira Long, in the *Dictionary of New Zealand Biography*, noted that 'The watercolours display consummate skill, combining accuracy of observation with a balanced and graceful composition.'[6] The quality of King's artistry was widely recognised, but although she later exhibited at the Sydney International Exhibition of 1879, no other examples of her work have survived. Apart from the aesthetic appeal of King's illustrations, the images went some way to satisfying the public's appetite for new and unfamiliar flora. This was a time when plants from all over the globe were being brought into cultivation in Europe, either for medicinal, agricultural or other useful properties, or simply for their horticultural qualities. While the earlier exploring expeditions had given the world a taste of what New Zealand's plants had to offer, they were still to be fully exploited for any of their benefits.

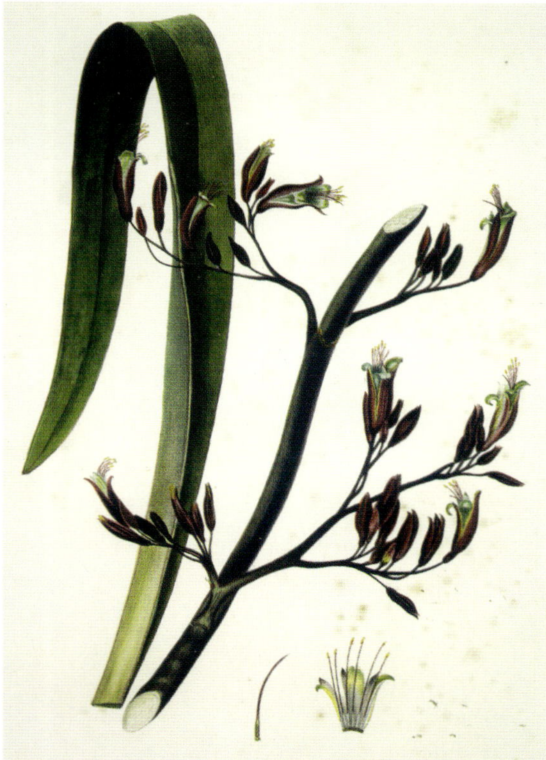

A hand-coloured lithograph of *Phormium tenax* (New Zealand flax, harakeke) based on a painting by Martha King, published in *Adventure in New Zealand*, 1845.

By the 1850s, immigration from Britain was increasing, and in the 1860s the gold rush reached its peak, with hopefuls from all over the world coming to New Zealand to seek their fortune. John Buchanan, like Walter Fitch, had been trained as a calico designer, but the decline of the textile trade in Britain caused him to look overseas for his future. Arriving in Dunedin in 1852, he spent a short time gold prospecting, before being employed by James Hector to assist with his geological survey of inland Otago, the start of a very successful partnership between the two men. Buchanan was employed as draughtsman and botanist, and for three years, as part of a small hard-working team, he collected and drew plants, and sketched a succession of landscapes, maps, geological sections and fossils. Most of his delightful plant watercolours are now held at Kew, with a few also at the Museum of New Zealand Te Papa Tongarewa.

When Hector moved to Wellington in 1865 to become the first Director of the Colonial Museum, Buchanan went with him. Hector founded the New Zealand

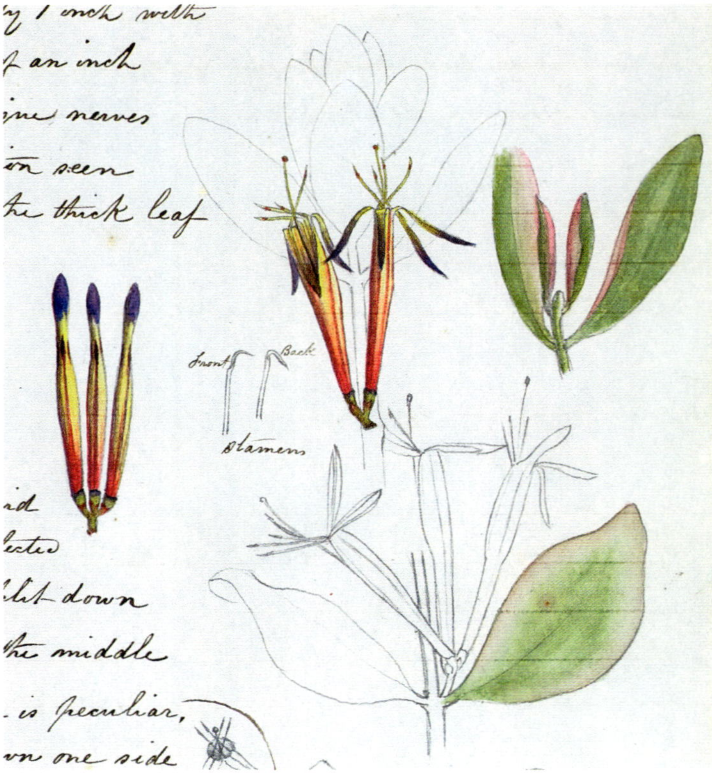

A watercolour detail of a mistletoe from
John Buchanan's notebook of New Zealand flora, now held
at the Auckland War Memorial Museum.

Institute in 1867, later to become the Royal Society of New Zealand, and for eighteen years Buchanan was given responsibility for producing lithographic plates in the *Transactions of the New Zealand Institute*, many of them illustrating plant species discovered and described during this time of burgeoning botanical exploration. While working at the Colonial Museum, he also kept notes and paintings of native plants in a notebook now at the Auckland War Memorial Museum. The experience Buchanan gained during this period ultimately led to the publication in 1880 of *The Indigenous Grasses of New Zealand* – the first major botanical work produced and published entirely in New Zealand. In the late nineteenth century, expansion of pastoral farming had created a need for better information about the identification and distribution of native and introduced grasses. The book was ordered by the government and considerable trouble was taken in its publication, as Hector recorded: 'The condition imposed, that the plates should be nature-printed, has rendered it necessary to publish the work in this large size [quarto], which is both inconvenient and expensive.'[7]

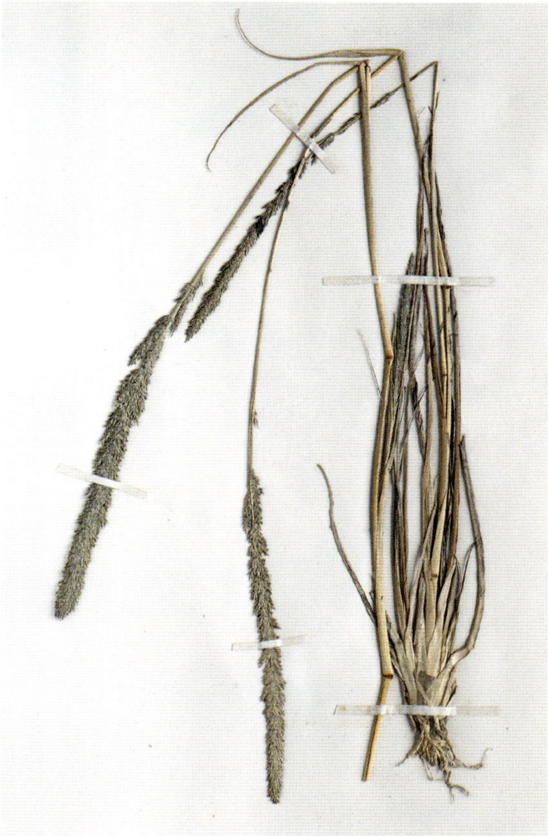

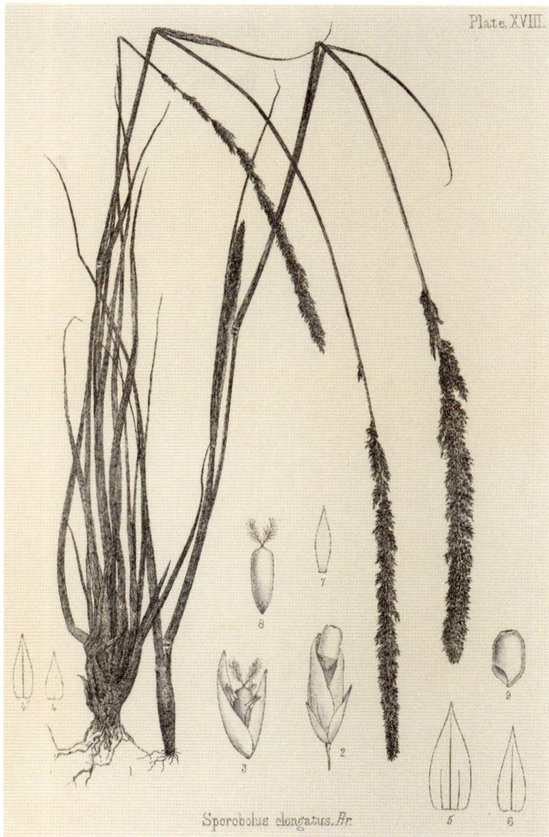

The inked specimen and nature-printed plate of *Sporobolus africanus* (ratstail), from John Buchanan's *The Indigenous Grasses of New Zealand*, 1880.

Nature-printing was a technique in which specimens were dried, lightly inked, impressed on a lithographic stone and printed at life-size. Buchanan supplemented this method with detailed drawings of the floral parts, and a second lithographic stone provided a tinted background. The plates are highly informative, elegantly designed and a novel form of botanical illustration that works well for flat specimens such as grasses and ferns. The inked grasses used for the book are held by Te Papa.

By the late 1880s, the flora of New Zealand was becoming increasingly well known, both nationally and internationally. Using Joseph Hooker's work as a basis, resident botanists such as William Colenso, John Buchanan, Thomas Kirk, Thomas Cheeseman and Donald Petrie were discovering and documenting new species at a furious rate. This mirrored the Australian scene, where the seven-volume *Flora Australiensis* (1878), written at Kew by the British botanist George Bentham, stimulated intensive investigation by local Australian botanists. In New Zealand, monochrome plates were available in the

Transactions, and in publications such as Kirk's *Forest Flora of New Zealand* (1889), but what people really wanted were affordable colour illustrations. Engraving of copper plates by skilled craftsmen could provide subtle detail of floral characters and, when coloured by hand, the prints were stunningly beautiful and accurate, but unfortunately very expensive. Few people in New Zealand could afford books with the number of hand-coloured plates in Hooker's *Flora Antarctica* and *Flora Novae-Zelandiae*.

Lithography was simpler, quicker and cheaper, and by the mid-1800s had largely replaced engraving. The technique involved drawing an image on a flat, specially prepared printing stone. When the stone was moistened, oil-based ink was taken up only by the drawn or impressed areas, and repelled by the untouched surfaces, so that it could be used to print a mirror-image of the original drawing. As with engravings, the outline drawings could be hand-coloured. However, by using different, carefully matched stones, it was possible to print several layers of colour onto one image to achieve a fully coloured lithograph, or chromolithograph. Chromolithographs had an appealing softness to them but could only employ a restricted range of pigments, and resulted in somewhat dull images with a yellow-brown tinge, lacking the sharpness and fresh colours of hand-coloured lithographs.

Amateur botanists were the first to adopt the new process of chromolithography. Two significant publications appeared in the late 1880s, at the same time as Walter Buller was using the technique to print the second edition of *A History of the Birds of New Zealand* (1888). Georgina McKellar arrived with her family in New Zealand in 1852, soon marrying Charles Hetley and moving to a farm in Taranaki. Charles died less than a year later and Georgina moved with her young son to Auckland. In Taranaki she had made sketches of her farms, and in 1879 she exhibited with the Auckland Society of Artists, developing her skills to win first prize at the New Zealand Industrial Exhibition in Wellington in 1885 for paintings of indigenous plants. Following a lecture given by Cheeseman at the Auckland Museum in 1881 about a botanical trip to Nelson, and with considerable encouragement from him, Kirk, Buchanan and other botanists, Hetley was inspired 'to do for the flora of New Zealand what Dr Buller has done for its birds'.[8] By 1884, she had started work on a book of native plants, *The Native Flowers of New Zealand Illustrated in Colours*, embarking (with backing from the government and the Union Steamship Company) on a trip around New Zealand to obtain live specimens.

Hetley felt intense disappointment, even anger, as she saw at first hand how the New Zealand forest was being cleared for agricultural development: 'Some evil disposed men had wantonly set fire to the forest all along the road, burning whole sides of mountains and destroying some of the most magnificent scenery'.[9] But her effort at actually venturing into the field was frequently rewarded with displays of New Zealand plants at

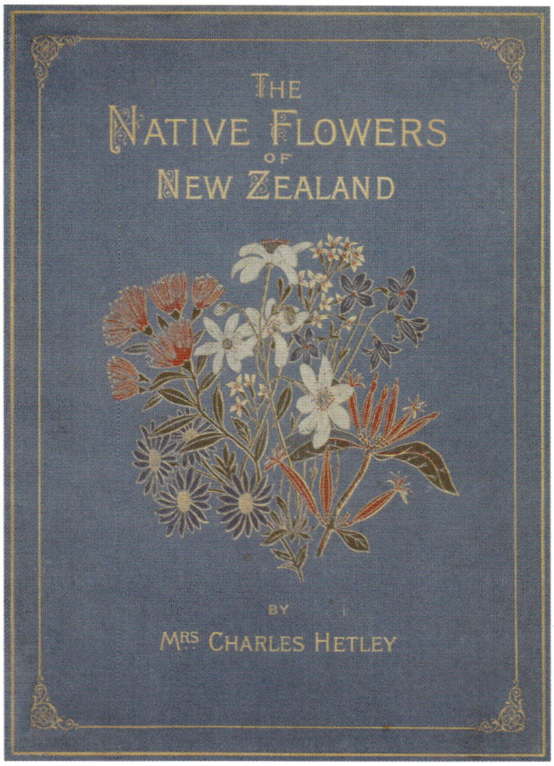

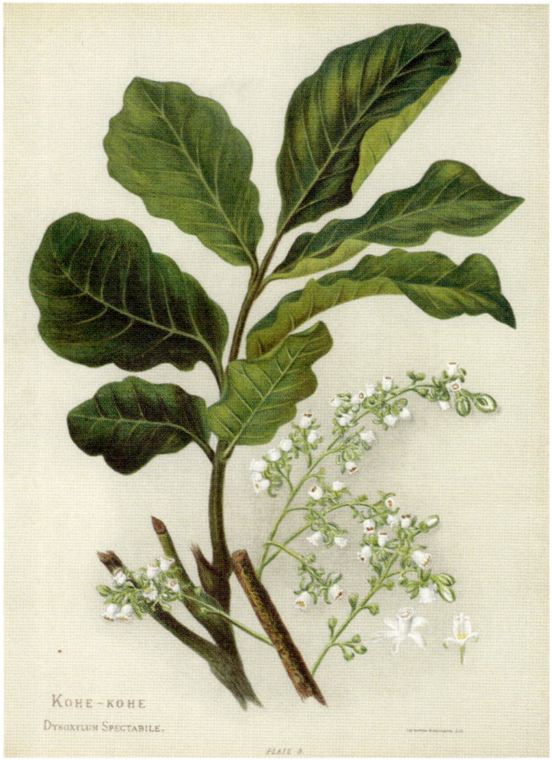

The cover of Georgina Hetley's *The Native Flowers of New Zealand Illustrated in Colours*, 1888, and a chromolithograph from the book, by Leighton Brothers, from a watercolour by Hetley of *Dysoxylum spectabile* (kohekohe).

their colourful best, and revealing this beauty to her compatriots ultimately brought her considerable satisfaction:

> Lace-bark trees ... were in full blossom with bunches of cherry-like flowers, and close alongside them were trees of crimson rata ... also bushes of Veronica, a mass of white flowers ... My sketches created some surprise in Auckland, as even people who were born in New Zealand had no idea we had such lovely flowers.[10]

Like Buller, she went to England to seek a publisher, receiving assistance along the way from authorities at Kew, and the chromolithographs were ultimately produced in 1888 by Leighton Brothers, who did colour printing for the *Illustrated London News.* The plates are refreshingly life-like, albeit not of the quality of modern colour printing. The plates also had the distinction of being published in a French edition a year later, one reviewer reflecting that they 'brought home to us, as never before, the gorgeousness of the New Zealand flora'.[11]

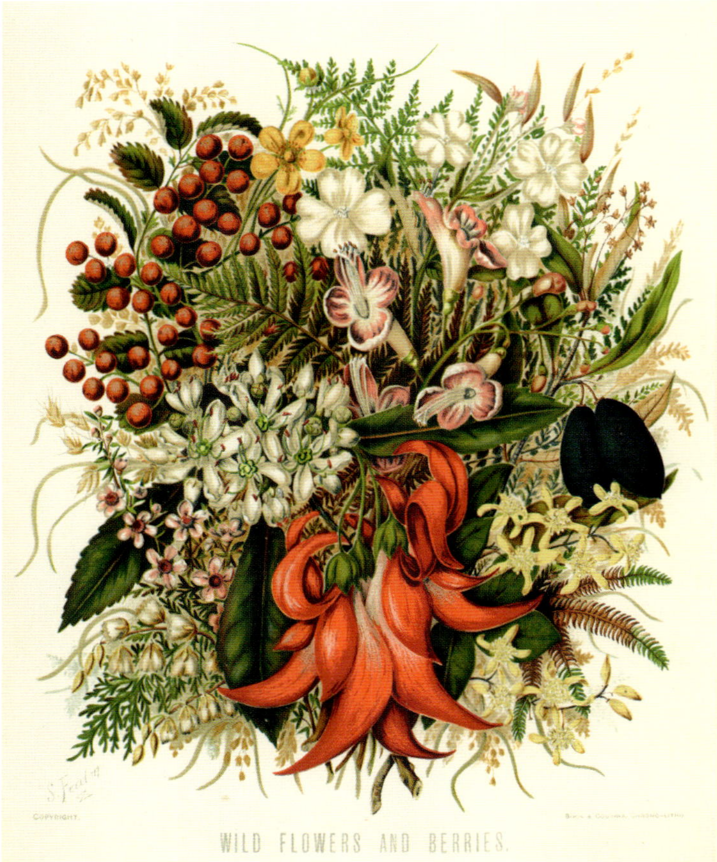

The frontispiece of *The Art Album of New Zealand Flora*, 1889, featuring *Clianthus puniceus* (kakabeak), by Sarah Featon.

In the same year, *The Art Album of New Zealand Flora* was published by Edward and Sarah Featon, a book comparable in style and substance to Hetley's, and with strikingly similar aims:

> It has been repeatedly asserted that there are no flowers in New Zealand, and by very many the statement is generally believed to be true. The authors of this work deem it to be their privilege to prove how fallacious and incorrect such beliefs are … the flowers of our adopted land cannot vie with the beauteous floral displays of more tropical climes, but … they display a beauty of form and character which in many instances are peculiarly their own.[12]

The book was the first containing full-colour illustrations to be produced and published entirely in New Zealand. The text was written by Edward, who served in the military, worked for many years in the District Land Office at Gisborne, and had a wide interest in art and literature. The forty plates were printed from original watercolours by Sarah, who Edward married in 1870, and who had a life-long interest in painting and handicrafts. Sarah used brighter colours than Hetley, and these often succeed better as chromolithographs, but her work lacks some of the botanical detail evident in Hetley's paintings. Printed by William Bock and Alfred Cousins, the book was undoubtedly appealing, but unfortunately it was released at a time of economic depression and was a commercial failure, resulting in the dissolution of the firm of Bock and Cousins. Nevertheless, contemporary records show that the book was well received and, like Hetley's, introduced native plants to a wider public, opening many eyes for the first time to New Zealand's floral diversity. A copy of *The Art Album*, in a specially made casket of native timbers, was presented to Queen Victoria by the New Zealand Government at the time of her Diamond Jubilee in 1897. It was intended to publish two more volumes, and ninety-three unpublished plates, together with the original paintings, are in the collection at Te Papa. Exactly why the last two volumes were never published is unclear, but the commercial failure of the first volume and the death of Edward Featon in 1909 are likely contributing factors.

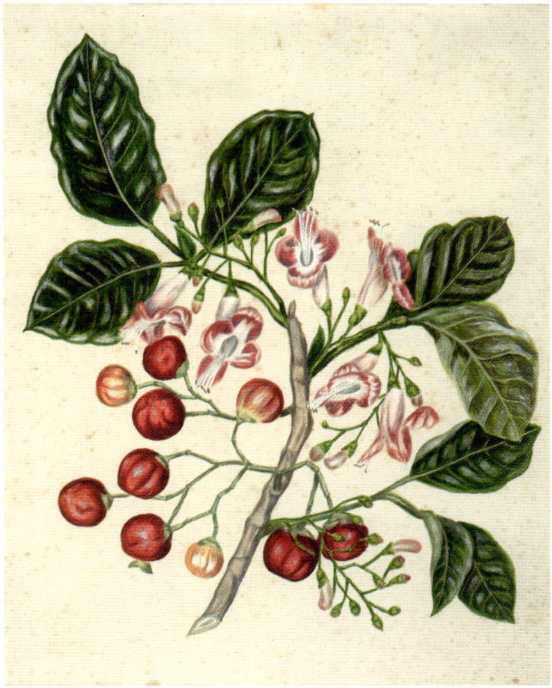

Sarah Featon's watercolour of *Vitex lucens* (pūriri), *c.* 1889, later reproduced as a chromolithograph in *The Art Album of New Zealand Flora*, 1889.

By the 1890s, Joseph Hooker's seminal *Handbook of the New Zealand Flora* had become outdated because of the sheer number of new species discovered in the intervening period. Thomas Kirk was commissioned by the government to write a new Flora (treatise on plants), but he died in 1898, leaving his *Students' Flora of New Zealand and the Outlying Islands* to be published posthumously a year later, unillustrated and missing several families.

Thomas Cheeseman was then asked to step in, and his *Manual of the New Zealand Flora* was published in 1906 with descriptions of more than 1500 species. This time a volume of illustrations was carefully planned, although it was another eight years before they finally appeared, just before the outbreak of World War I. Cheeseman's objective

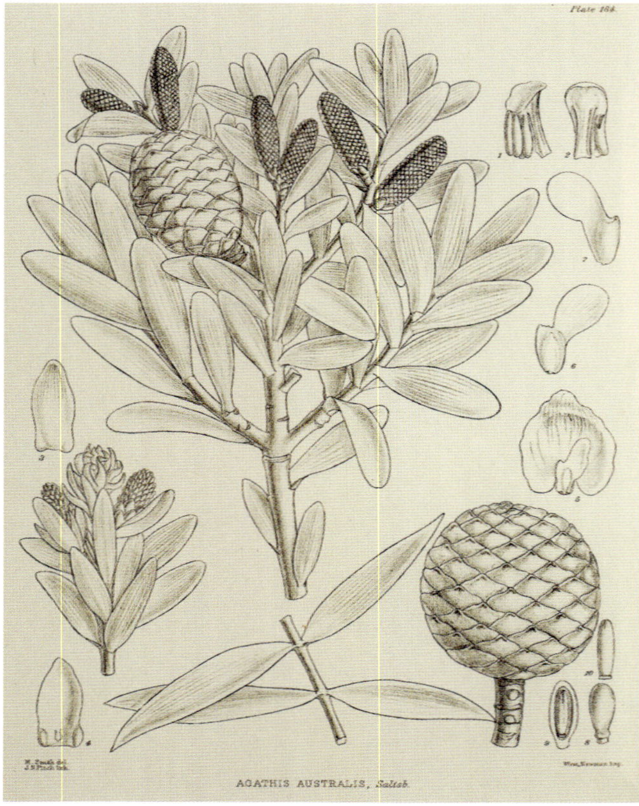

Agathis australis (kauri) from Thomas Cheeseman's *Illustrations of the New Zealand Flora*, 1914, drawn by Matilda Smith and lithographed by John Fitch.

was 'to issue plates of an educational character, so selected as to present an accurate and comprehensive idea of the main features of the flora, and so designed to be of real use in the study and identification of the plants of the Dominion'.[13] To achieve this he felt obliged to search outside New Zealand for the necessary artistic and lithographic skills, overlooking, among others, his own sister Emma, who had produced exquisite watercolours of orchids (now held by Auckland Museum). Matilda Smith, botanical artist at Kew and cousin of Joseph Hooker, was commissioned to prepare the illustrations and John Fitch, nephew of Walter, the lithographs. Smith had been trained by Hooker, and prepared plates for *Curtis's Botanical Magazine* after Walter's resignation in 1877. She contributed around 2300 plates to the *Botanical Magazine* as well as another 1500 plates for Hooker's *Icones Plantarum*. For Cheeseman's *Flora* she drew 268 species in 251 uncoloured lithographic plates, easily the largest and most comprehensive selection of New Zealand plant illustrations then completed by one artist. She never visited New

Zealand, but regularly received pressed specimens, photographs and spirit-preserved flowers and fruit. Her skill was in making life-like drawings from dried specimens. *Illustrations of the New Zealand Flora* remained for several decades the authoritative visual guide to the seed plant flora. It set new standards, recording when and where the species was first discovered, its known range and the source of the plant illustrated by Smith. No subsequent New Zealand Flora has provided such a comprehensive range of illustrations to accompany the written descriptions, and possibly none may ever do so again.

By 1914, nearly half of New Zealand's native seed plants and ferns had been illustrated in various publications. But the medium of choice was changing once again and photography was the developing technology that eclipsed botanical illustration during much of the twentieth century. *New Zealand Native Plant Studies* by WC Davies (1956, in black and white), *The Native Trees of New Zealand* by JT Salmon (1980, in colour) and *New Zealand's Native Trees* by JW Dawson and R Lucas (2011, in colour) are state-of-the-art examples of the way botanical photography has evolved over the last half-century. Today, digital imaging and online publication are rapidly replacing hard-copy printing, and it is in this format that an electronic *Flora of New Zealand* is currently being produced by botanists at Landcare Research, Te Papa and the National Institute of Water and Atmospheric Research (NIWA). This online resource will replace the five-volume *Flora of New Zealand*, a series initiated by HH Allan in 1961, covering all the native and introduced seed plants and ferns, which has been the standard reference work for New Zealand botanists over the last fifty years, but which has now become somewhat outdated.

Despite the gradual ascendance of photography, many botanical artists continued to make contributions to documenting the New Zealand flora. Paintings or drawings were able to emphasise the key characters, or combinations of characters, essential for making an accurate identification in a way that eluded photography until digital imaging became available. The public, too, recognised that paintings had an individual beauty and aesthetic quality that could rival even the best photographs.

After World War II, two of New Zealand's very best botanical artists ensured the continuity of plant illustration alongside photography. The first of these was Nancy Adams. Her natural artistic talent was evident from a very early age and blossomed when she became the official artist at Botany Division, Department of Scientific and Industrial Research (DSIR) in 1950. Her 'line drawings of incomparable accuracy and uncluttered style'[14] became evident in major botanical works of the 1950s and early 1960s, and are

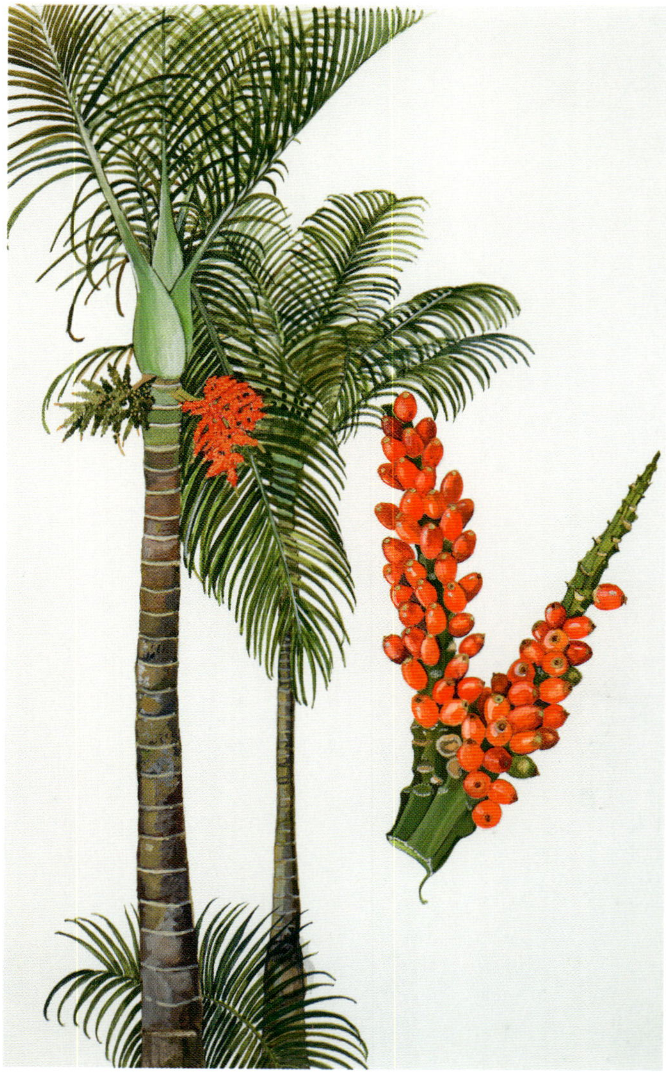

Nancy Adams' original watercolour of *Rhopalostylis sapida* (nīkau), published in *New Zealand Native Trees 2*, 1983.

best developed in *Trees and Shrubs of New Zealand*, jointly published in 1963 with Lindsay Poole and still in print after innumerable revisions. After leaving the DSIR, Adams continued to publish paintings and line drawings of plants in national park handbooks, ephemeral publications and a series of popular books on mountain flowers, trees and introduced plants, the latter something of a novelty at that time. Another landmark book, *New Zealand Alpine Plants*, was published with Alan Mark in 1973, with 118 colour plates. Appointment to the position of Assistant Curator of Botany at the Dominion

Museum in 1969 enabled her to develop her interest in marine algae, culminating twenty-five years later in the publication of *Seaweeds of New Zealand*, which has descriptions and 116 colour plates treating around 600 New Zealand species. No other botanical artist has come close to capturing this breathtaking sweep of native and introduced species from both marine and terrestrial habitats. As noted in her obituary: 'Nancy had a unique ability to capture the evocative and characteristic colours of New Zealand's native trees and shrubs, and she succeeded equally well with the seaweeds ... the indefinable shapes of the plants, are all captured in a way that words can never adequately express.'[15]

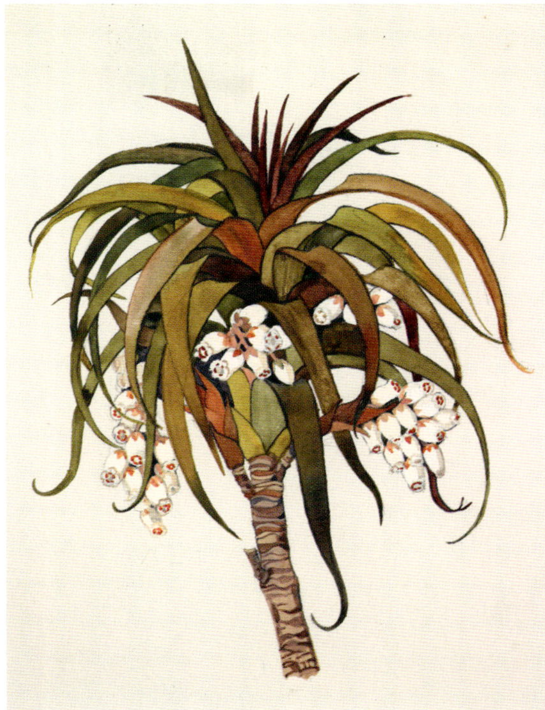

Dracophyllum menziesii, a watercolour painting by Nancy Adams, later published in Mark and Adams's *New Zealand Alpine Plants*, 1973.

The other dominant botanical artist of the postwar period in New Zealand is, of course, Audrey Eagle, whose work has focused specifically on our distinctive trees and shrubs. Starting out in 1954, she painted plants 'with the sole purpose of learning their botanical names',[16] but by 1968 the concept of a book with examples of every genus of tree and shrub in New Zealand had taken shape. She formed a working relationship with the leading botanists of the day at the DSIR, who provided her with material and advised her on taxonomic detail. In 1975, after more than twenty years of work, the meticulous *Eagle's Trees and Shrubs of New Zealand in Colour* was published, illustrating 228 species, all painted from live material, reproduced at life-size, and, following the standard set by Cheeseman and Smith, with detailed notes on identification, distribution and the source of the illustrated material. A contemporary reviewer, Sheila Natusch, was impressed at how true to life the illustrations were: 'Audrey Eagle has learned her drawing and painting from Mother Nature', and she further suggested that professional techniques could teach her nothing about 'the smell and feel of the bush'.[17]

A second book expanded Eagle's reputation as a sensitive and precise illustrator. It took only seven years to complete, appearing in 1982 and illustrating a further 405 species and varieties. The same reviewer found 'particularly appealing ... the delicacy

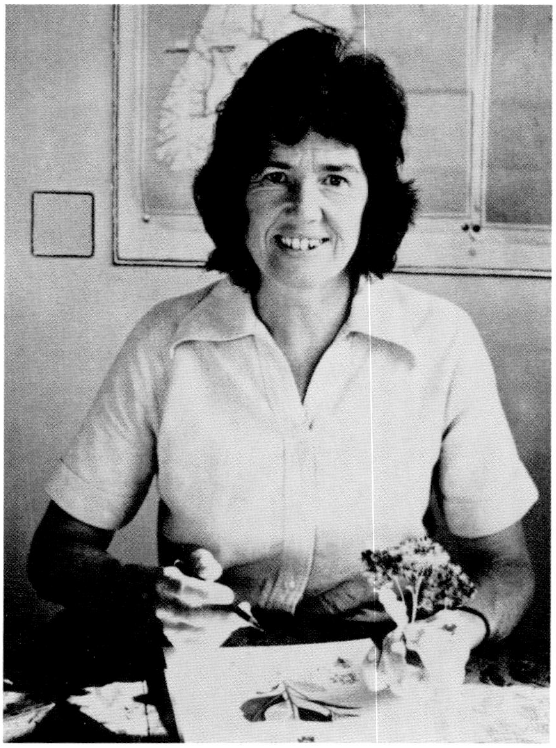

Audrey Eagle at work, 1970s.

of transition from near-white through lime-green to lemon-yellow in the *Clematis* species illustrated, and the silken translucency of *Ipomaea pes-caprae* . . . and all those gleaming fruit!'[18] Unquestionably, the two books were an outstanding contribution to New Zealand botany.

Their publication seemed to mark the end of Eagle's task of documenting the woody flora, especially when revised editions of both books in 1986 brought the nomenclature up to date. However, such was the pace of botanical research in the late twentieth century, led by a new generation of taxonomic botanists, that a third contribution became inevitable. In 2006, Te Papa Press published a ground-breaking two-volume edition incorporating all of Eagle's previous illustrations, together with 173 new paintings. *Eagle's Complete Trees and Shrubs of New Zealand* comprises illustrations of more than 800 species, subspecies, varieties and unnamed plants. In the opinion of the judges for the 2007 Montana New Zealand Book Awards, at which this book won the Montana medal for non-fiction, 'This monumental work is a magnificent tribute to her vision, her perseverance and to her consummate skill as a botanical artist.'[19]

Eagle's supreme discipline ensured comprehensive coverage of New Zealand's woody-stemmed plants, something that could only have been dreamed of by earlier botanical artists. Through her collaborations with many of New Zealand's best field botanists, she can rightly claim to be the only person in the world to have seen live material of virtually every woody plant in New Zealand, including one that is now extinct. With the confidence and knowledge that came from working with professional botanists, she included in her book many plants that still remain undescribed. *Eagle's Complete Trees and Shrubs of New Zealand* is the best example yet of description and illustration of every species for an entire section of the flora, and is an essential reference for professional and amateur botanists that will be consulted for years to come.

Some 163 of Eagle's finest illustrations from *Eagle's Complete Trees and Shrubs of New Zealand* have been carefully selected for this new book, covering the whole range of New

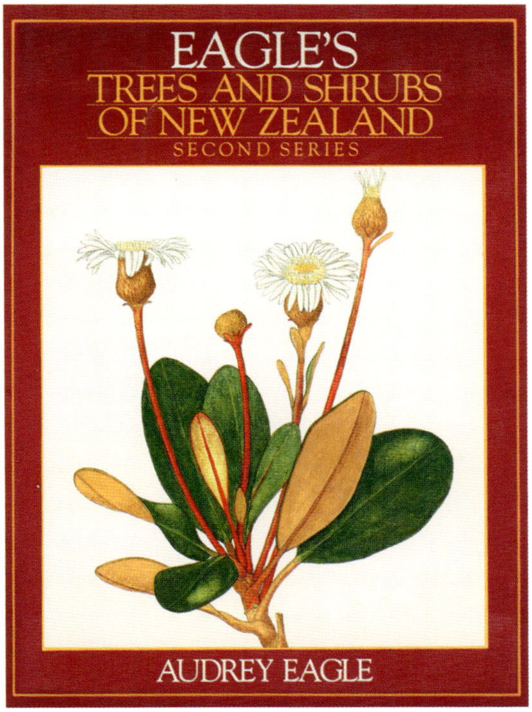

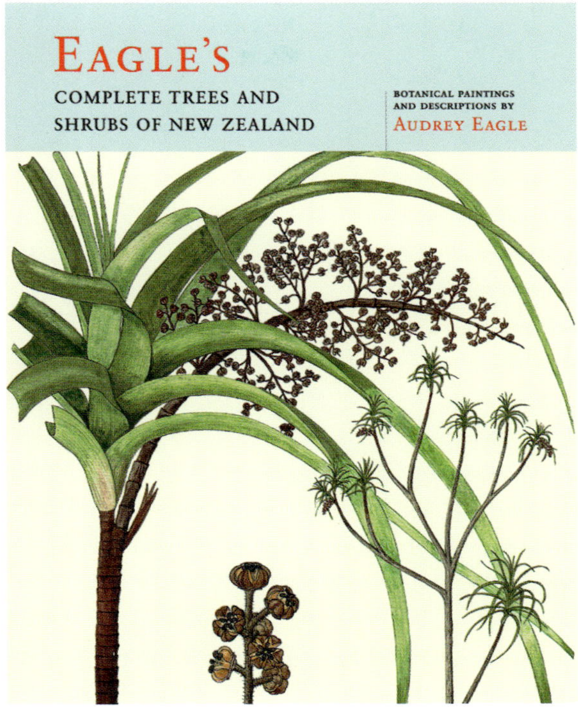

Audrey Eagle's *Eagle's Trees and Shrubs of New Zealand*, 1982
and *Eagle's Complete Trees and Shrubs of New Zealand*, 2006.

Zealand's woody flora. Beginning with the conifers, and moving through quintessentially New Zealand plants such as kōwhai, rātā, neinei, coprosma and hebe, to the highly distinctive flaxes, cabbage trees and nīkau, we can follow Eagle's own voyage of discovery as she painted familiar and not-so-familiar plants. Readers will detect her mastery of botanical detail over the course of almost sixty years of illustration, and will no doubt pause to admire their own favourite plants, or to appreciate the beauty of little-known plants such as those confined to offshore islands. We will surely all regret the loss of the mistletoe (*Trilepidea adamsii*), now extinct through the ravages of the introduced possum, and reflect on Eagle's painting of the last plant seen alive. But whatever our preferences, all these images should be savoured because botanical illustration of this quality, and on this scale, may never again be seen in New Zealand.

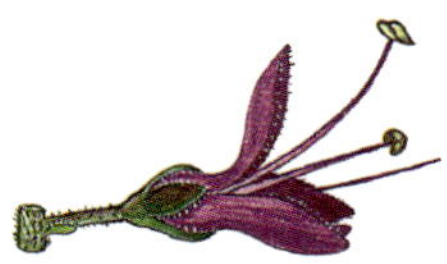

Plates

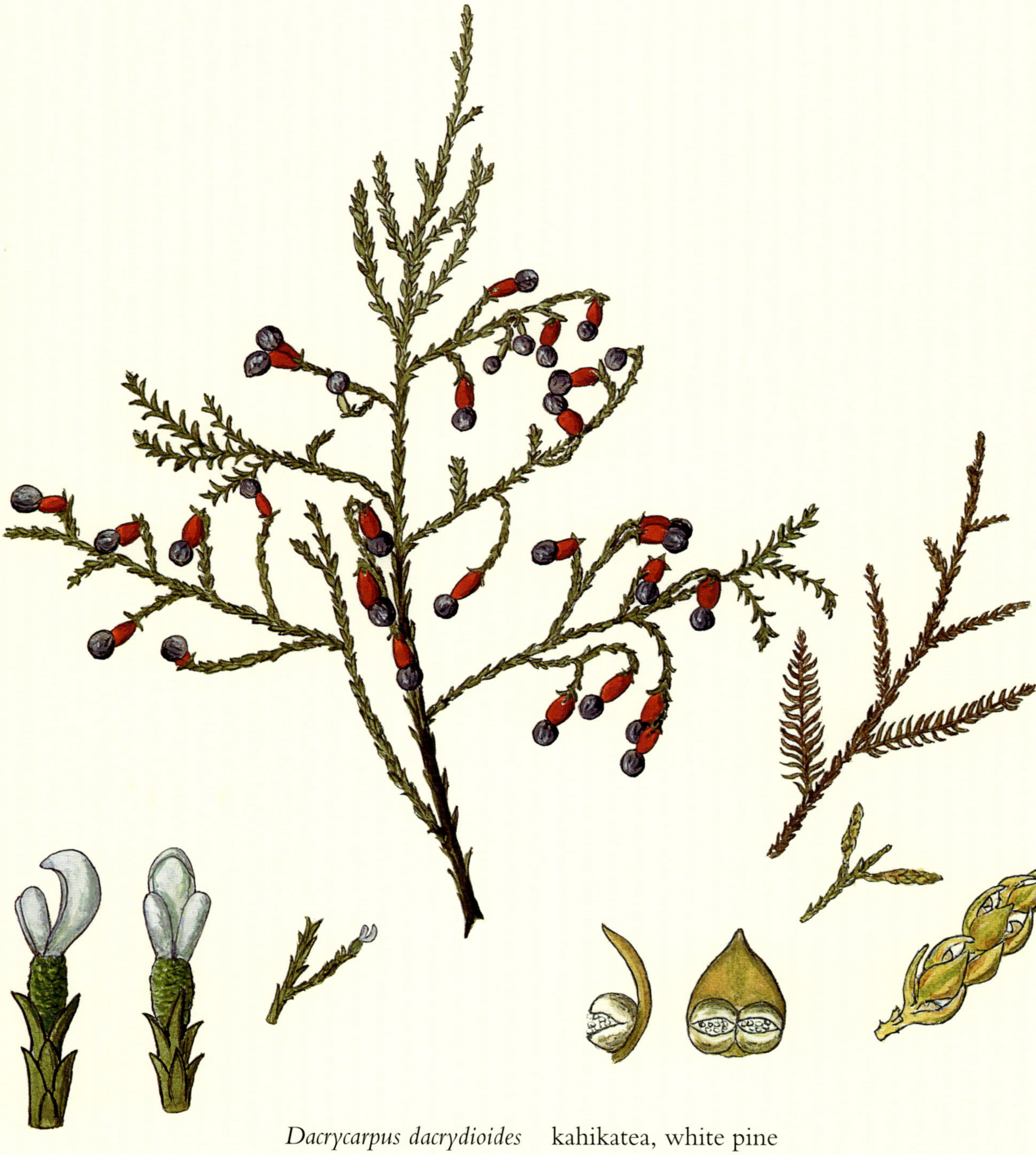

Dacrycarpus dacrydioides kahikatea, white pine

PLATE I

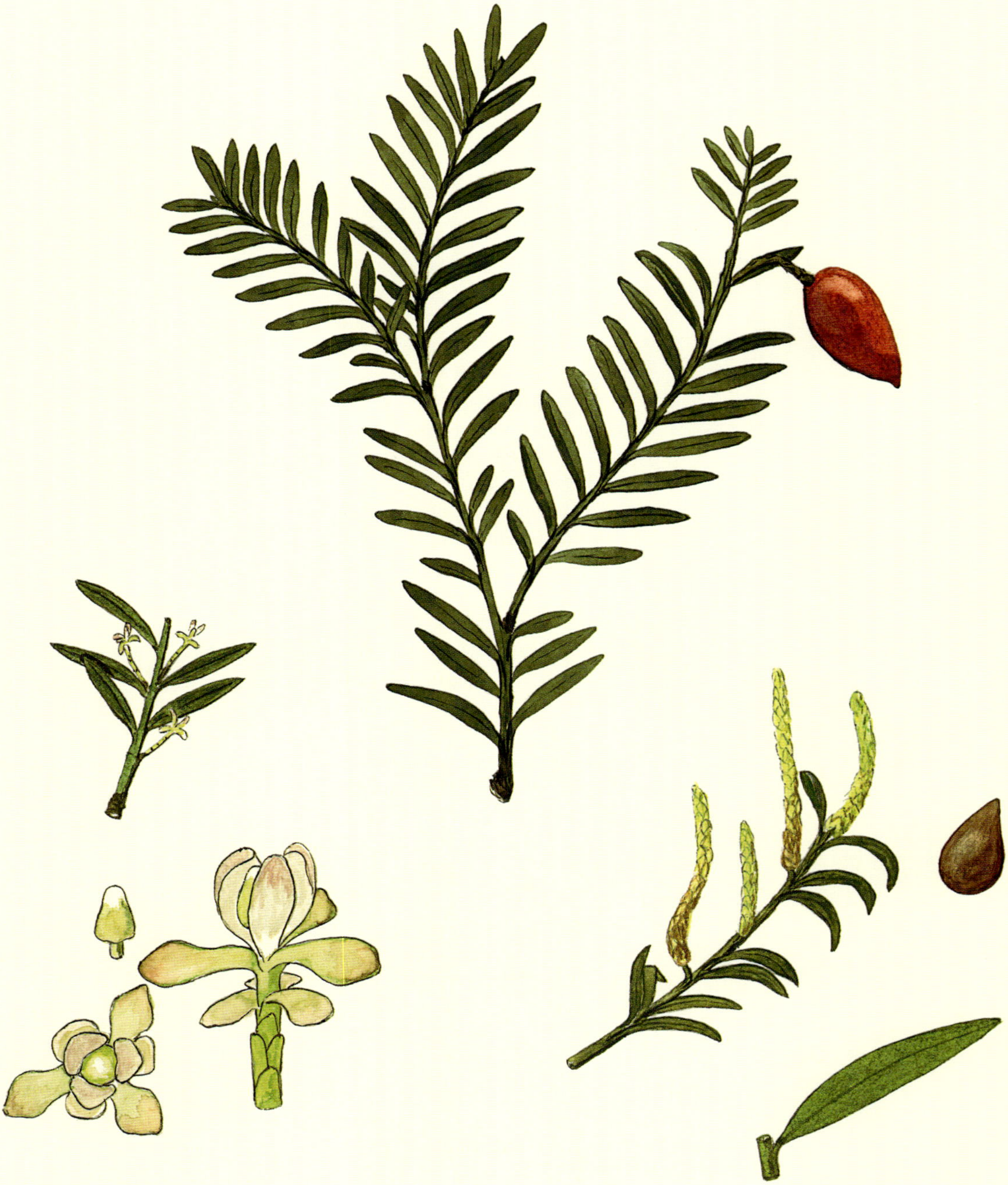

Prumnopitys ferruginea miro

PLATE 2

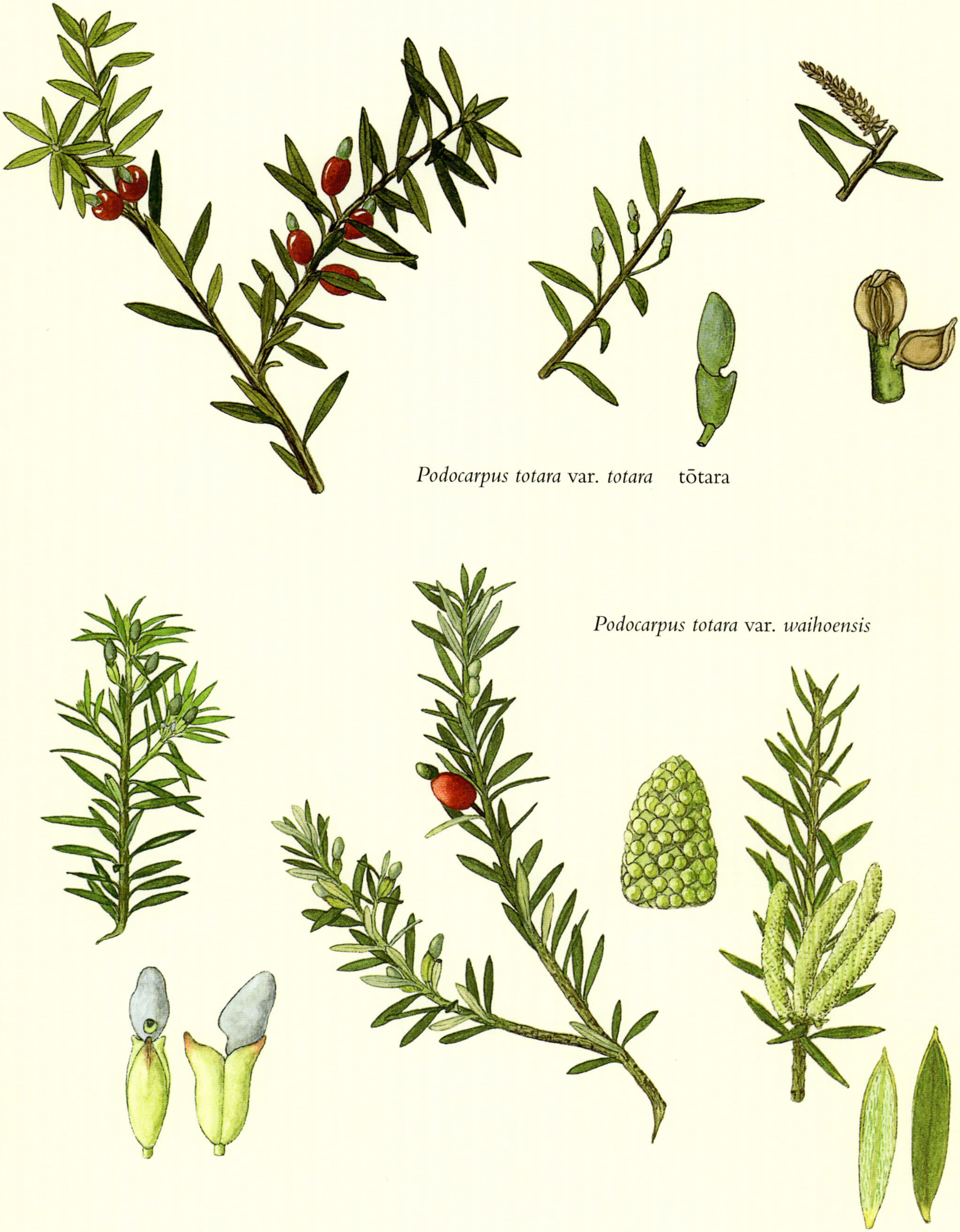

Podocarpus totara var. *totara* tōtara

Podocarpus totara var. *waihoensis*

PLATE 3

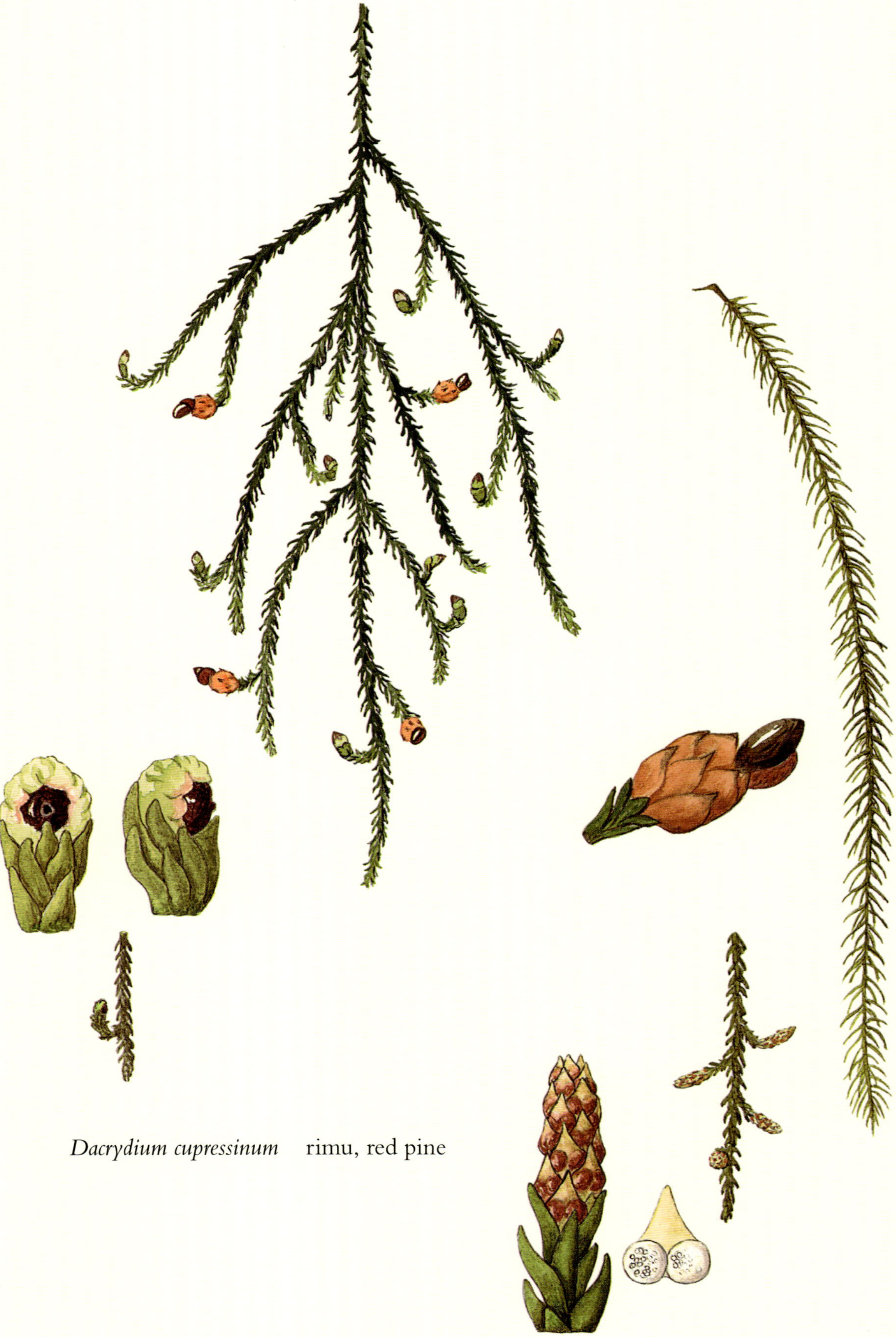

Dacrydium cupressinum rimu, red pine

PLATE 4

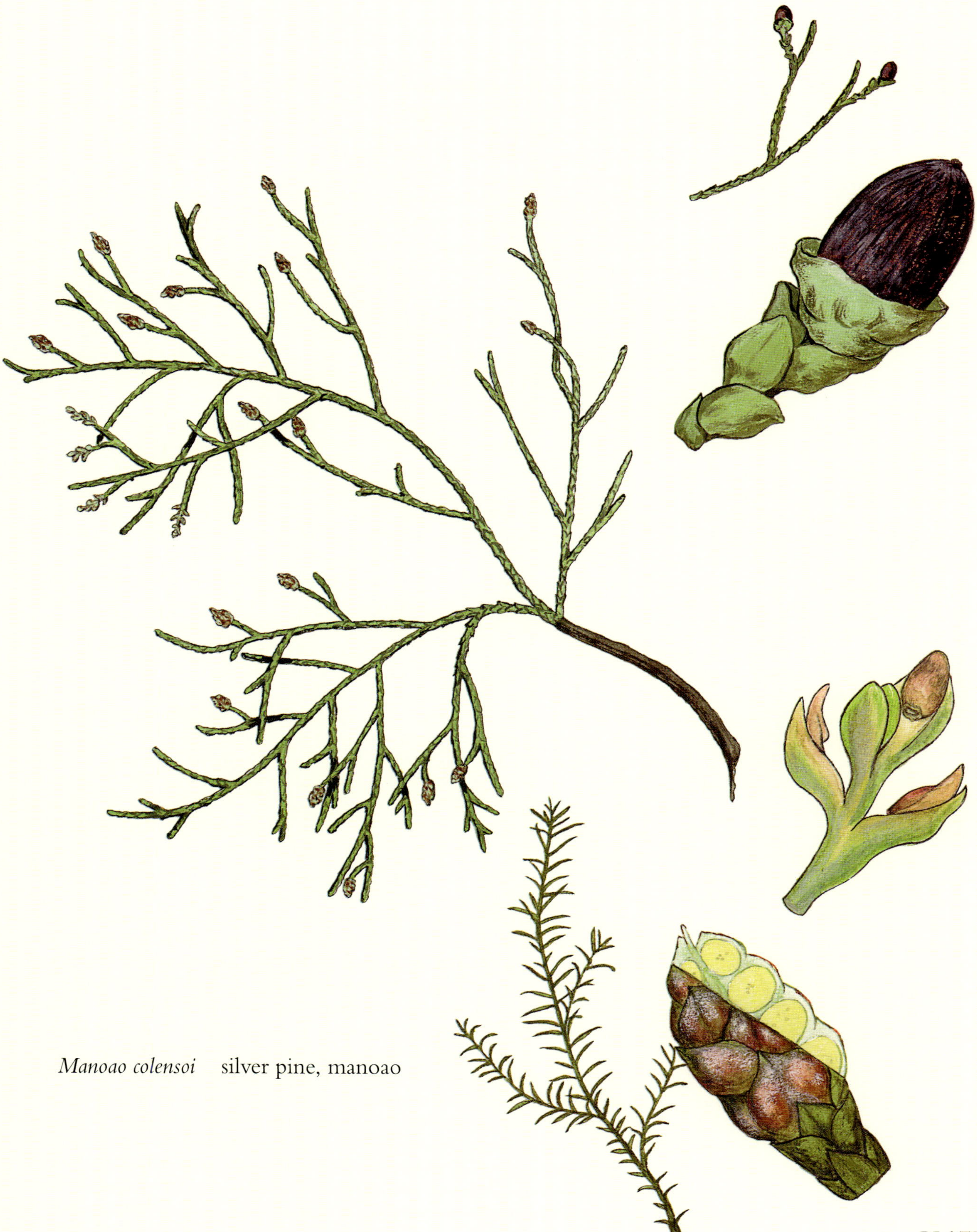

Manoao colensoi silver pine, manoao

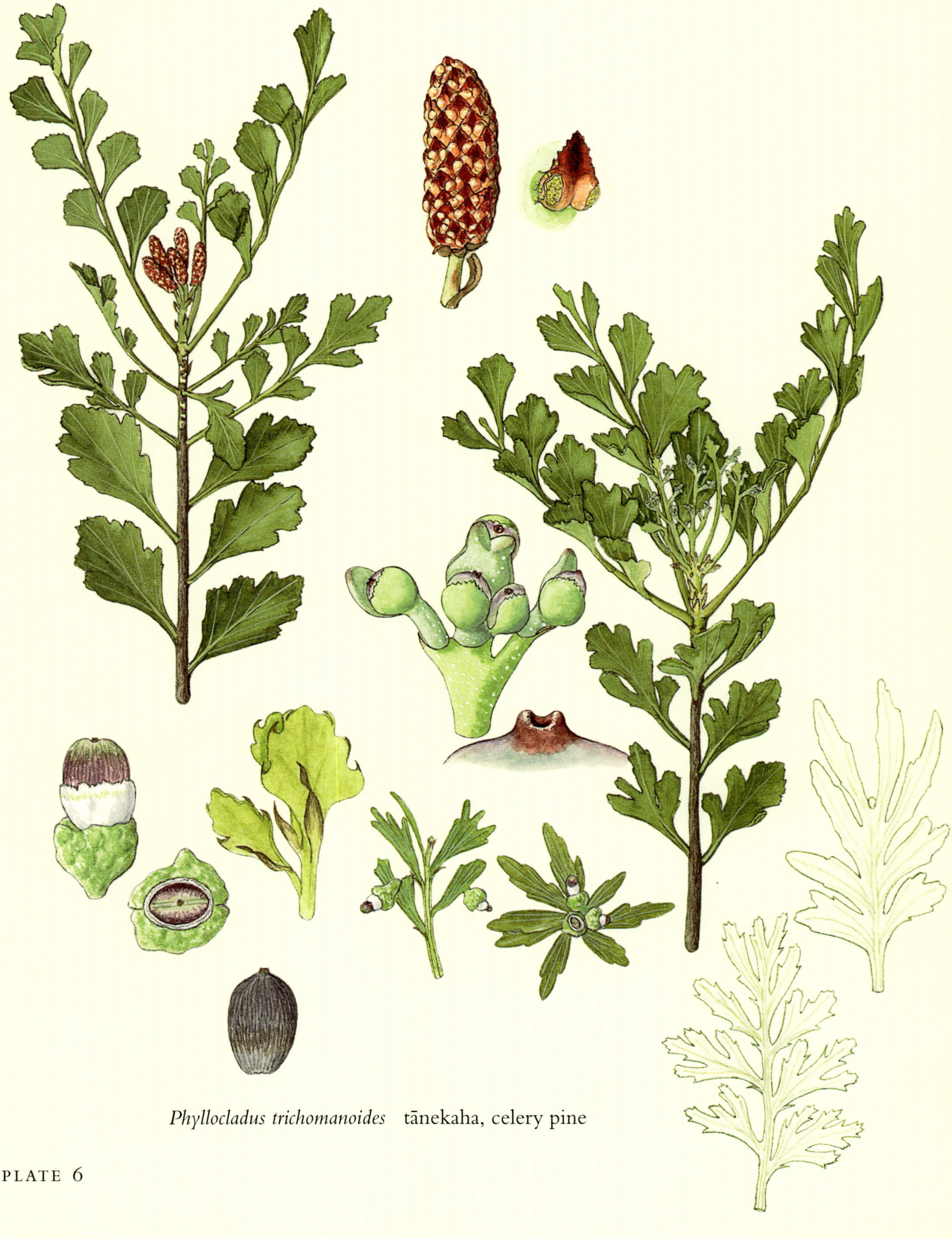

Phyllocladus trichomanoides tānekaha, celery pine

PLATE 6

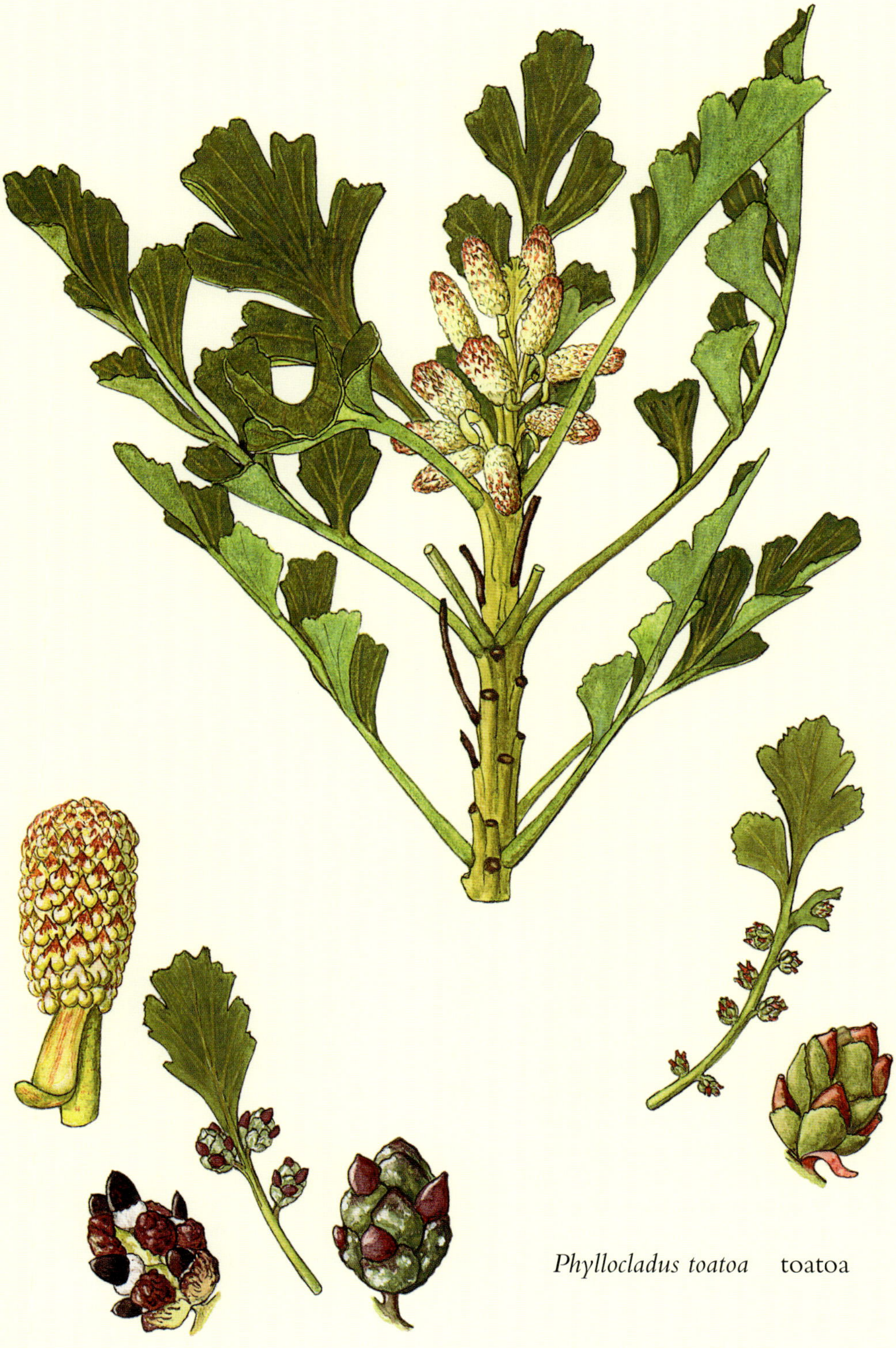

Phyllocladus toatoa toatoa

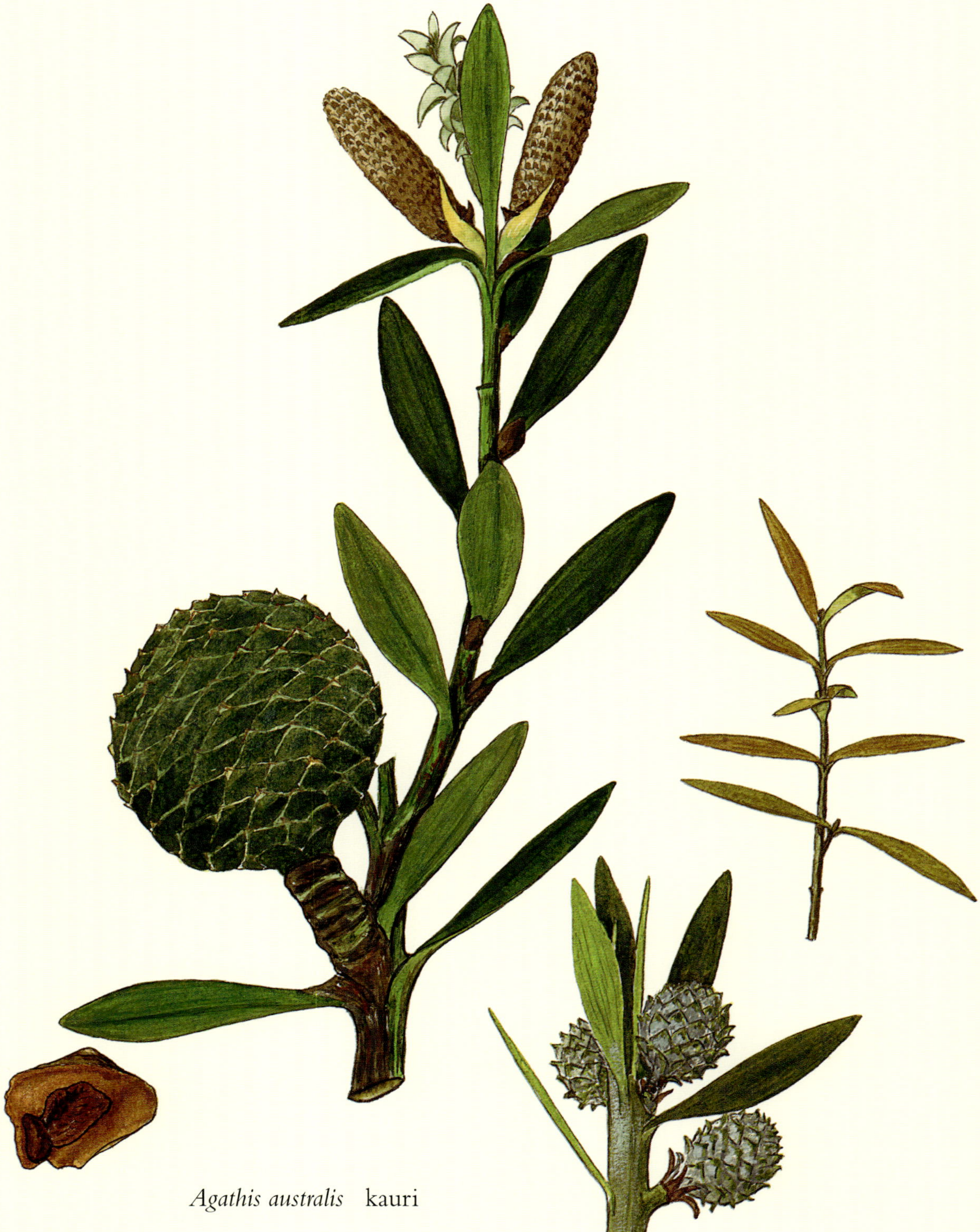

Agathis australis kauri

PLATE 8

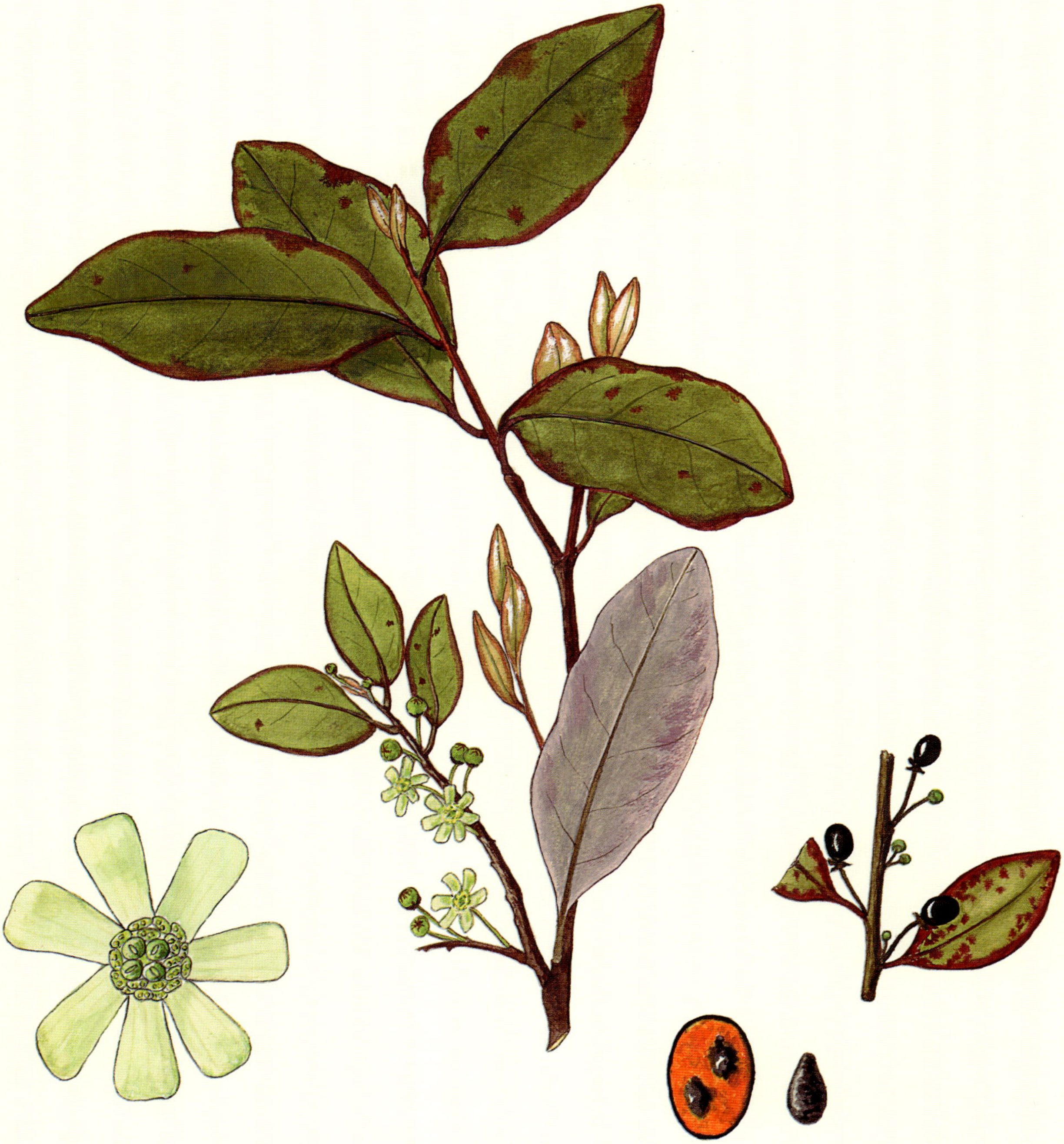

Pseudowintera colorata mountain horopito, pepper tree

PLATE 9

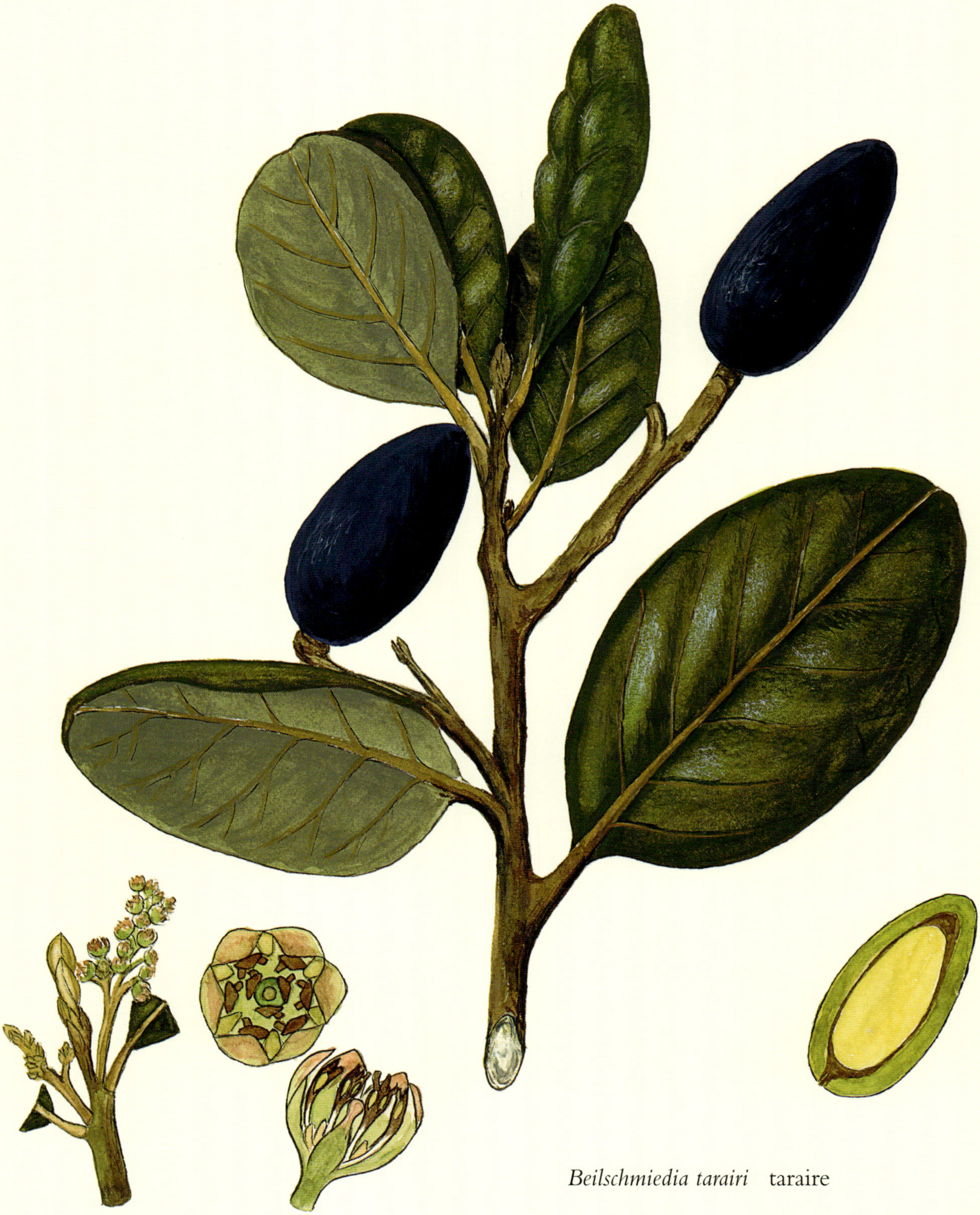

Beilschmiedia tarairi taraire

PLATE 10

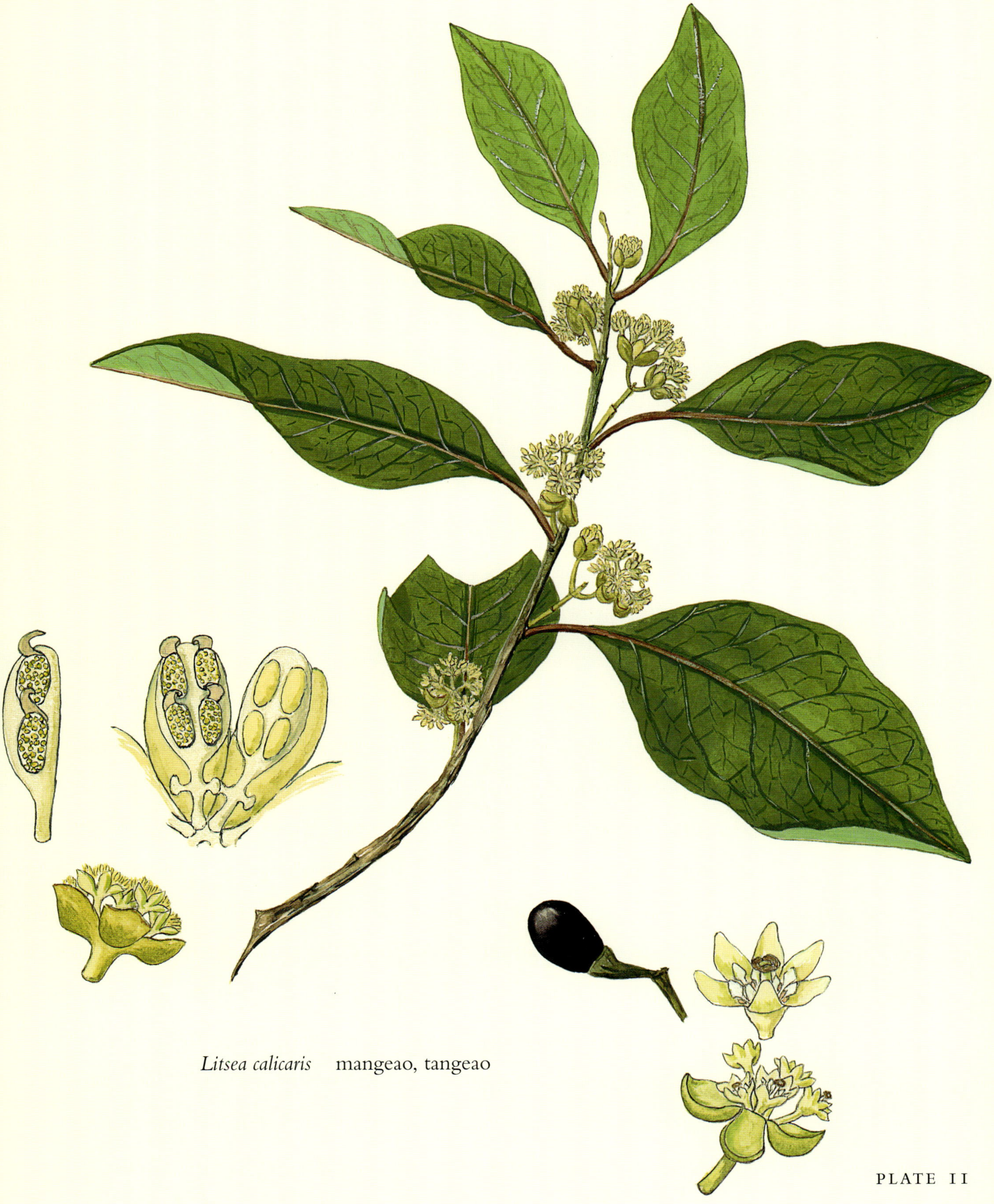

Litsea calicaris mangeao, tangeao

PLATE II

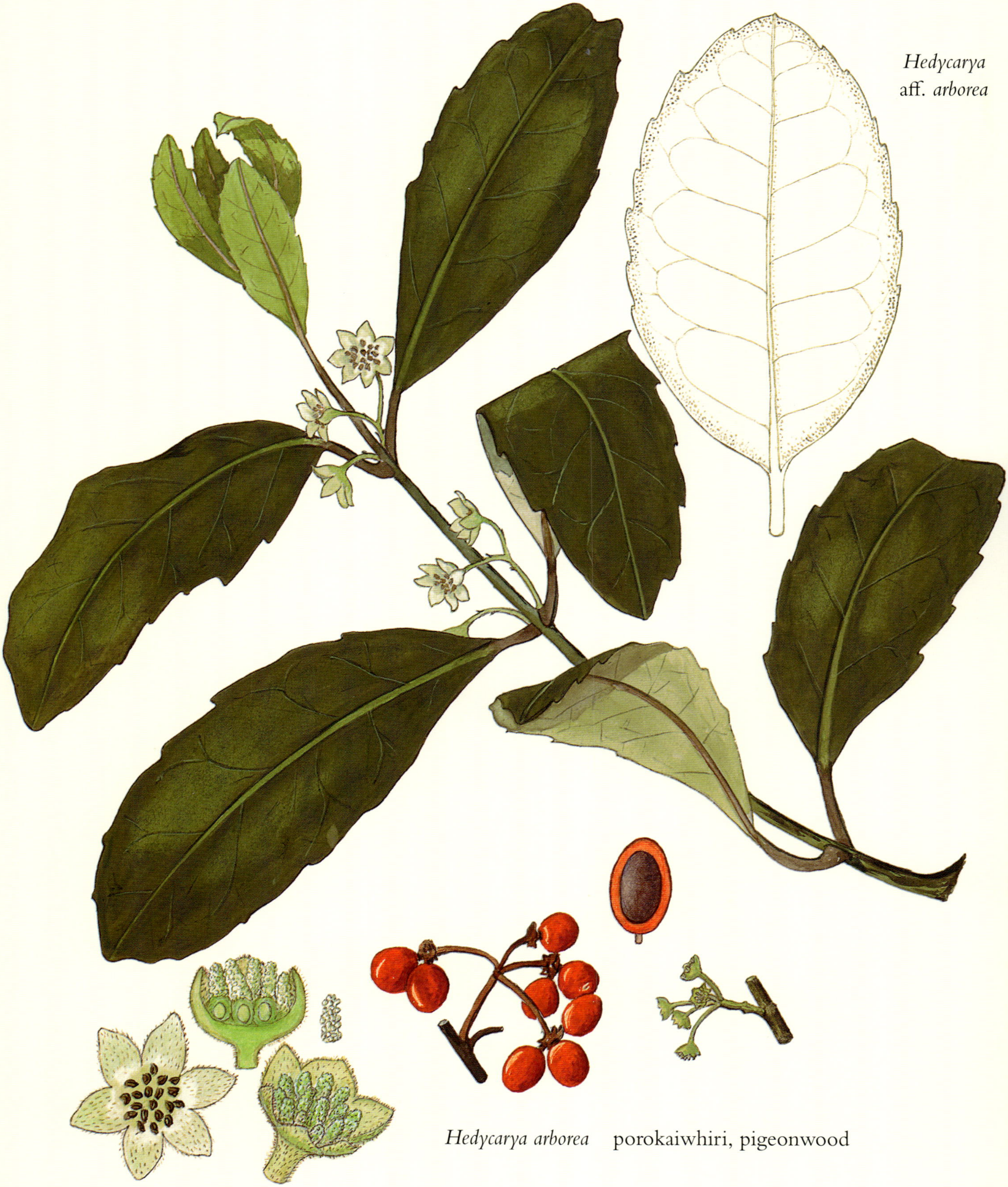

Hedycarya arborea porokaiwhiri, pigeonwood

PLATE 12

Clematis paniculata puawānanga, white clematis

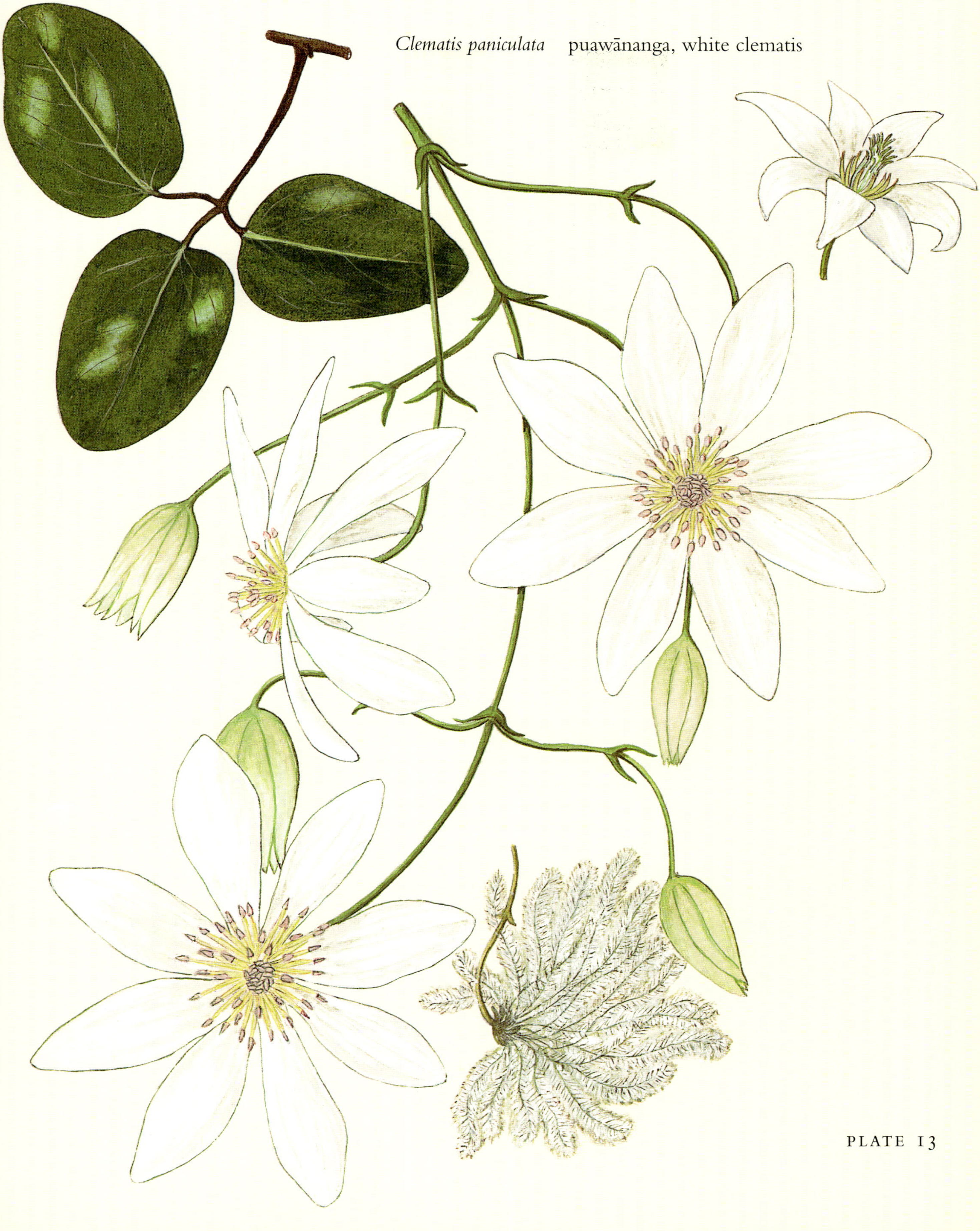

PLATE 13

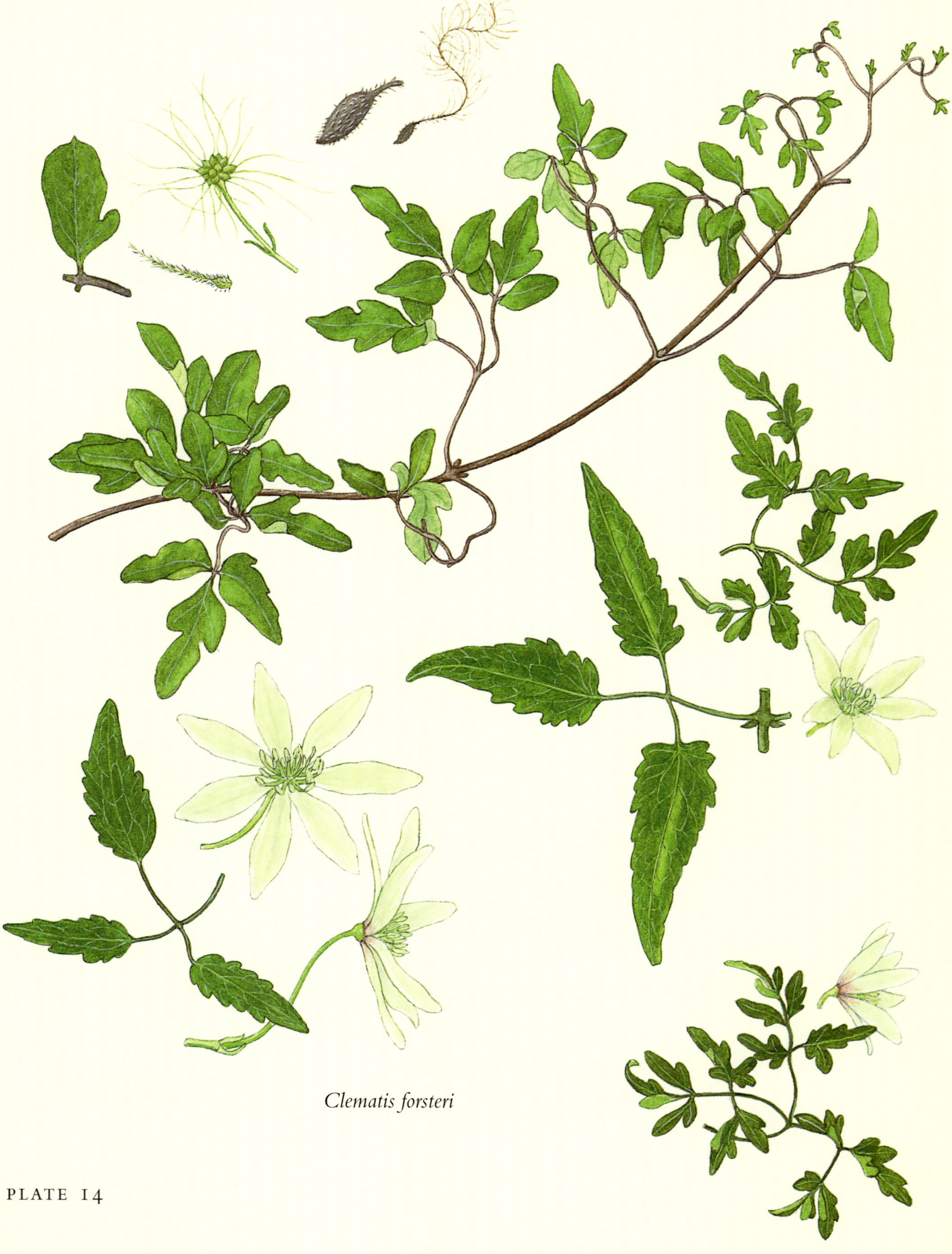

Clematis forsteri

PLATE 14

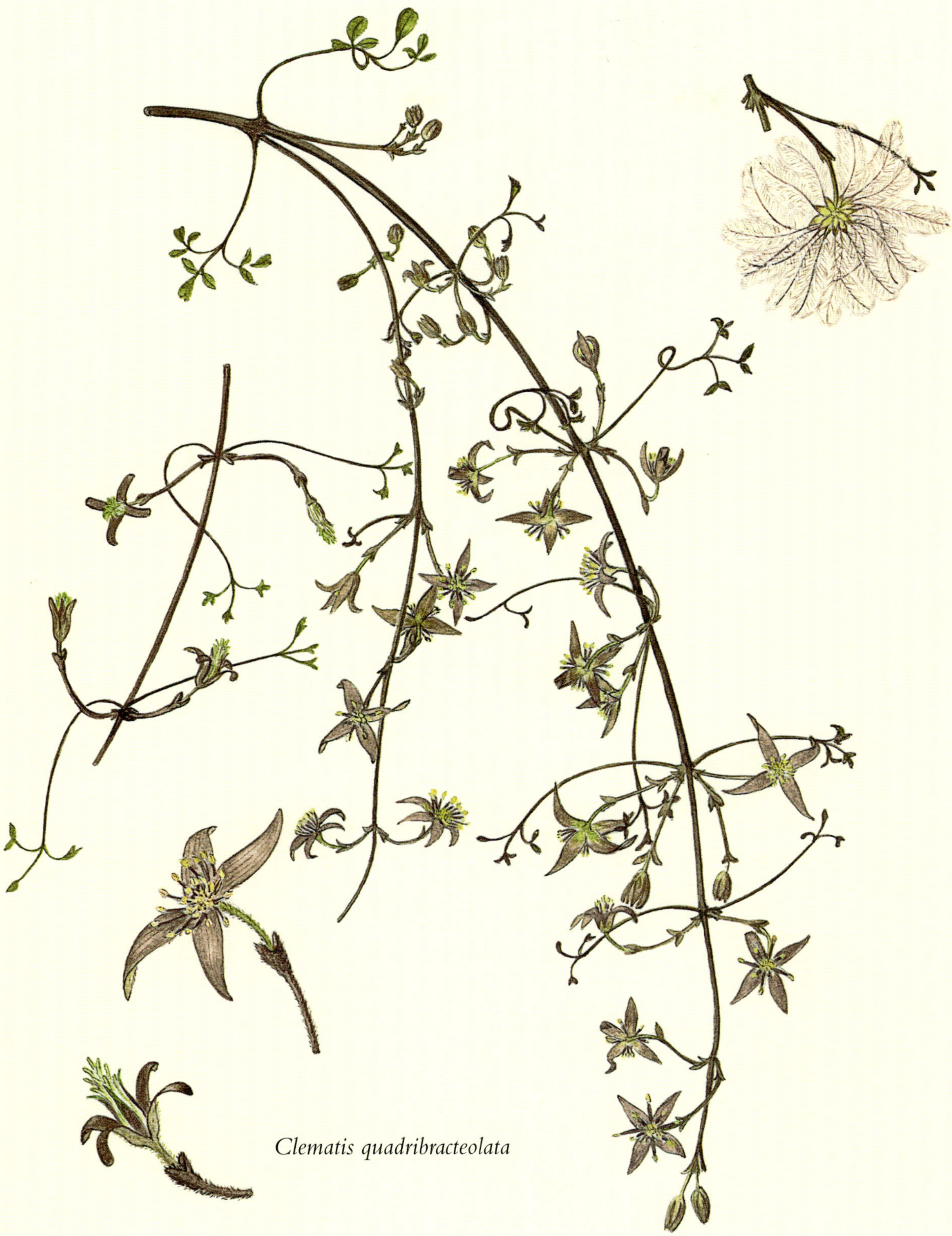

Clematis quadribracteolata

PLATE 15

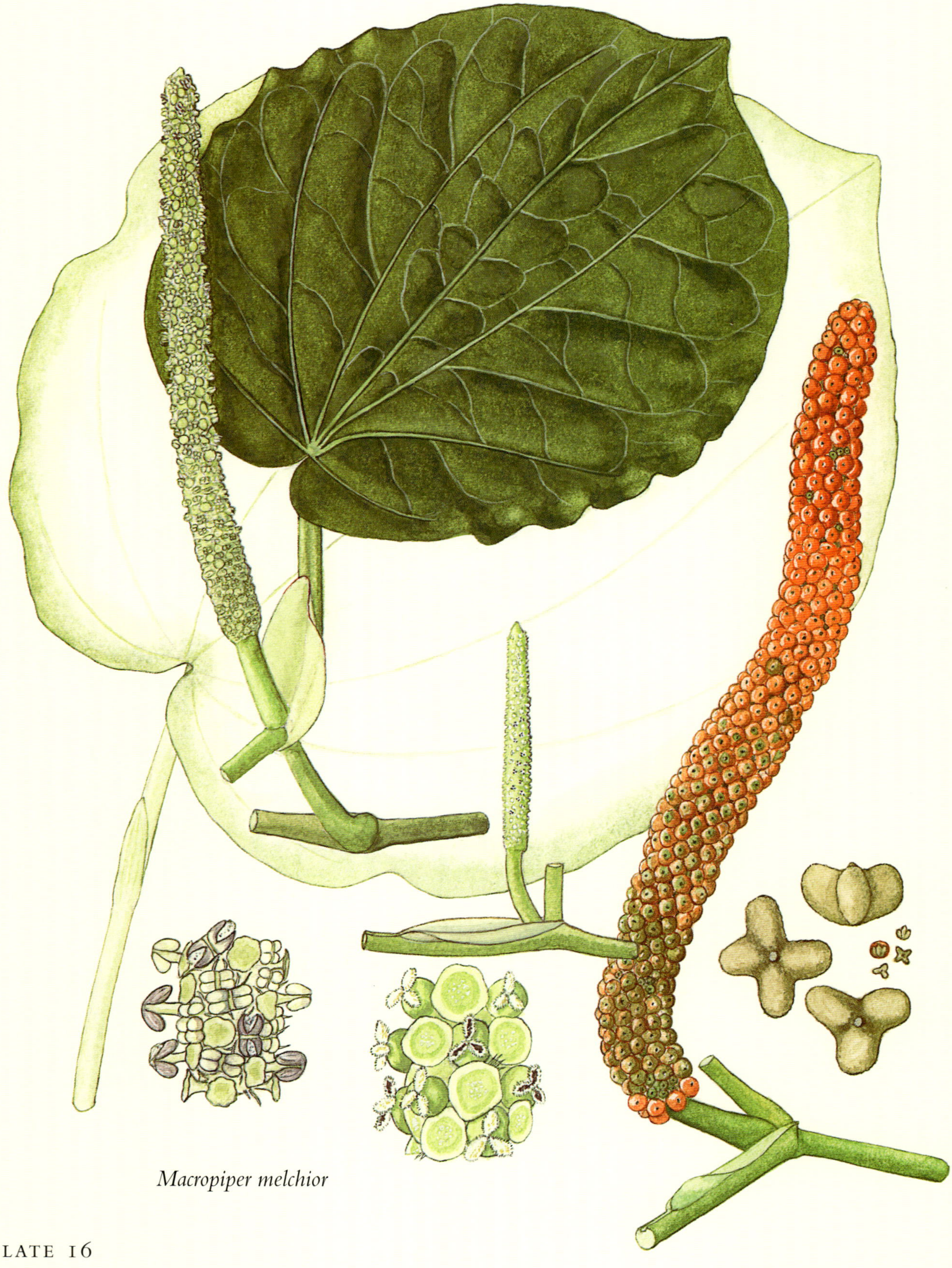

Macropiper melchior

PLATE 16

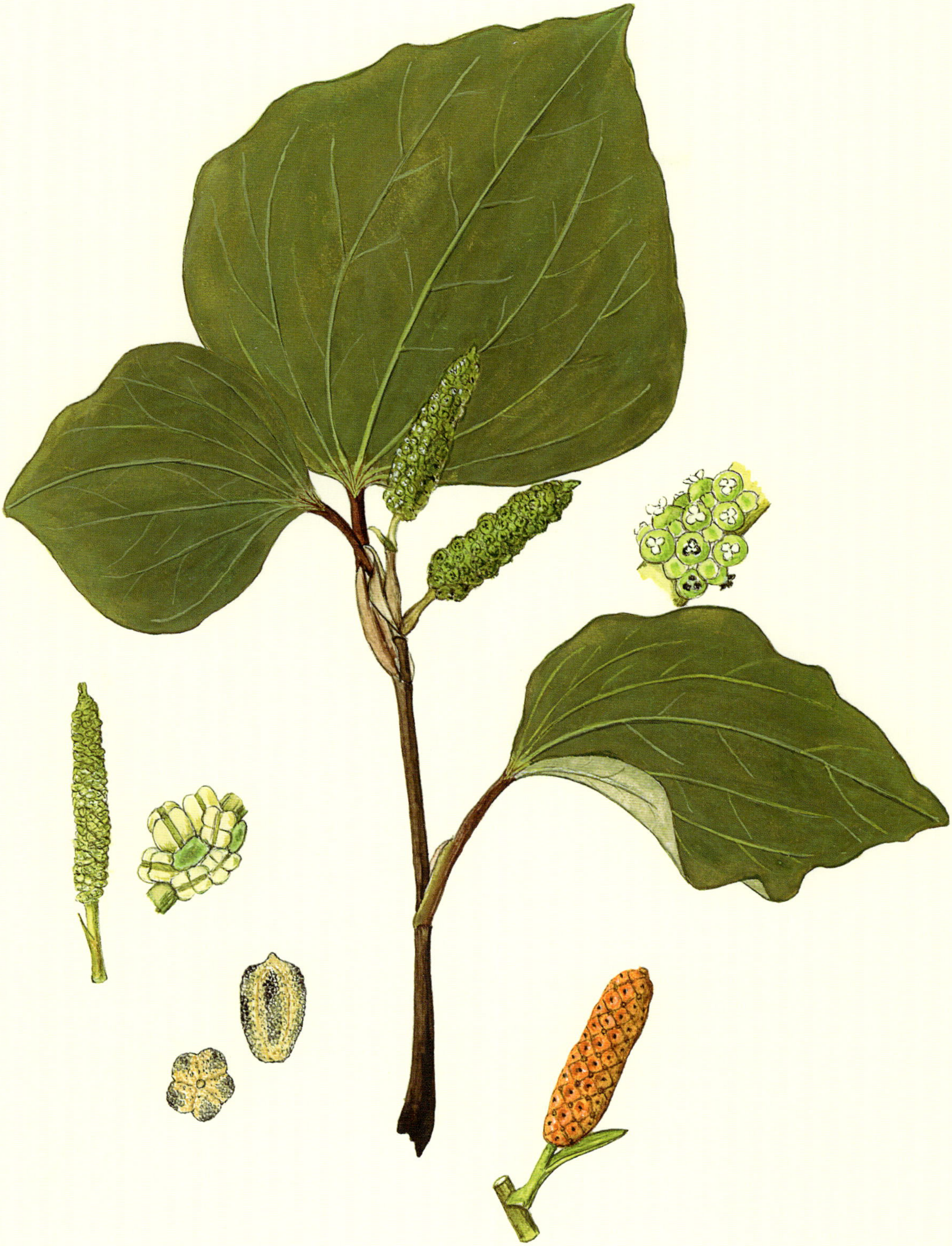

Macropiper excelsum subspecies *excelsum*   kawakawa, pepper tree

PLATE 17

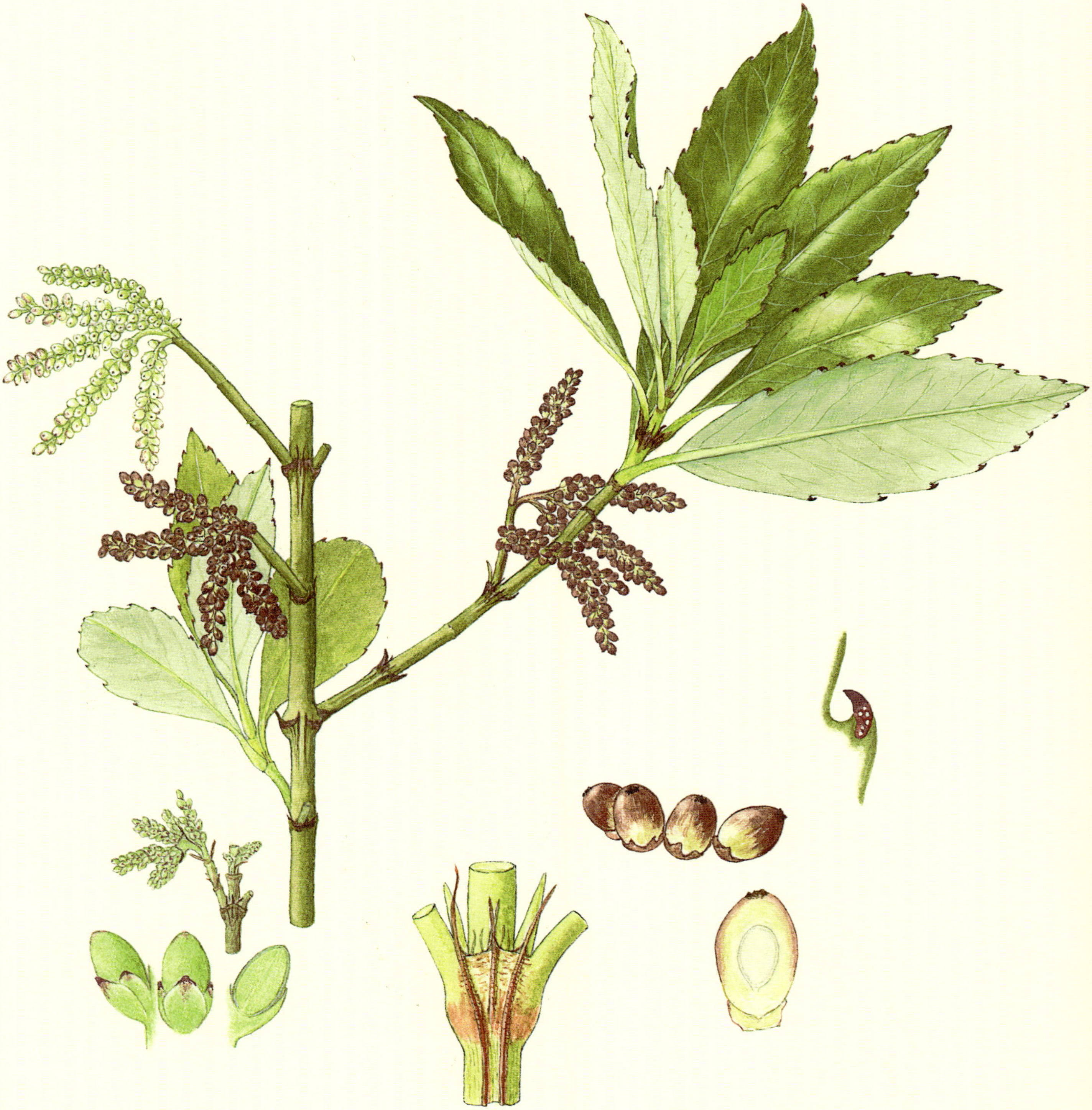

Ascarina lucida var. *lanceolata* hutu

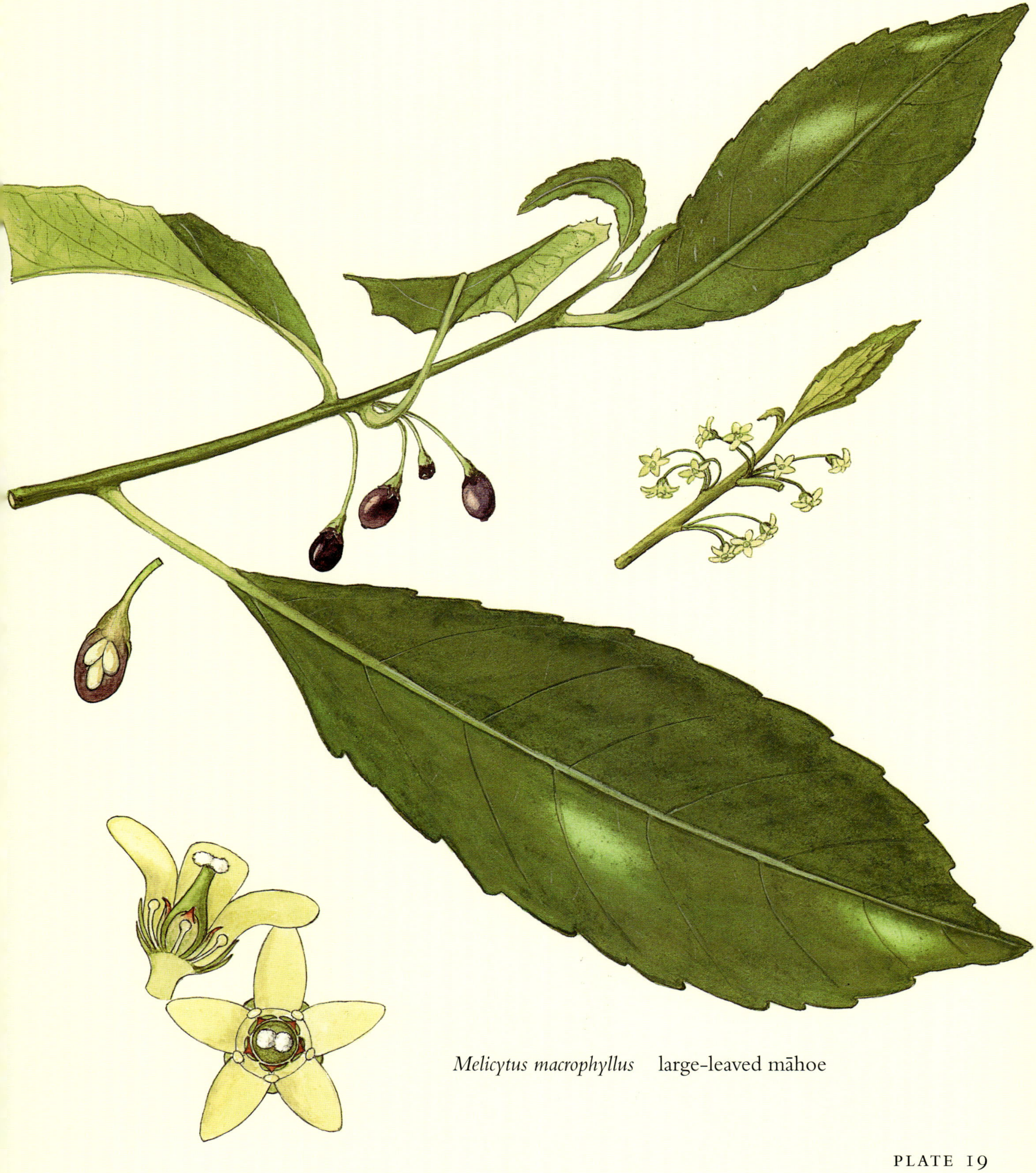

Melicytus macrophyllus large-leaved māhoe

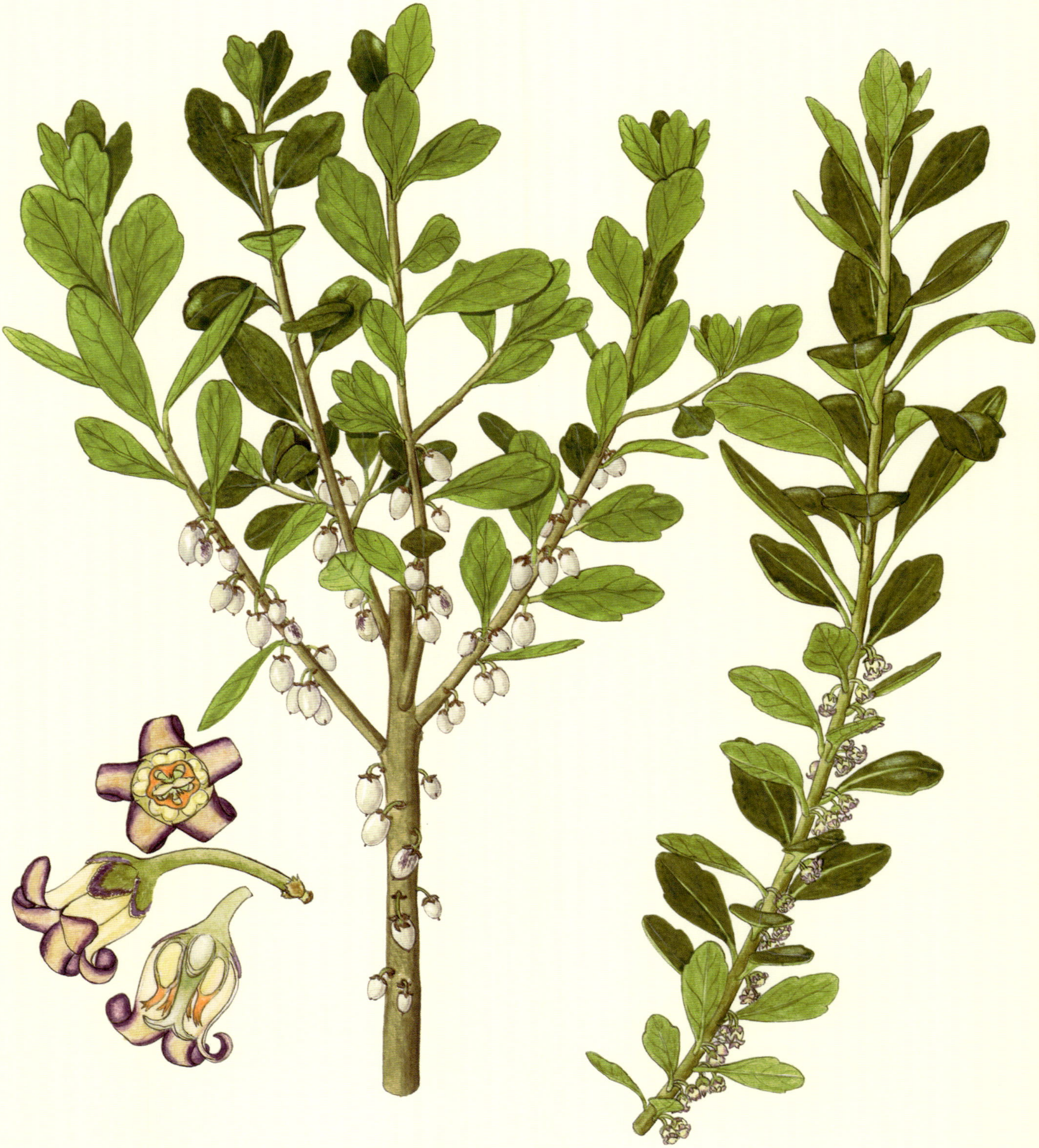

Melicytus aff. *obovatus*

PLATE 20

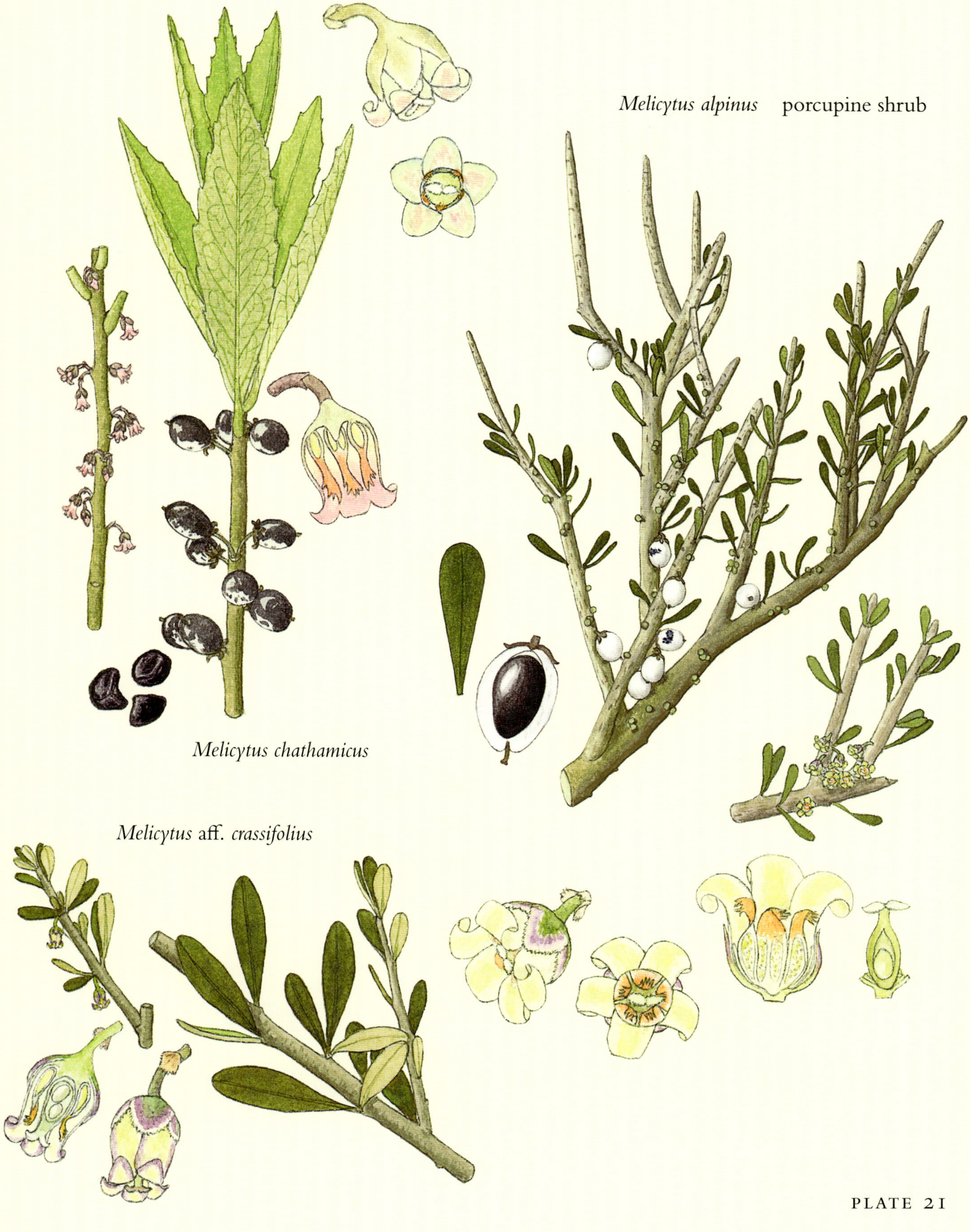

Melicytus alpinus porcupine shrub

Melicytus chathamicus

Melicytus aff. *crassifolius*

PLATE 21

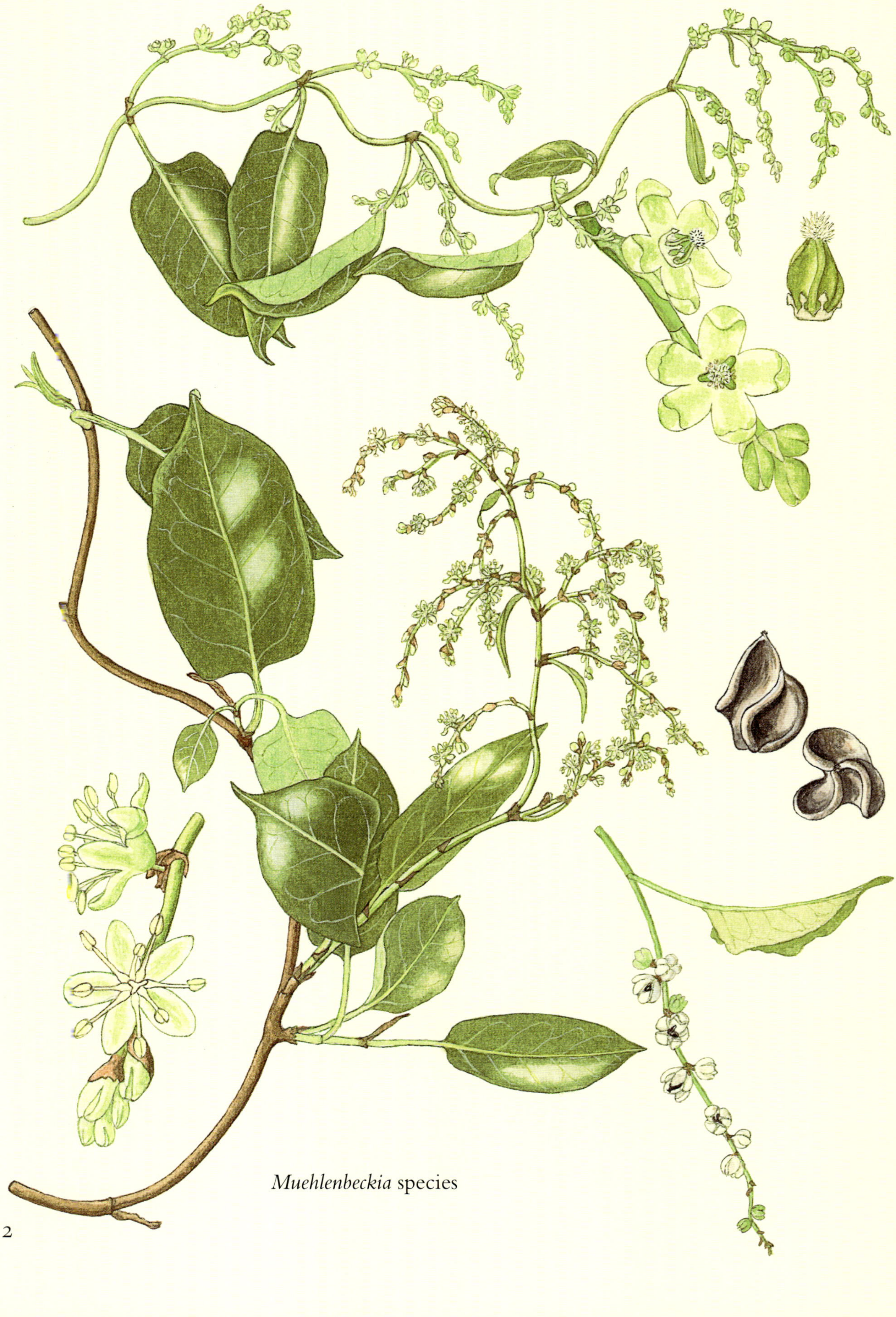

Muehlenbeckia species

PLATE 22

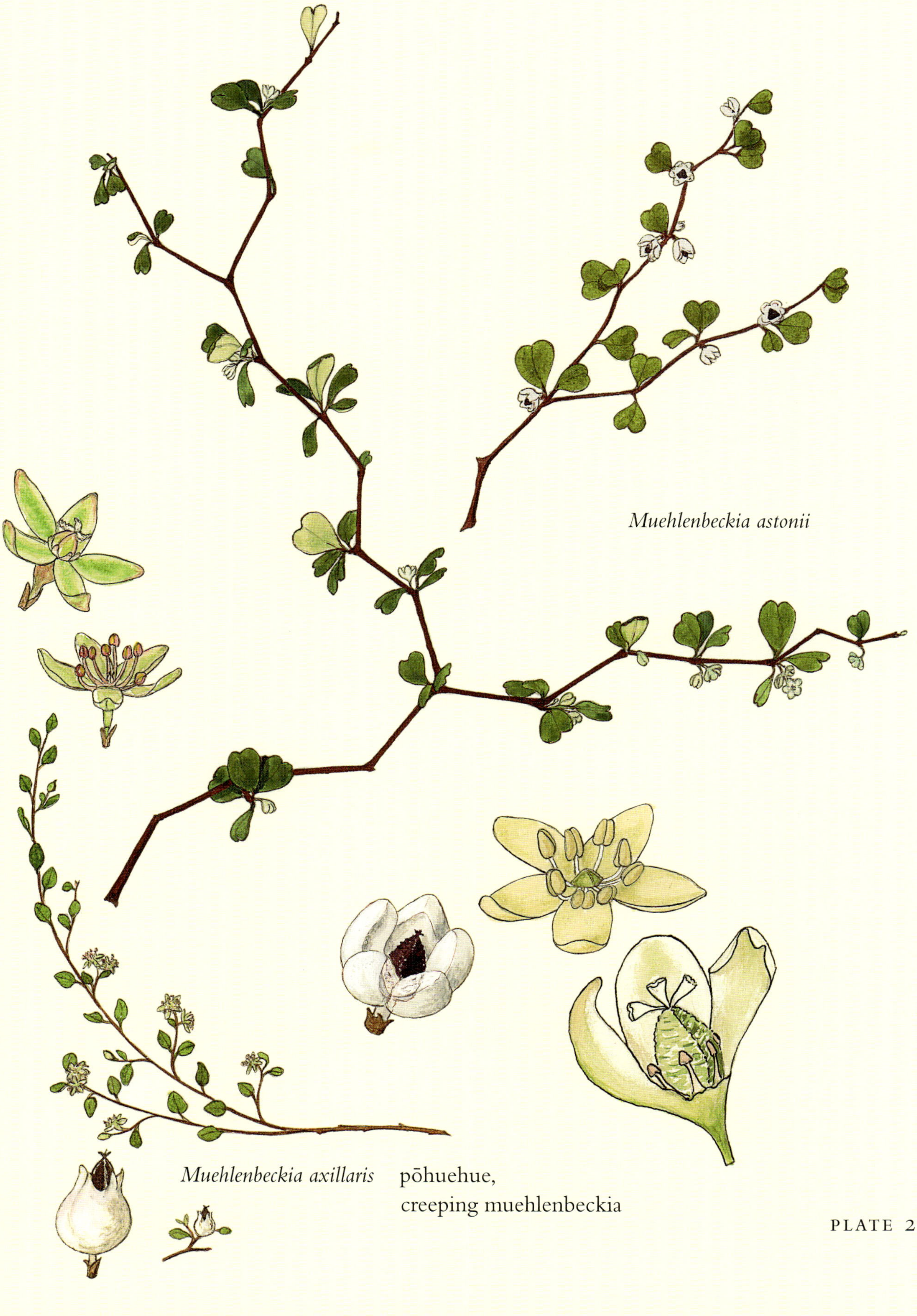

Muehlenbeckia astonii

Muehlenbeckia axillaris pōhuehue,
creeping muehlenbeckia

PLATE 23

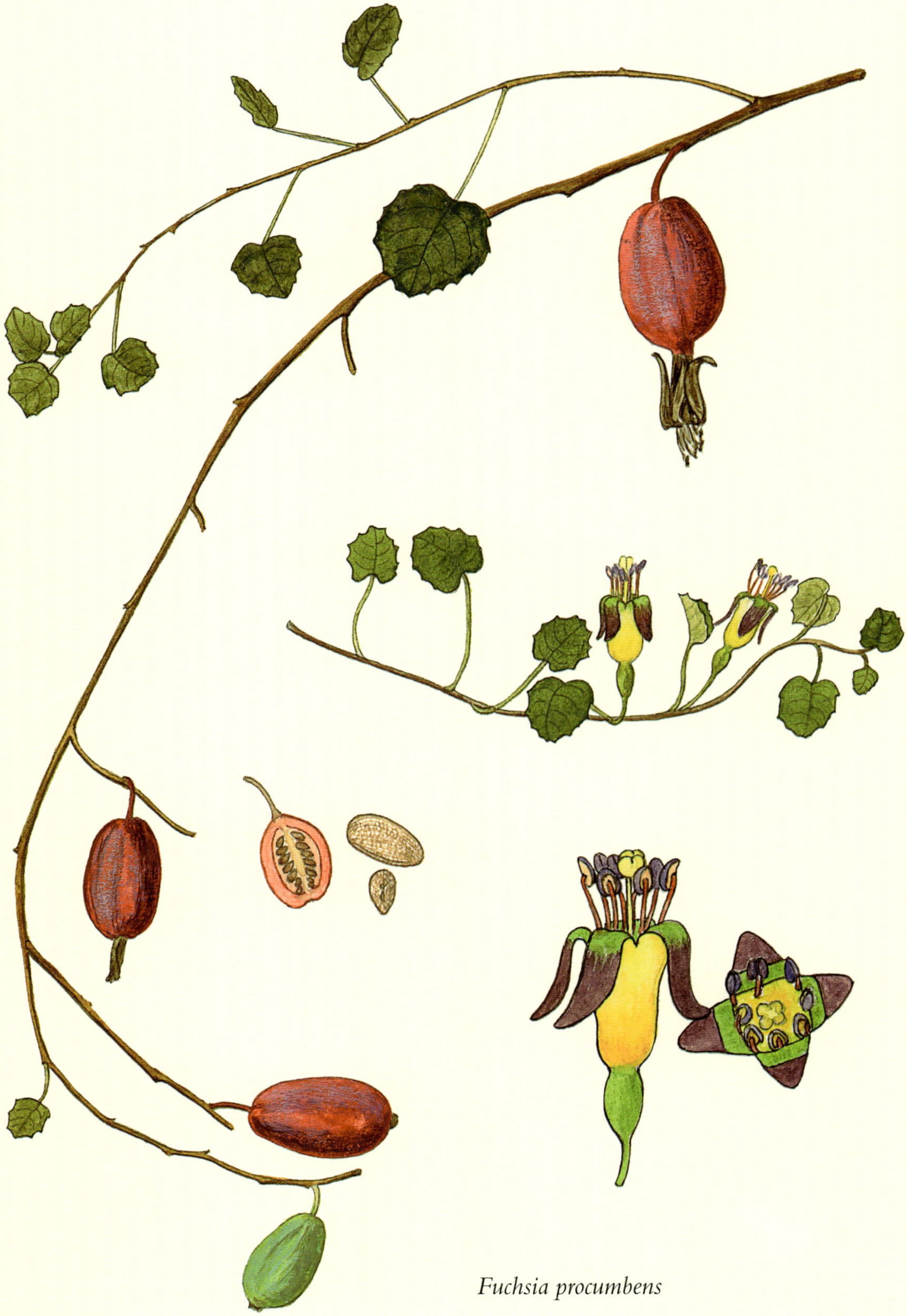

Fuchsia procumbens

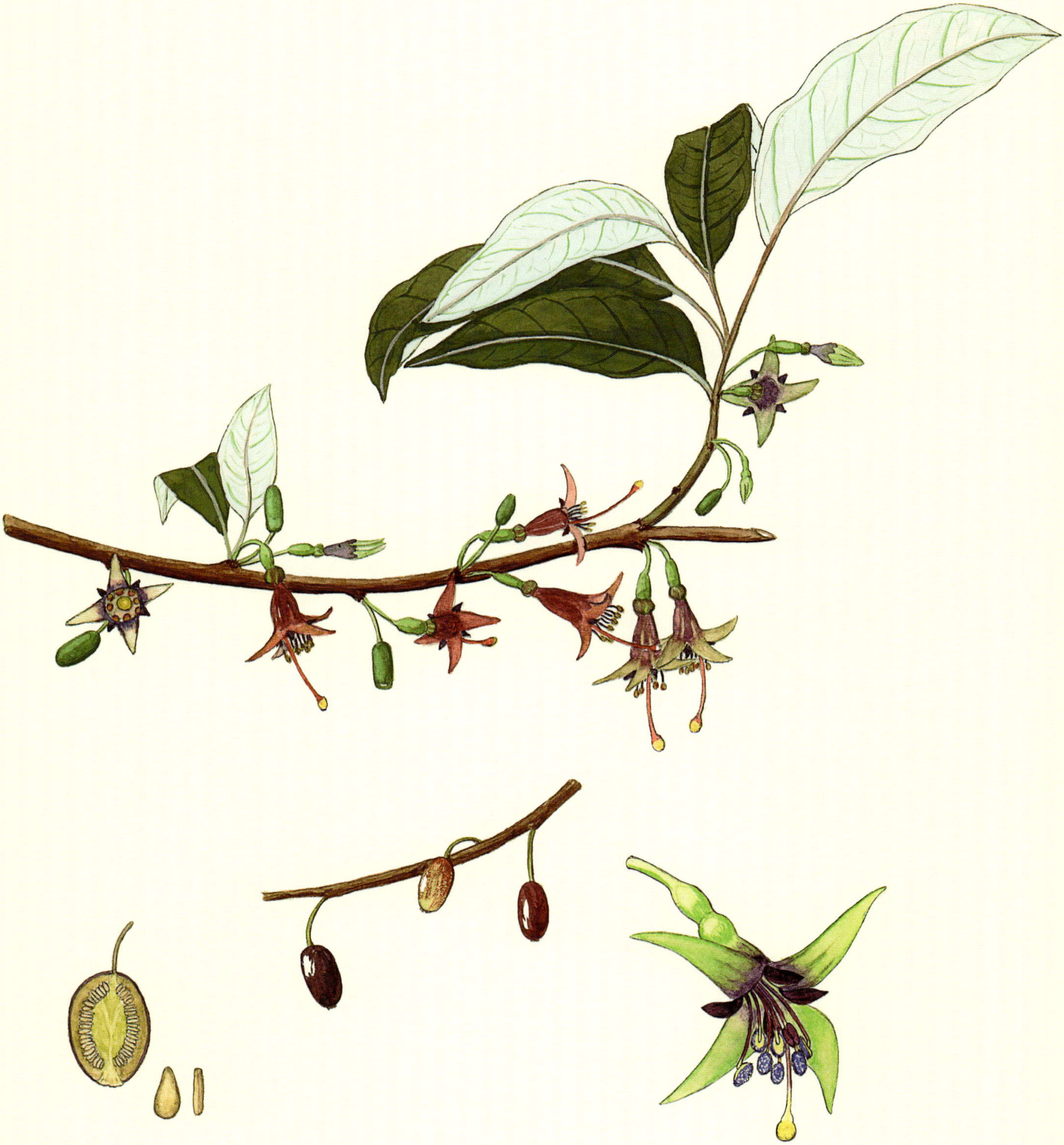

Fuchsia excorticata kōtukutuku, kōnini (berry)

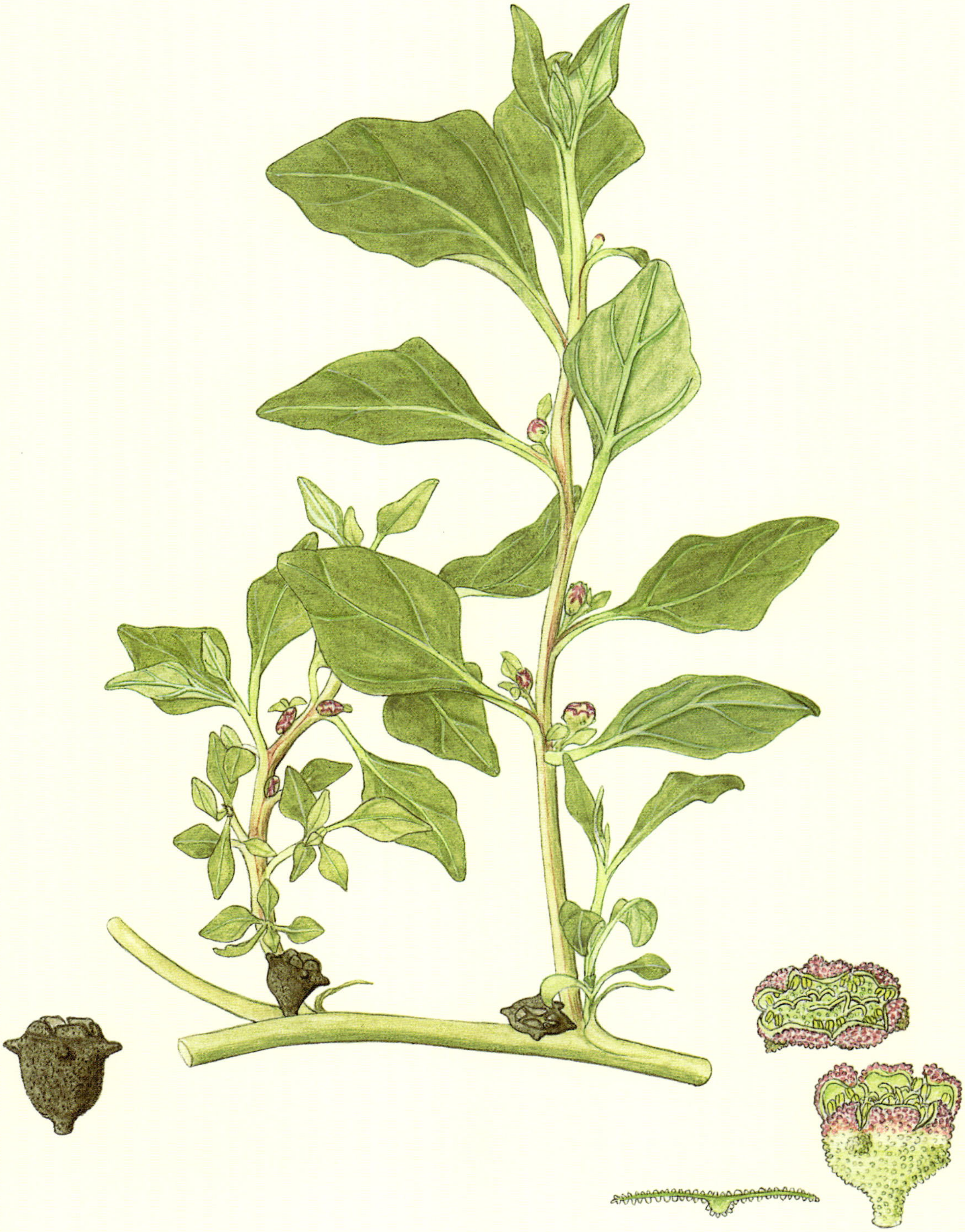

Tetragonia tetragonioides kōkihi, rengamutu, rengarenga, tūtae-ika-moana, New Zealand spinach

PLATE 26

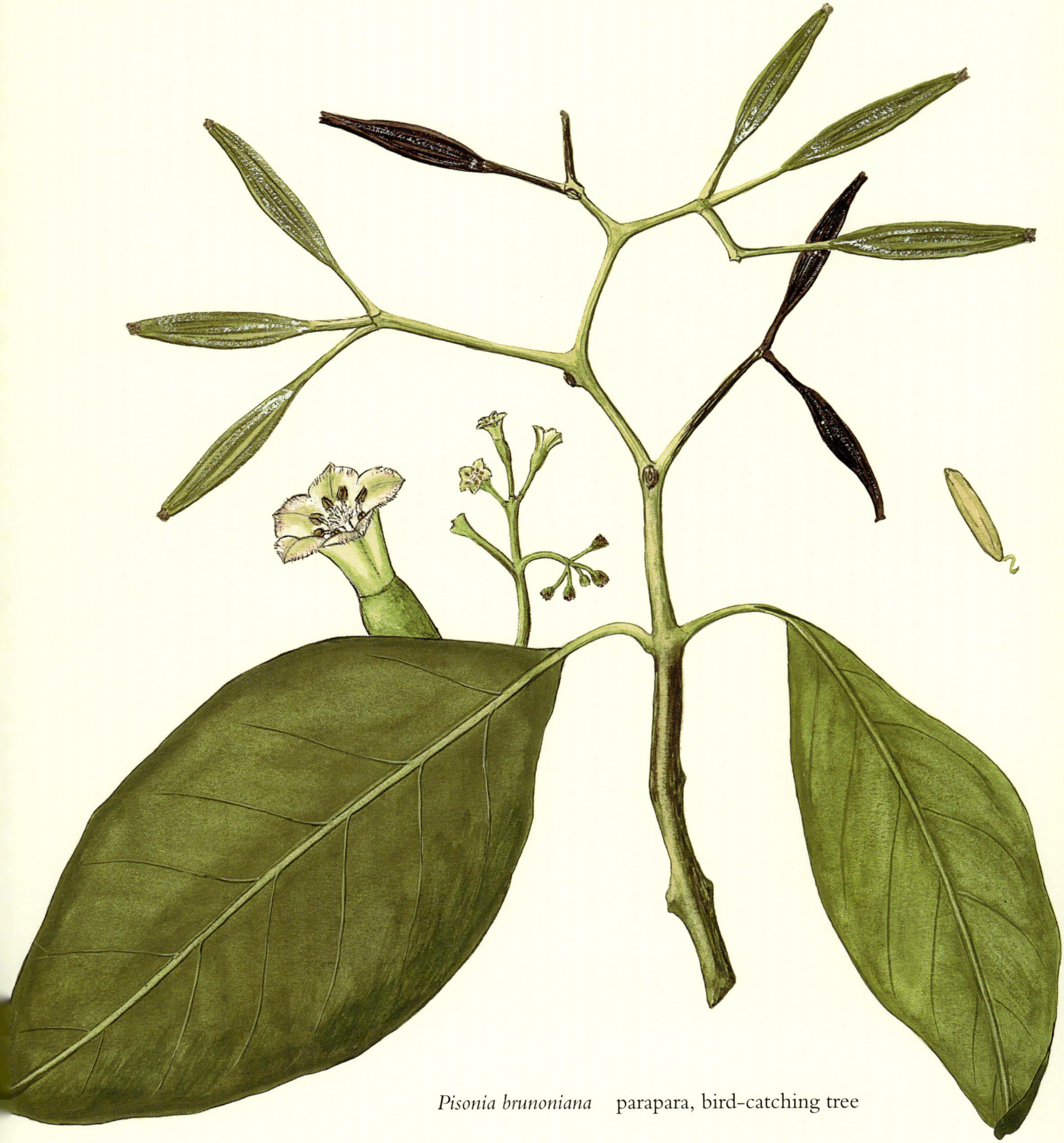

Pisonia brunoniana parapara, bird-catching tree

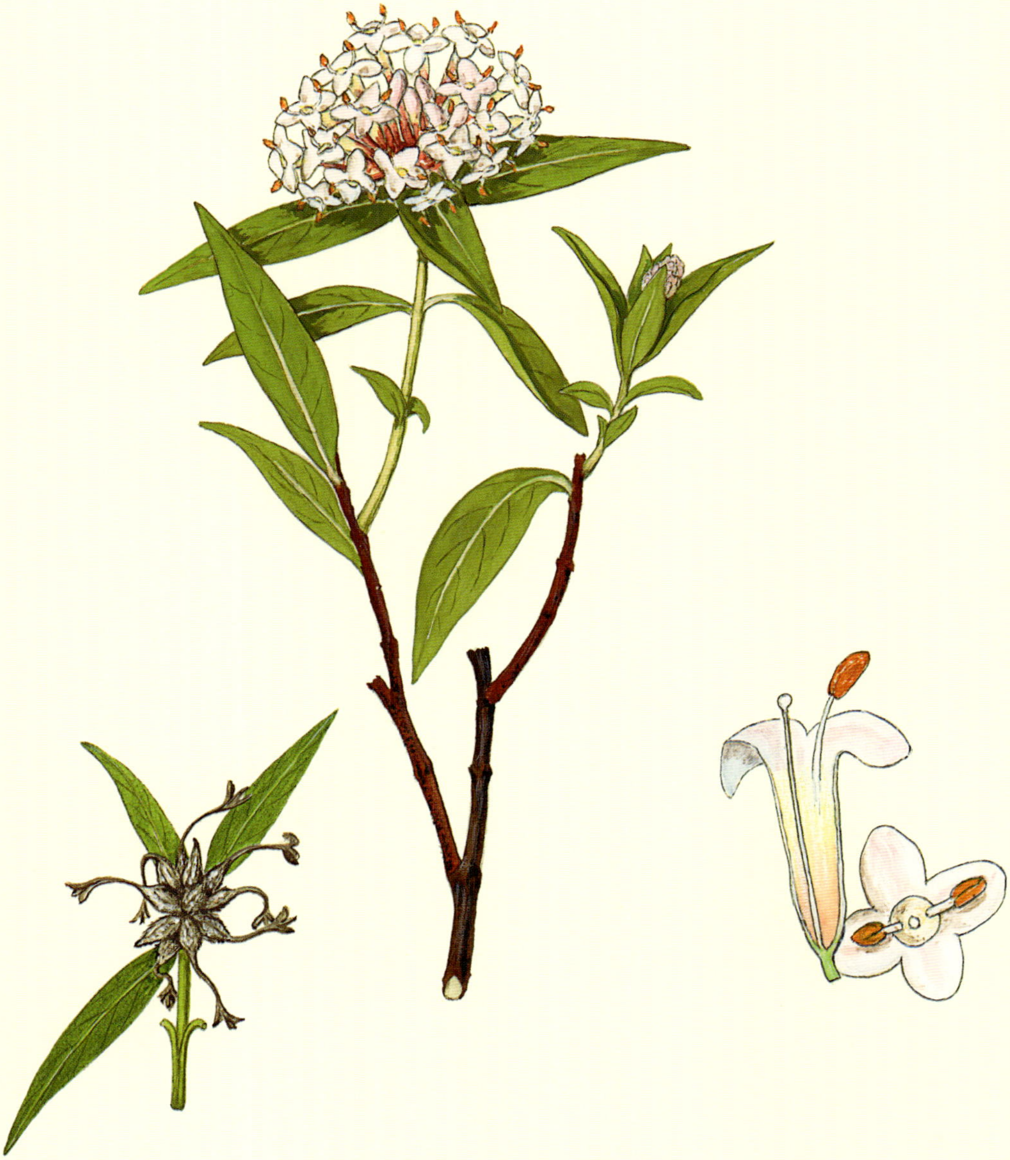

Pimelea longifolia tāranga

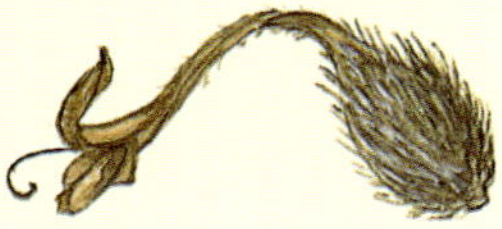

PLATE 28

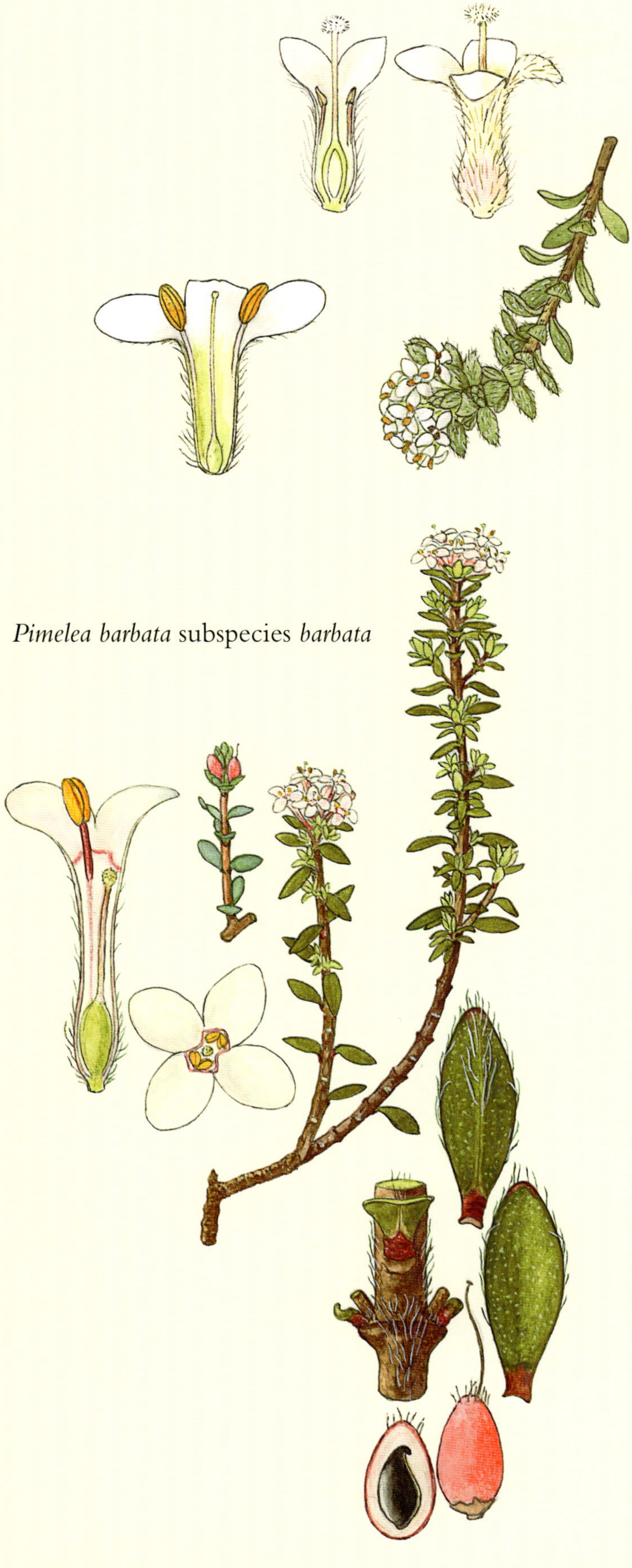

Pimelea barbata subspecies *barbata*

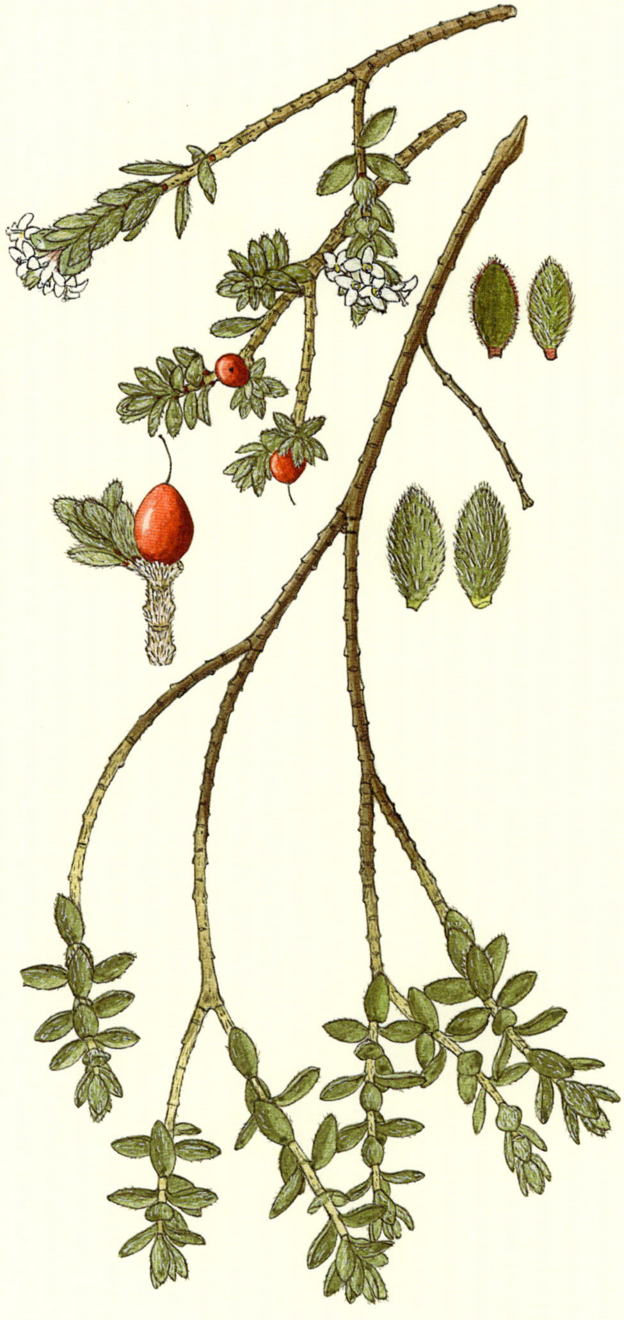

Pimelea species

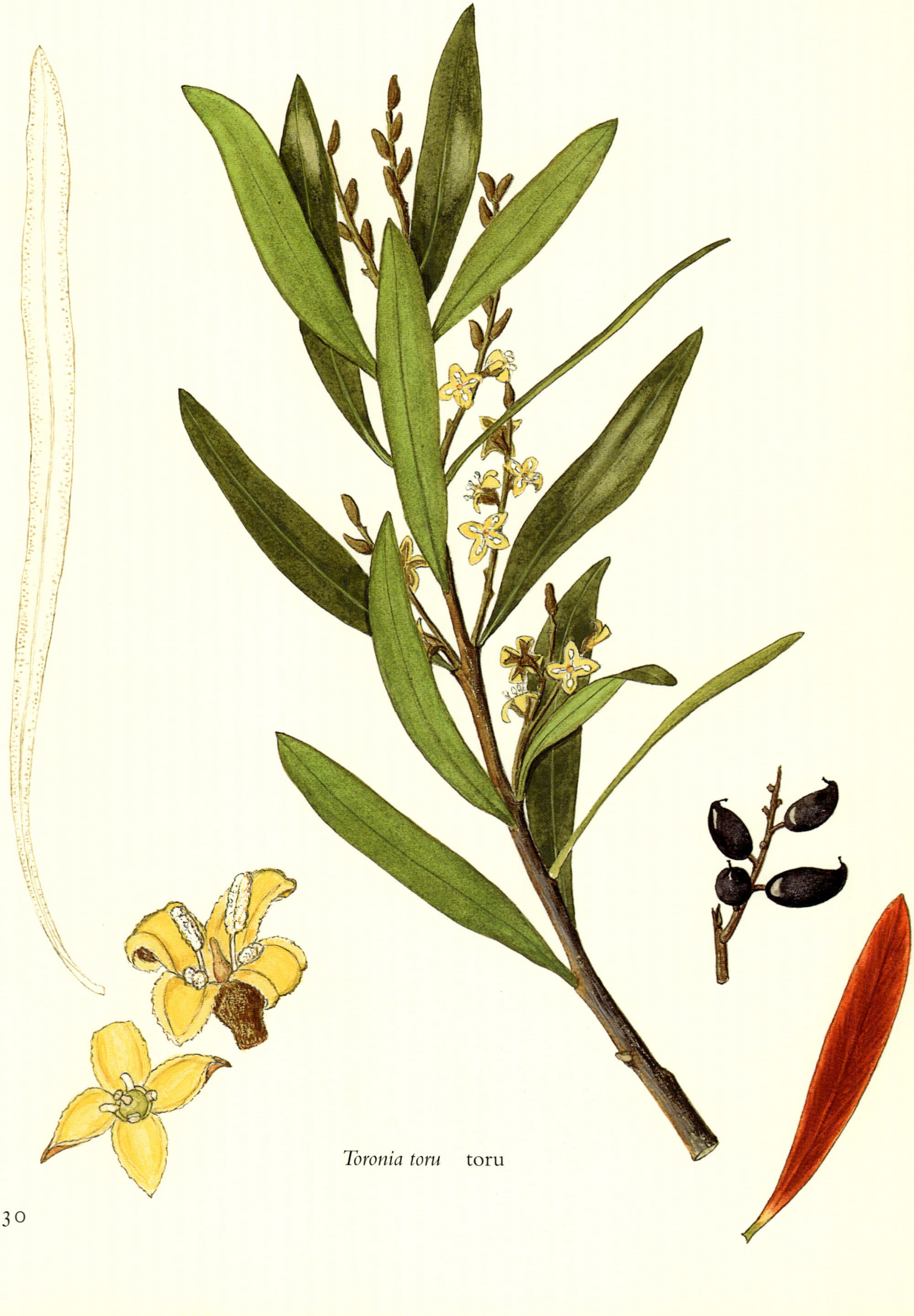

Toronia toru toru

PLATE 30

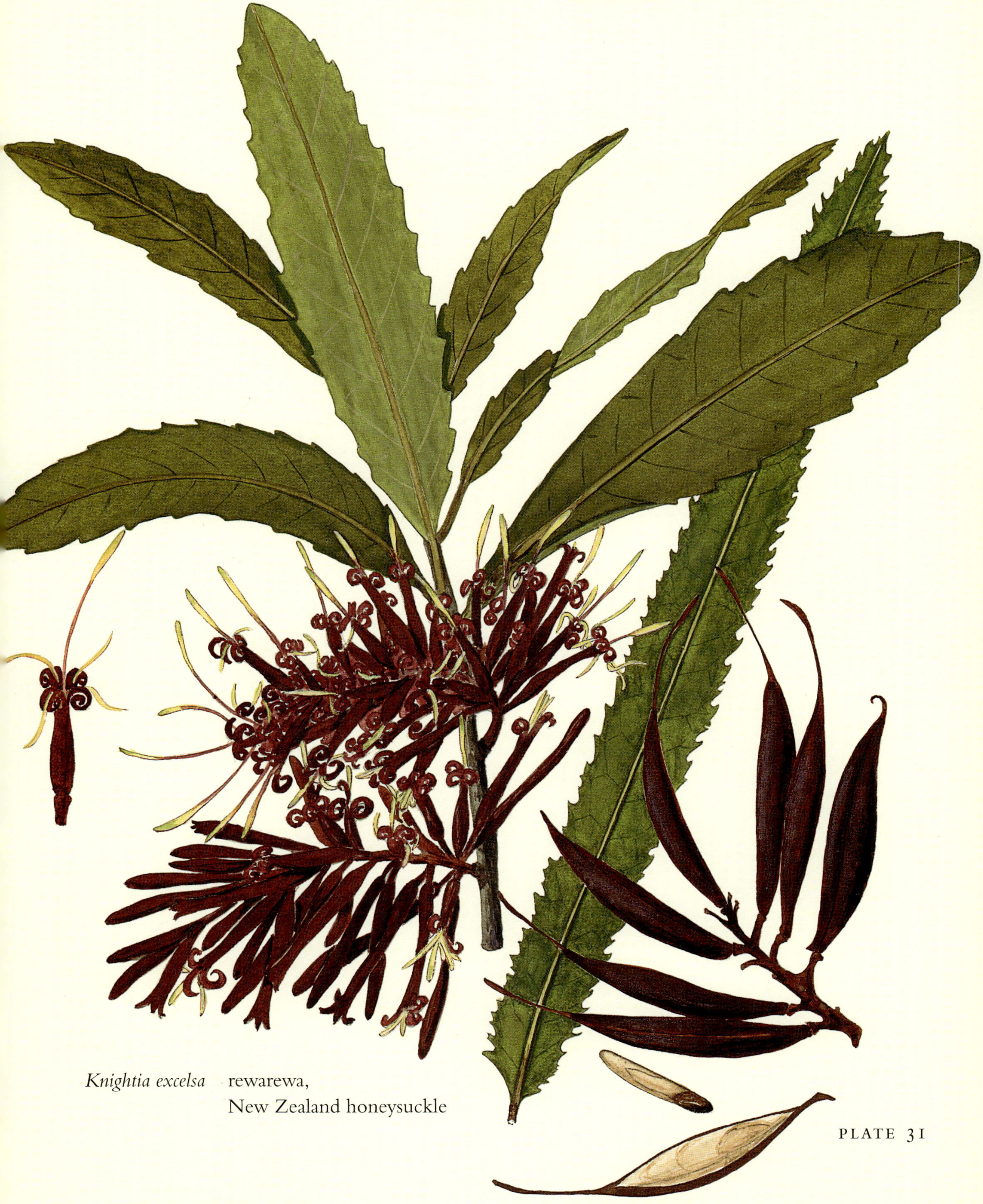

Knightia excelsa rewarewa,
New Zealand honeysuckle

PLATE 31

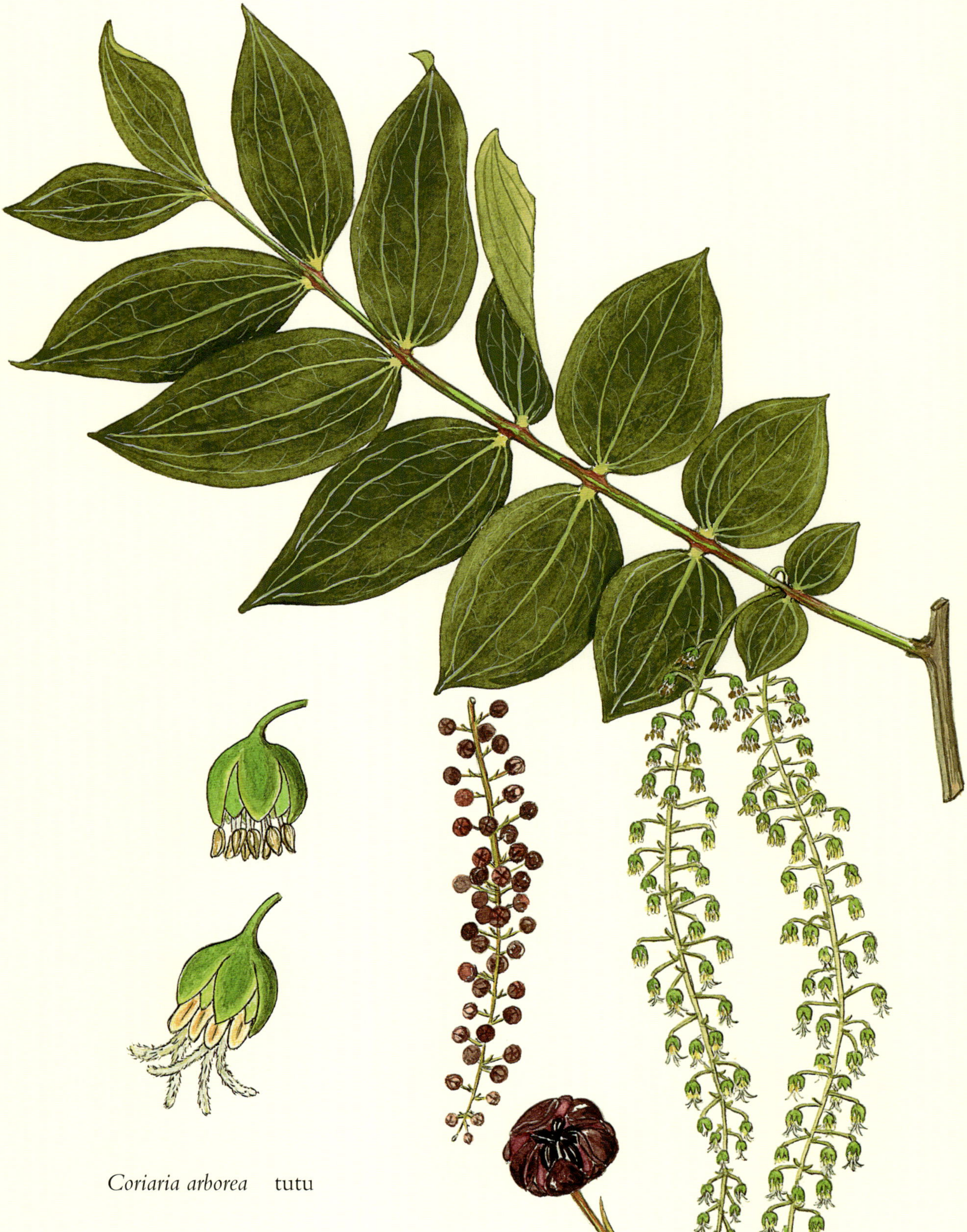

Coriaria arborea tutu

PLATE 32

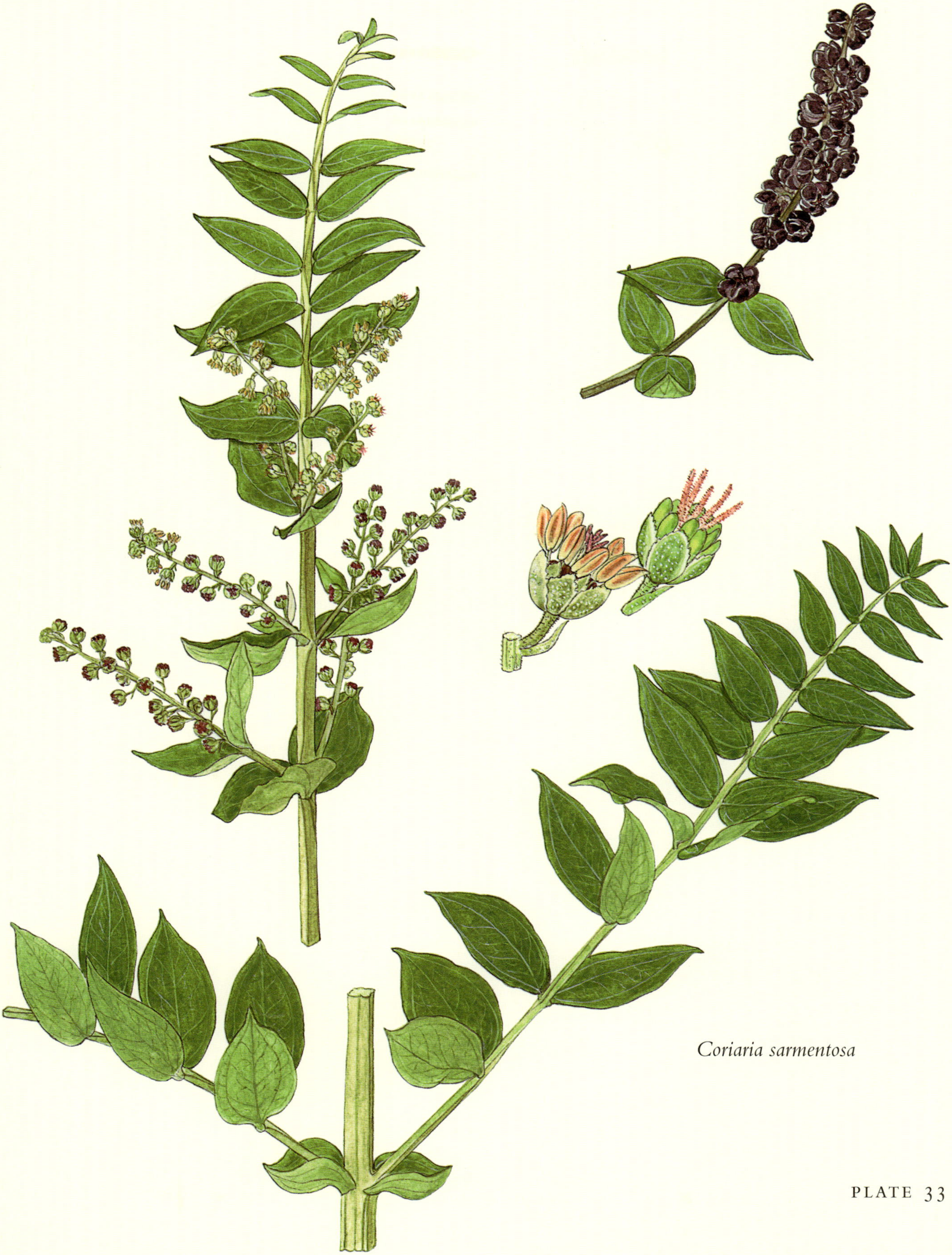

Coriaria sarmentosa

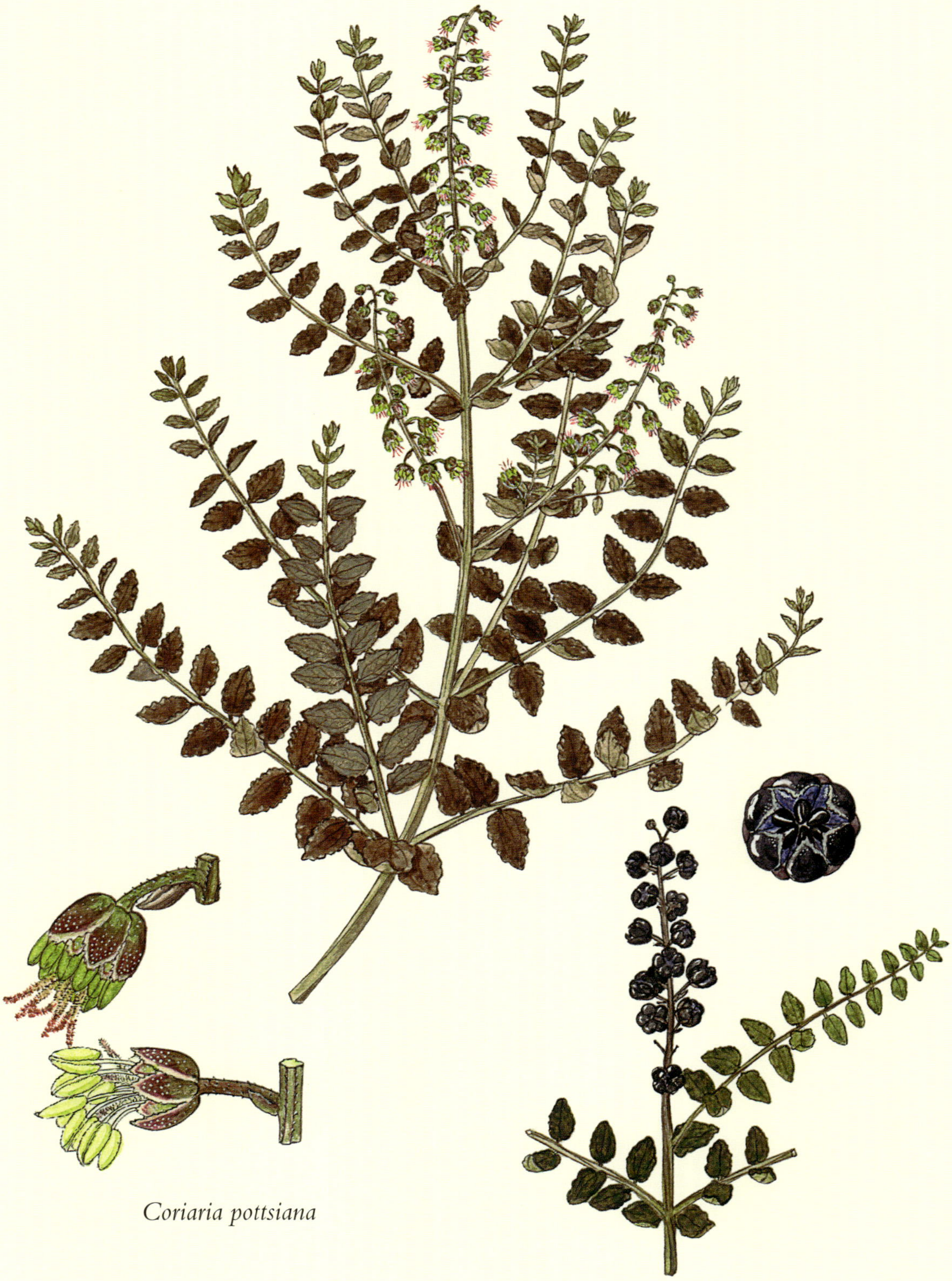

Coriaria pottsiana

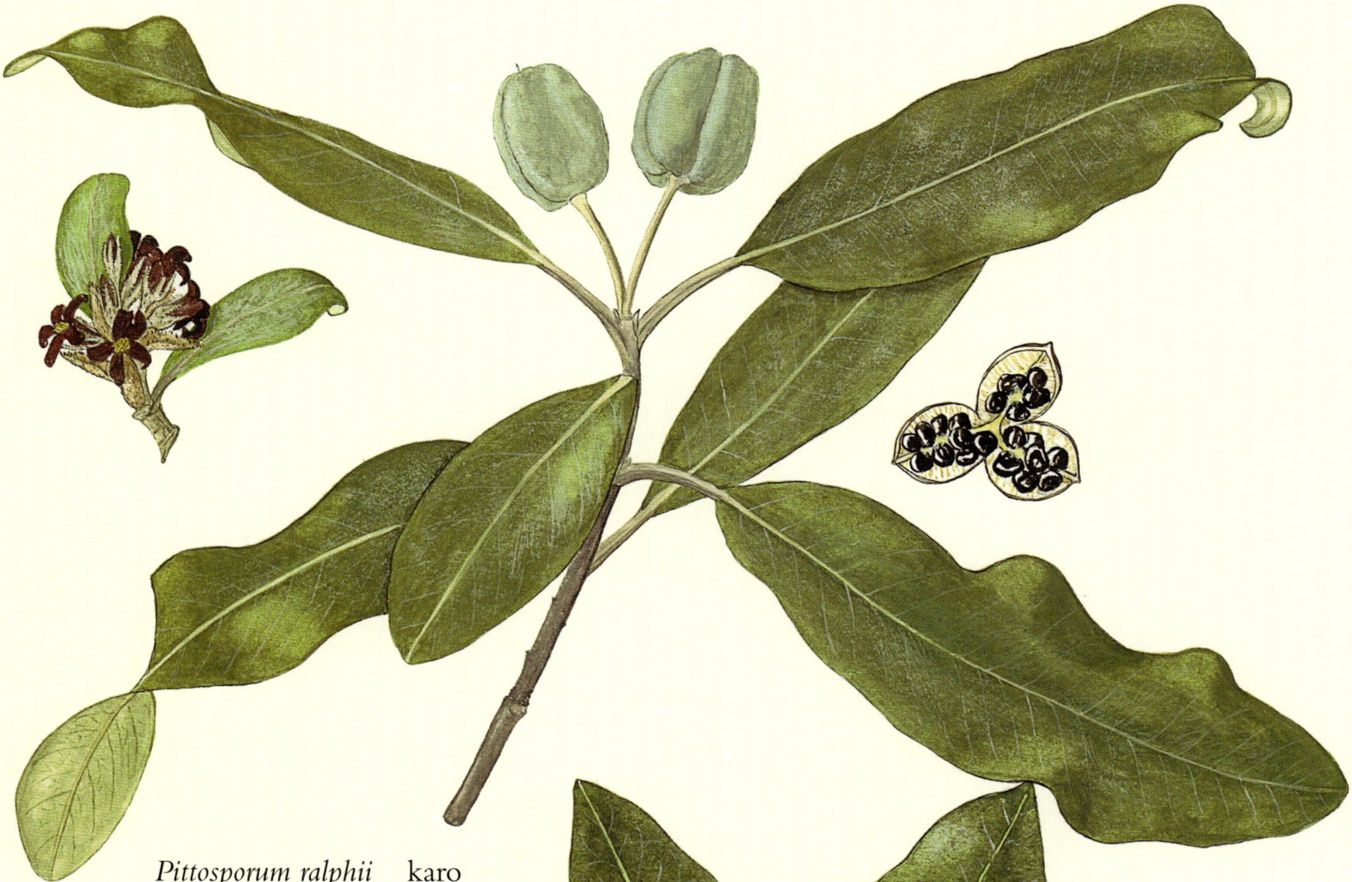

Pittosporum ralphii karo

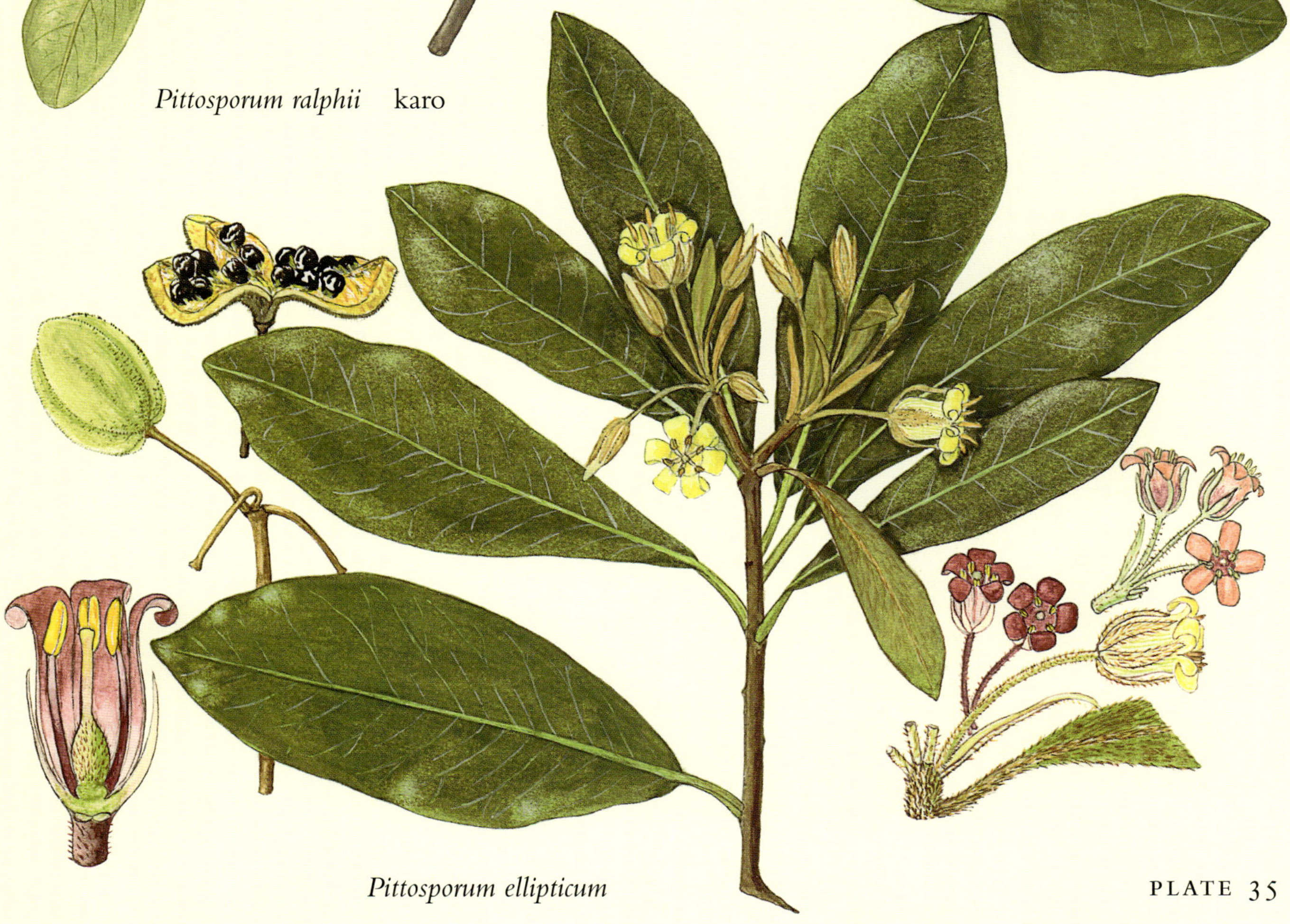

Pittosporum ellipticum

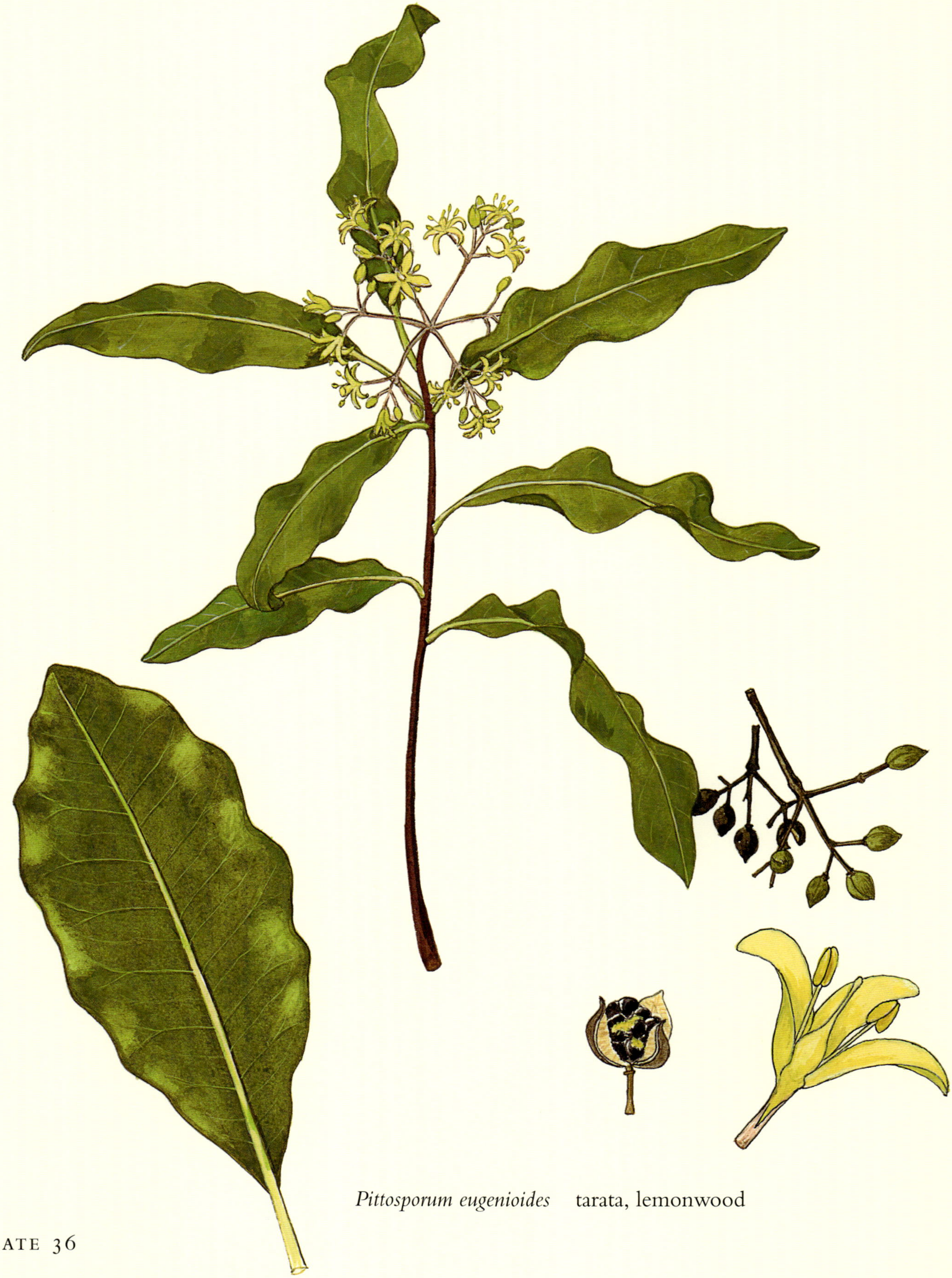

Pittosporum eugenioides tarata, lemonwood

PLATE 36

Pittosporum aff. *crassifolium*

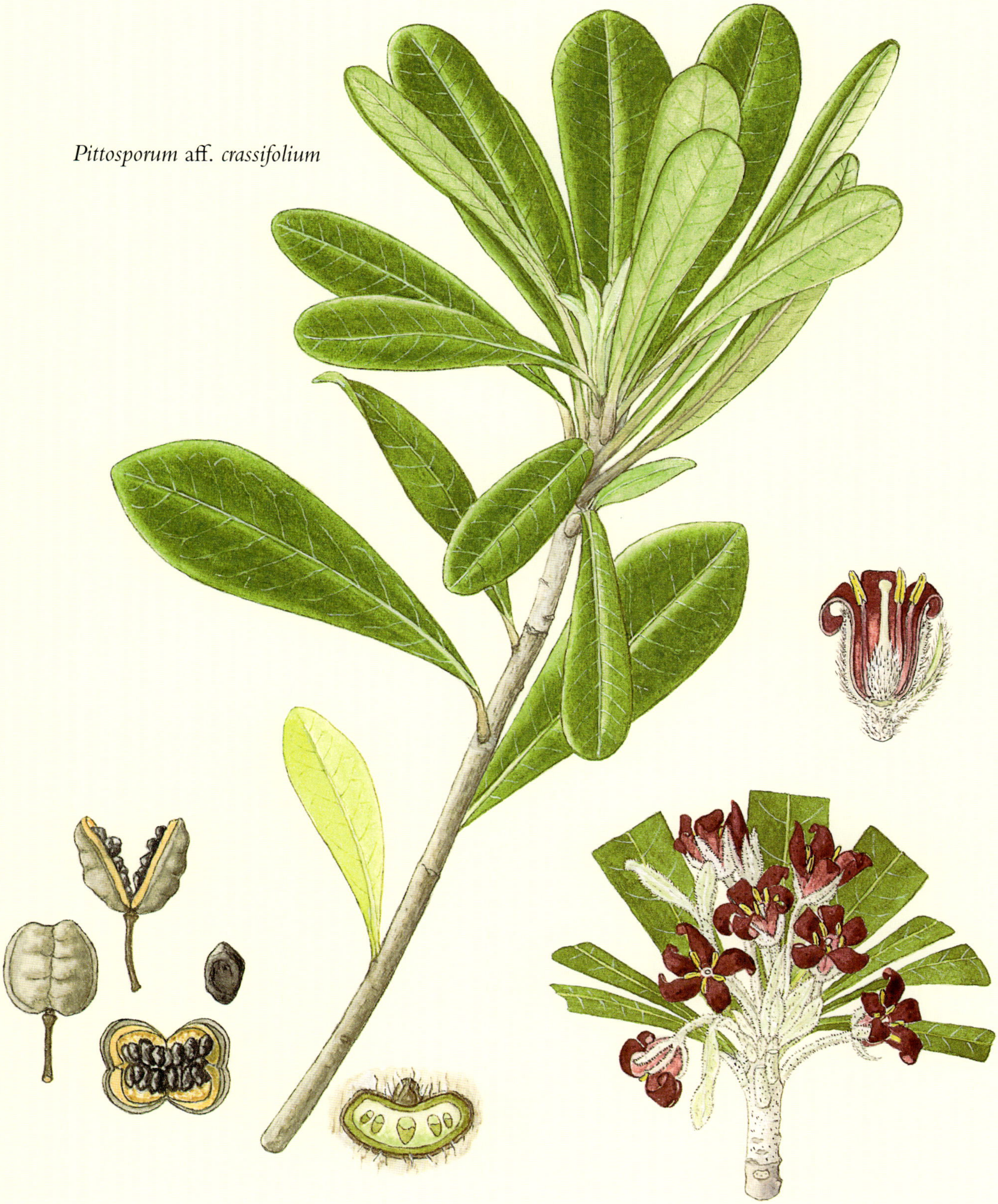

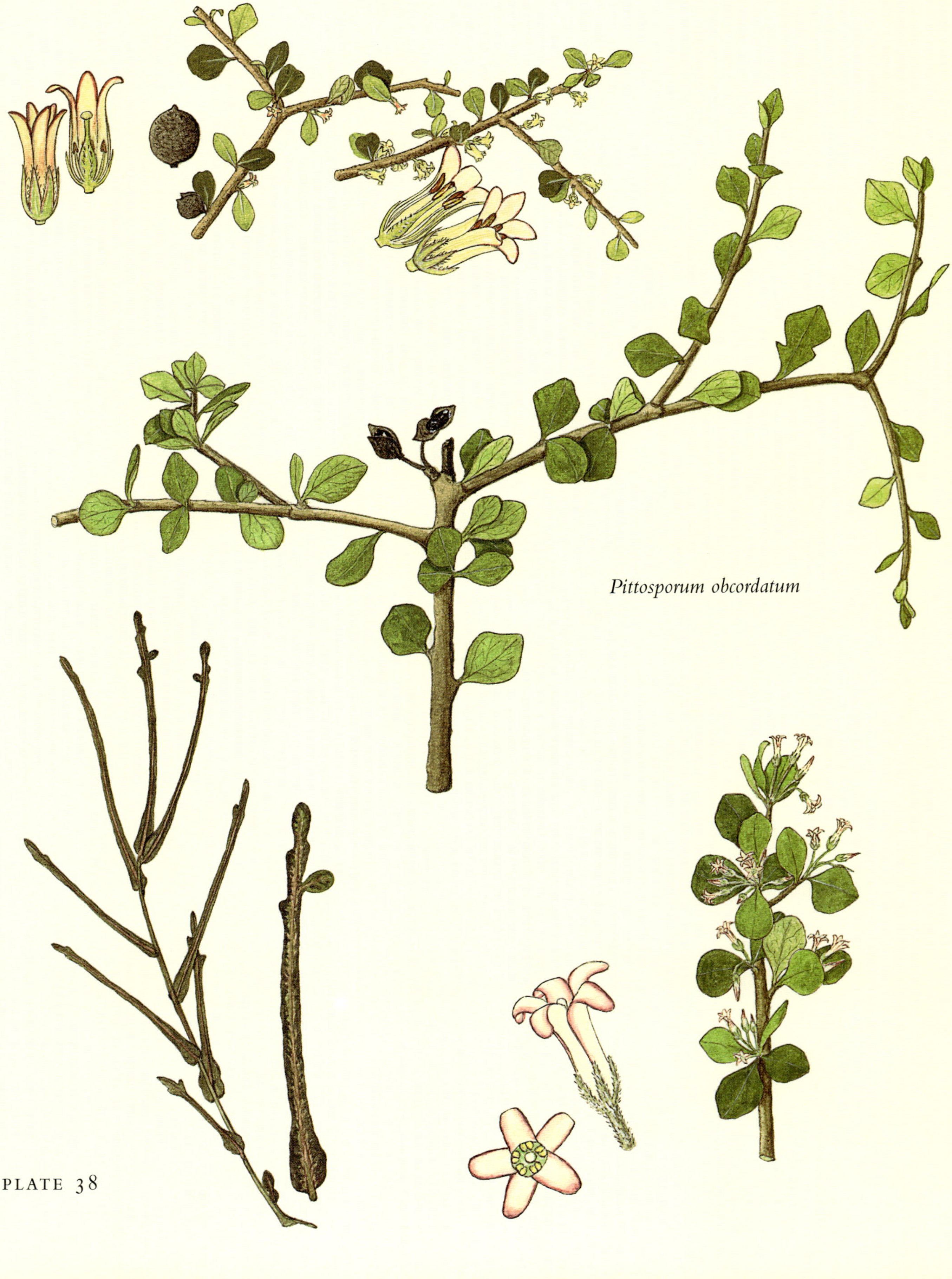

PLATE 38

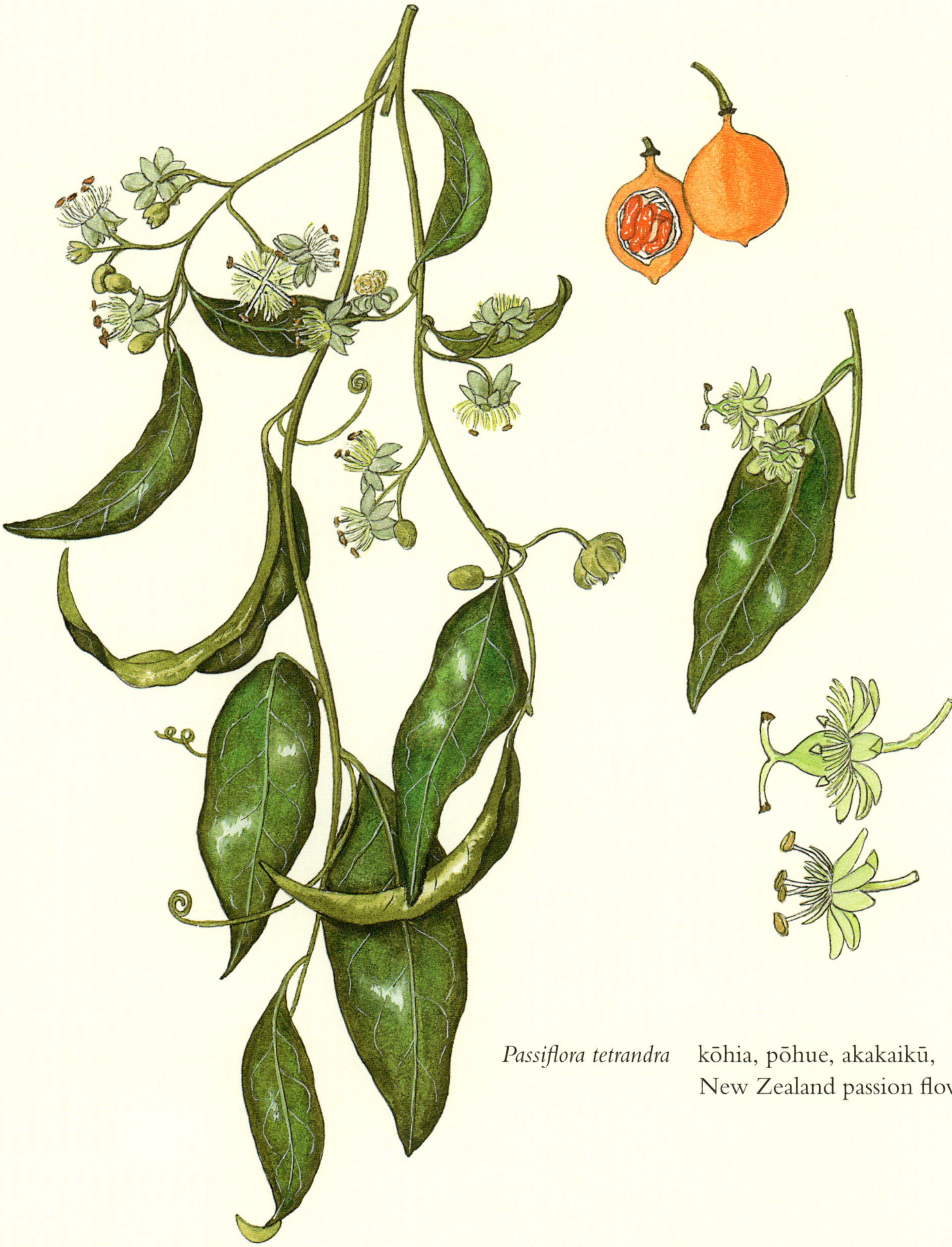

Passiflora tetrandra kōhia, pōhue, akakaikū,
New Zealand passion flower

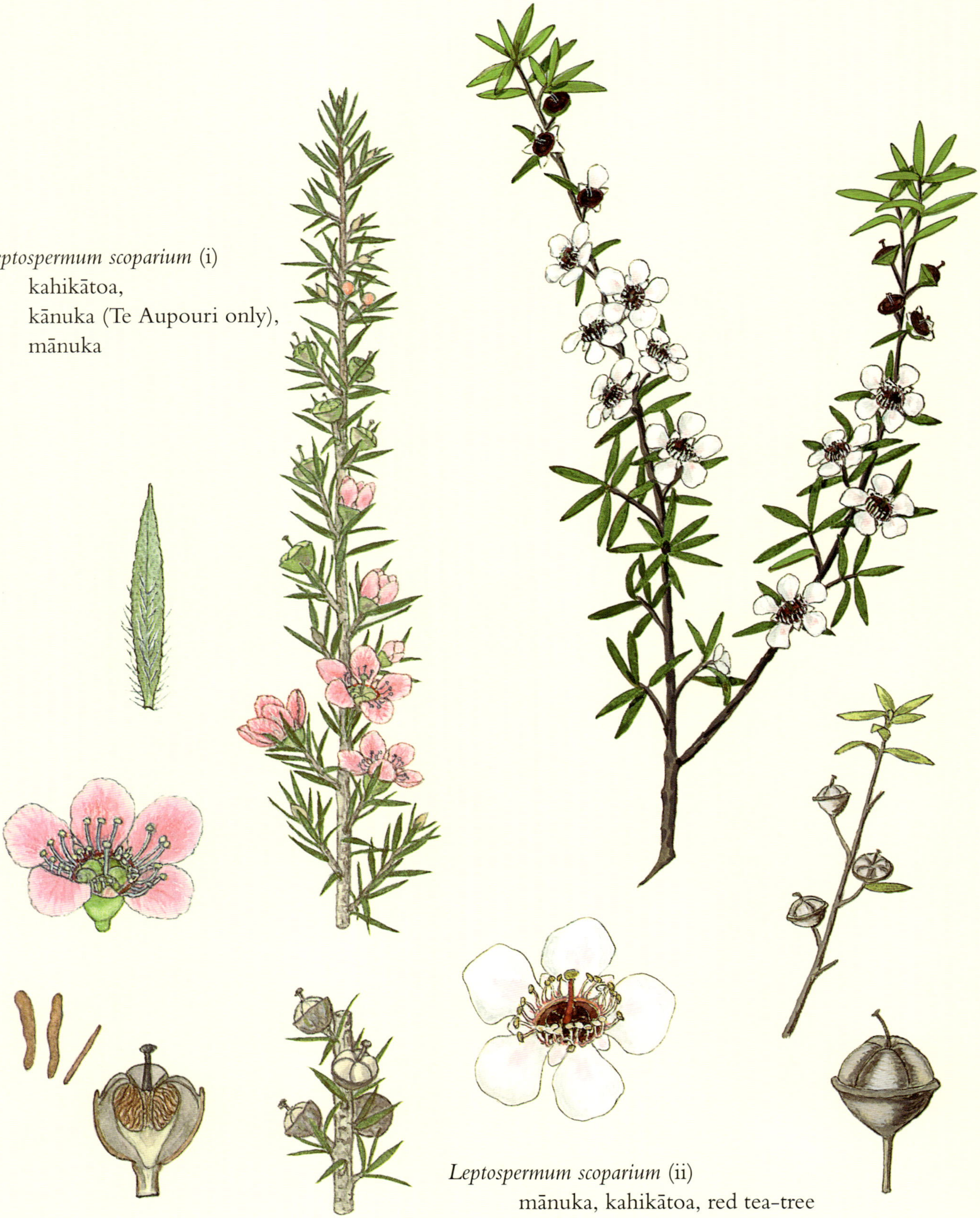

Leptospermum scoparium (i)
kahikātoa,
kānuka (Te Aupouri only),
mānuka

Leptospermum scoparium (ii)
mānuka, kahikātoa, red tea-tree

PLATE 40

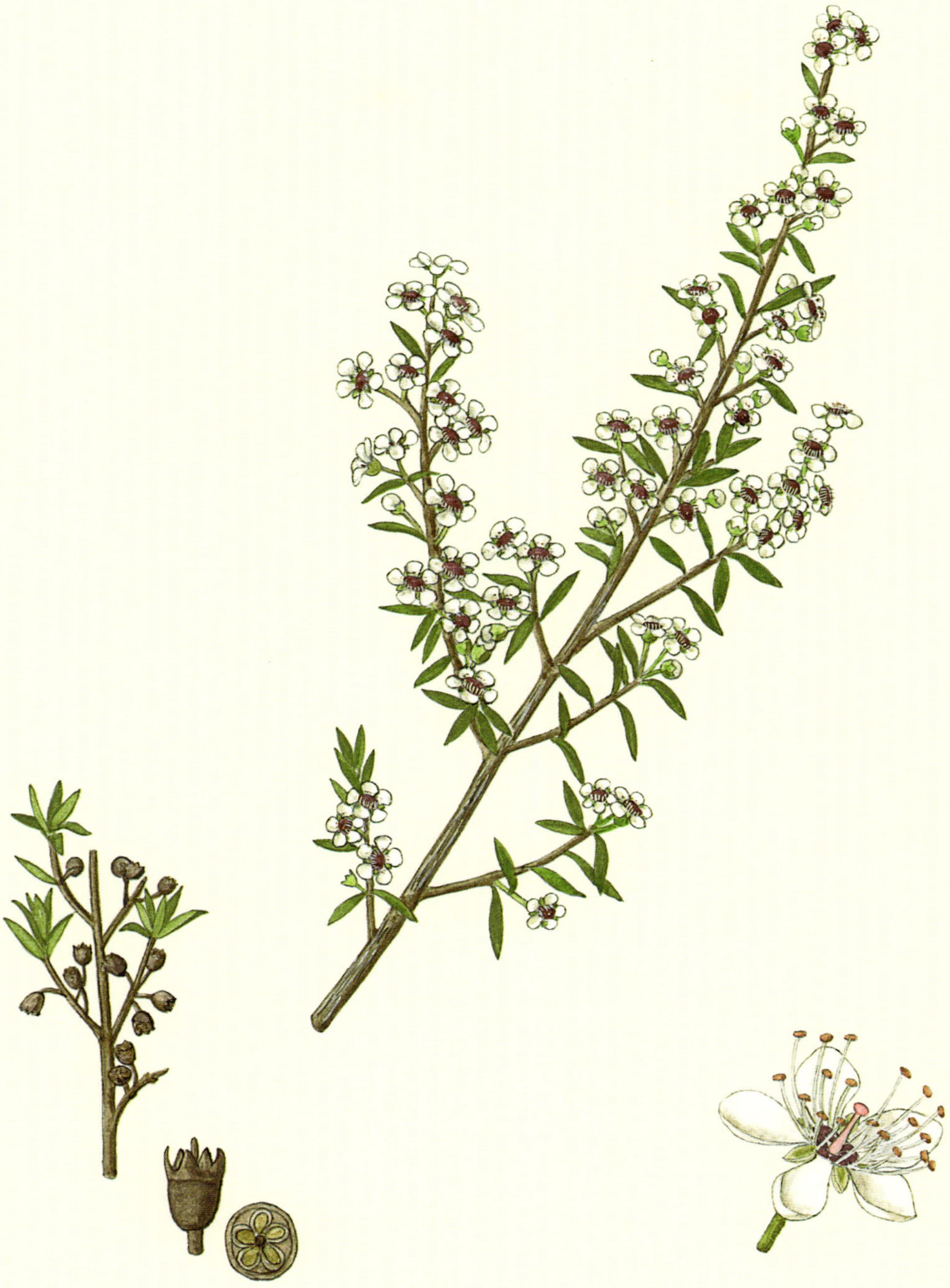

Kunzea aff. *ericoides*   mānuka, kānuka, white tea-tree

PLATE 41

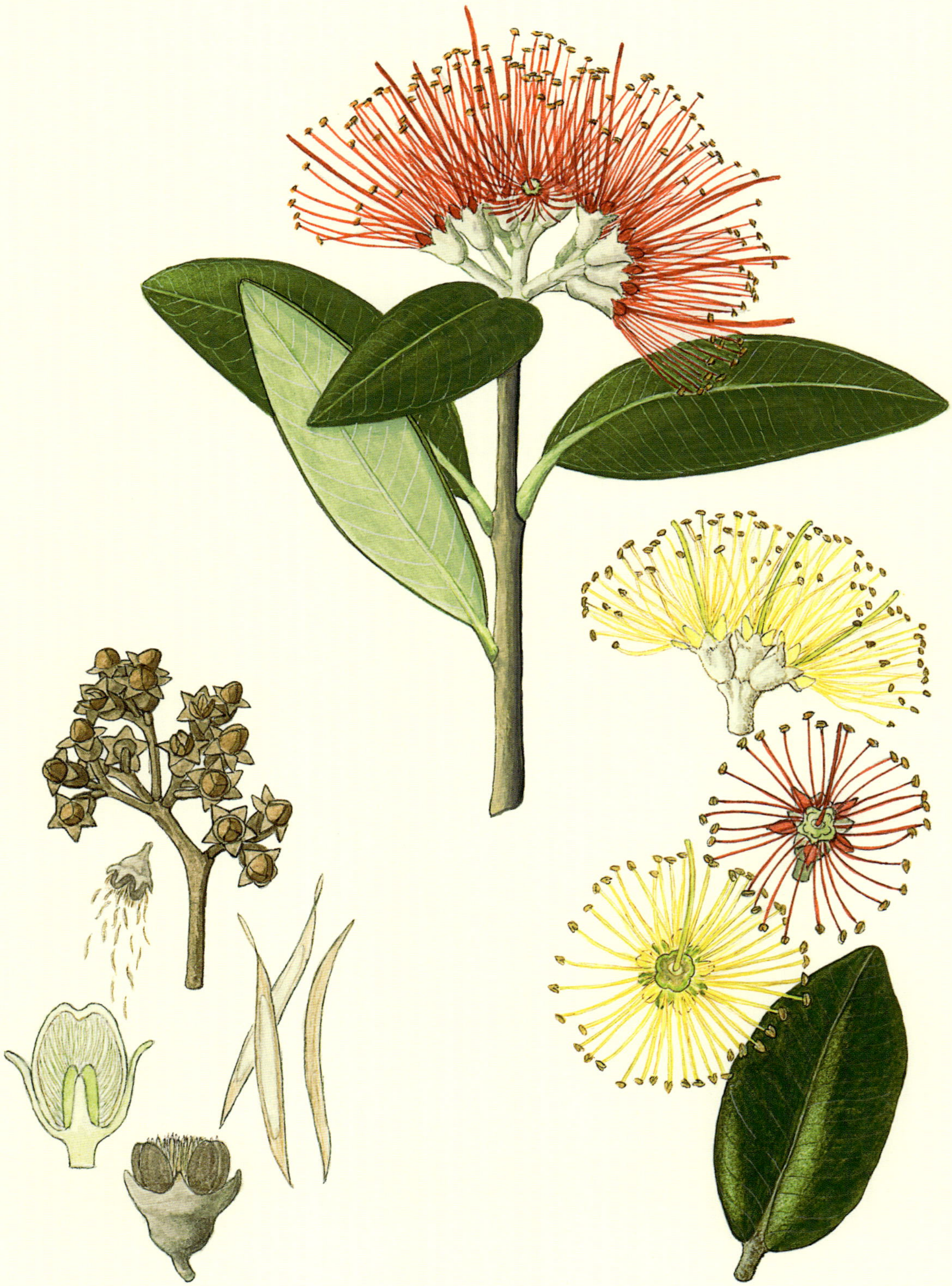

Metrosideros excelsa pōhutukawa, New Zealand Christmas tree

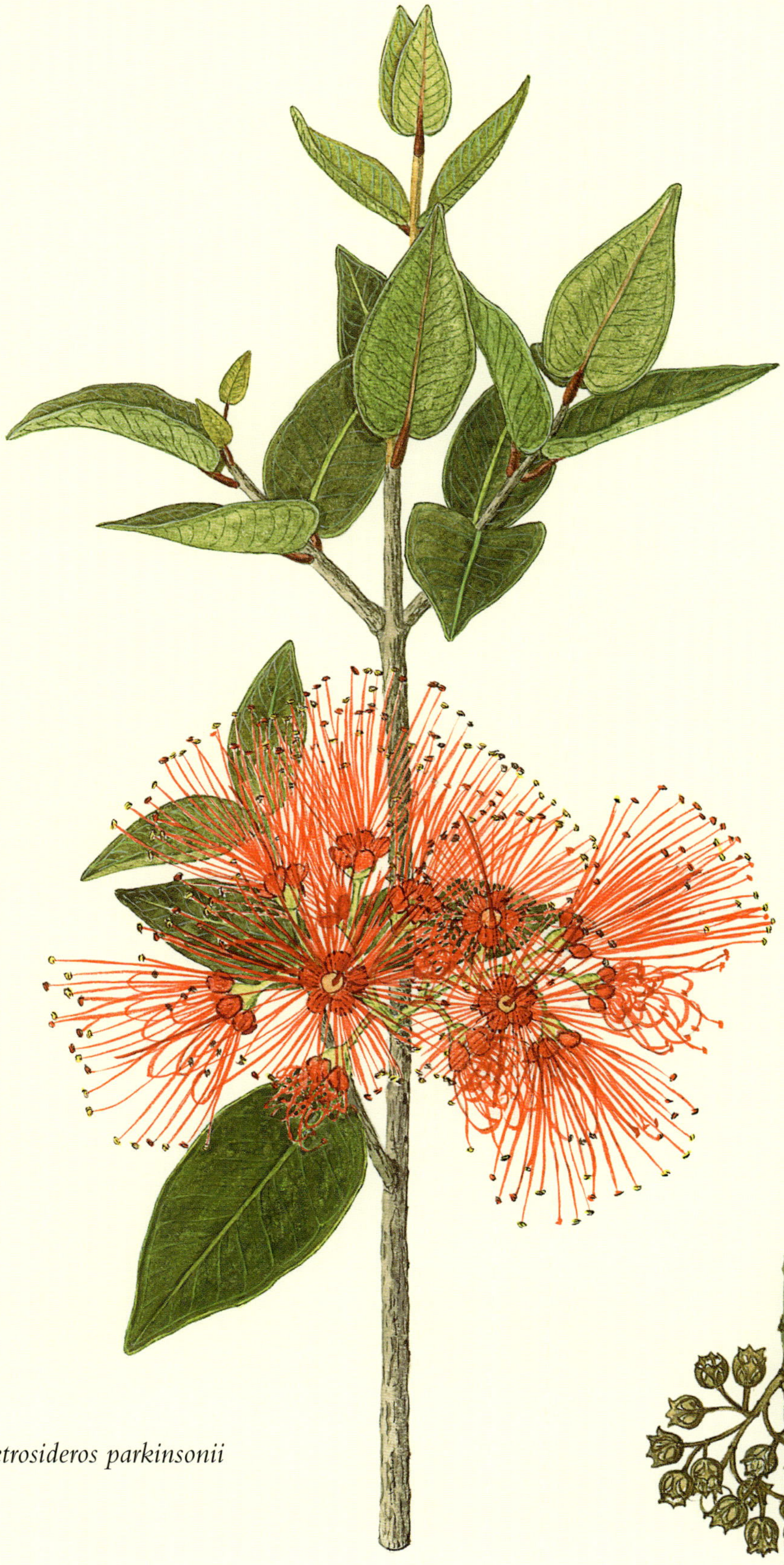

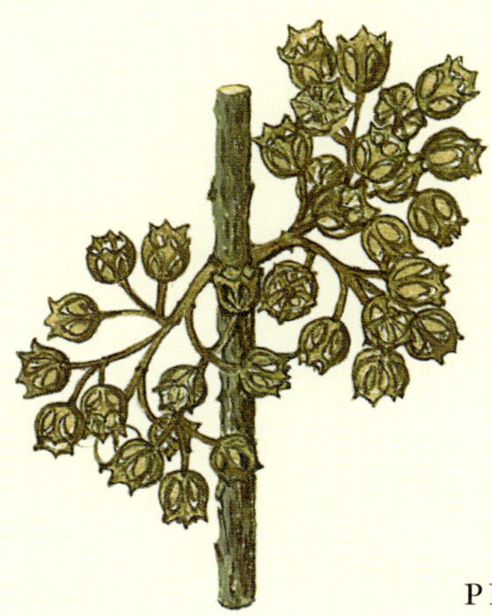

Metrosideros parkinsonii

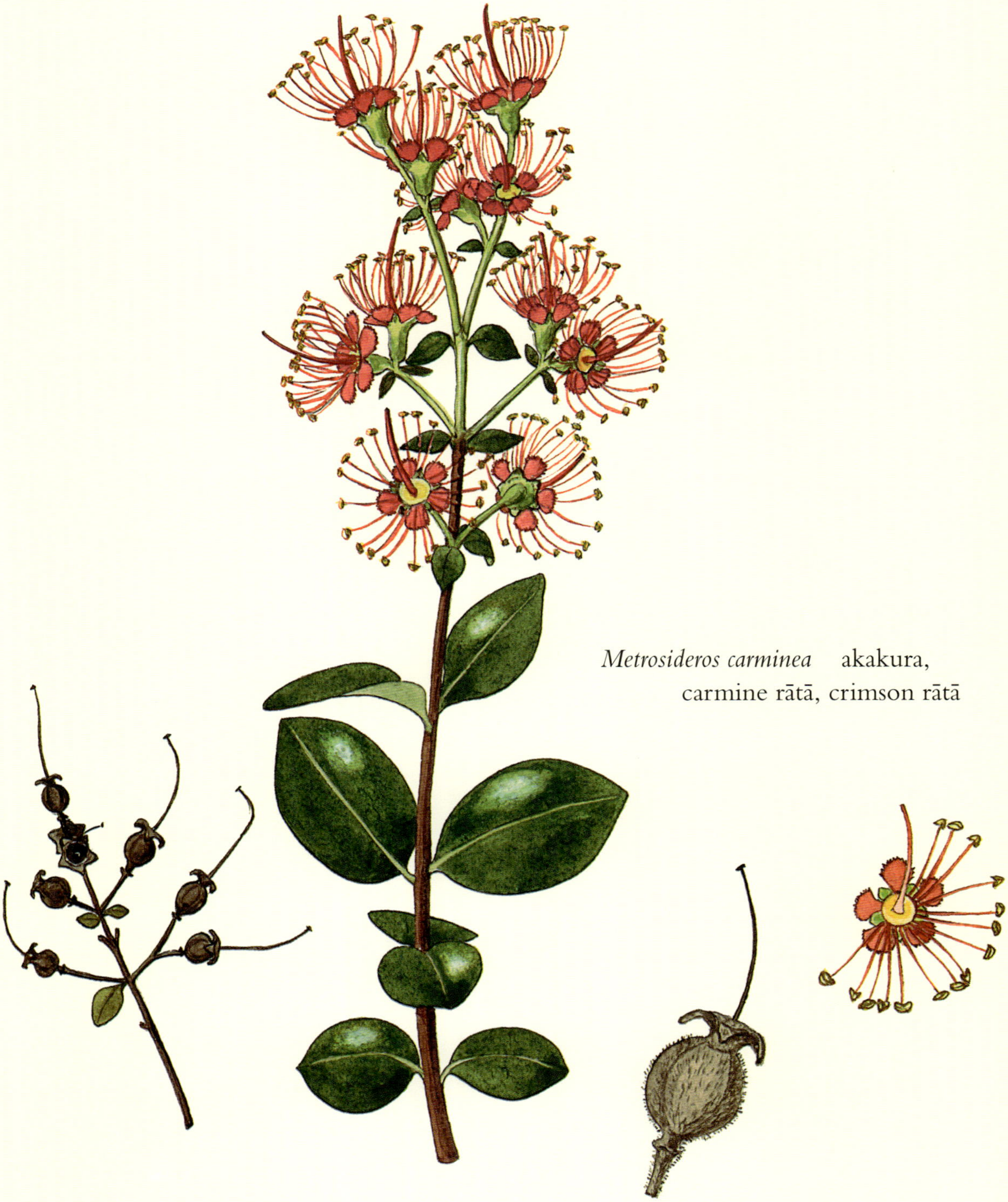

Metrosideros carminea akakura,
carmine rātā, crimson rātā

PLATE 44

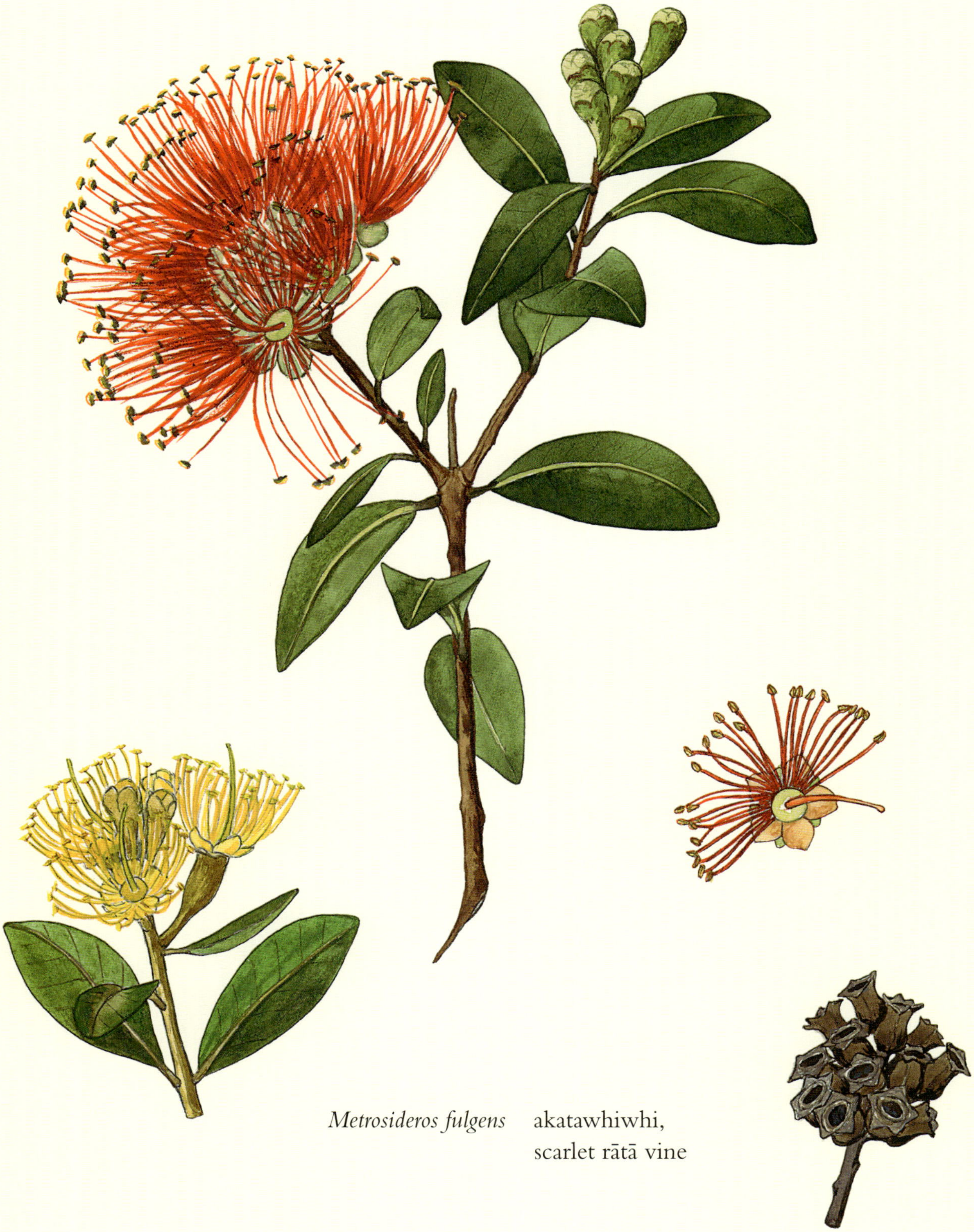

Metrosideros fulgens akatawhiwhi,
scarlet rātā vine

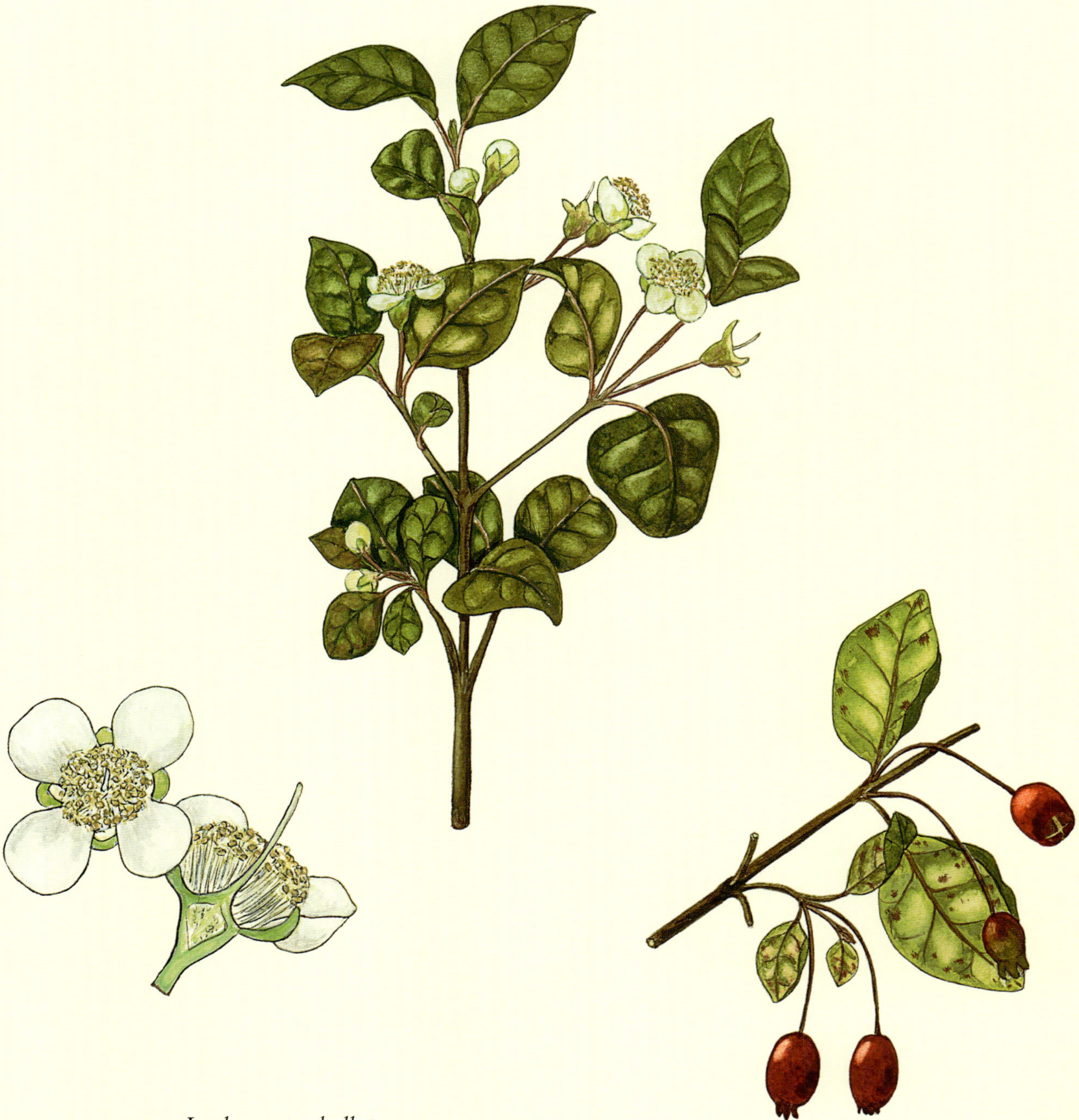

Lophomyrtus bullata ramarama

PLATE 46

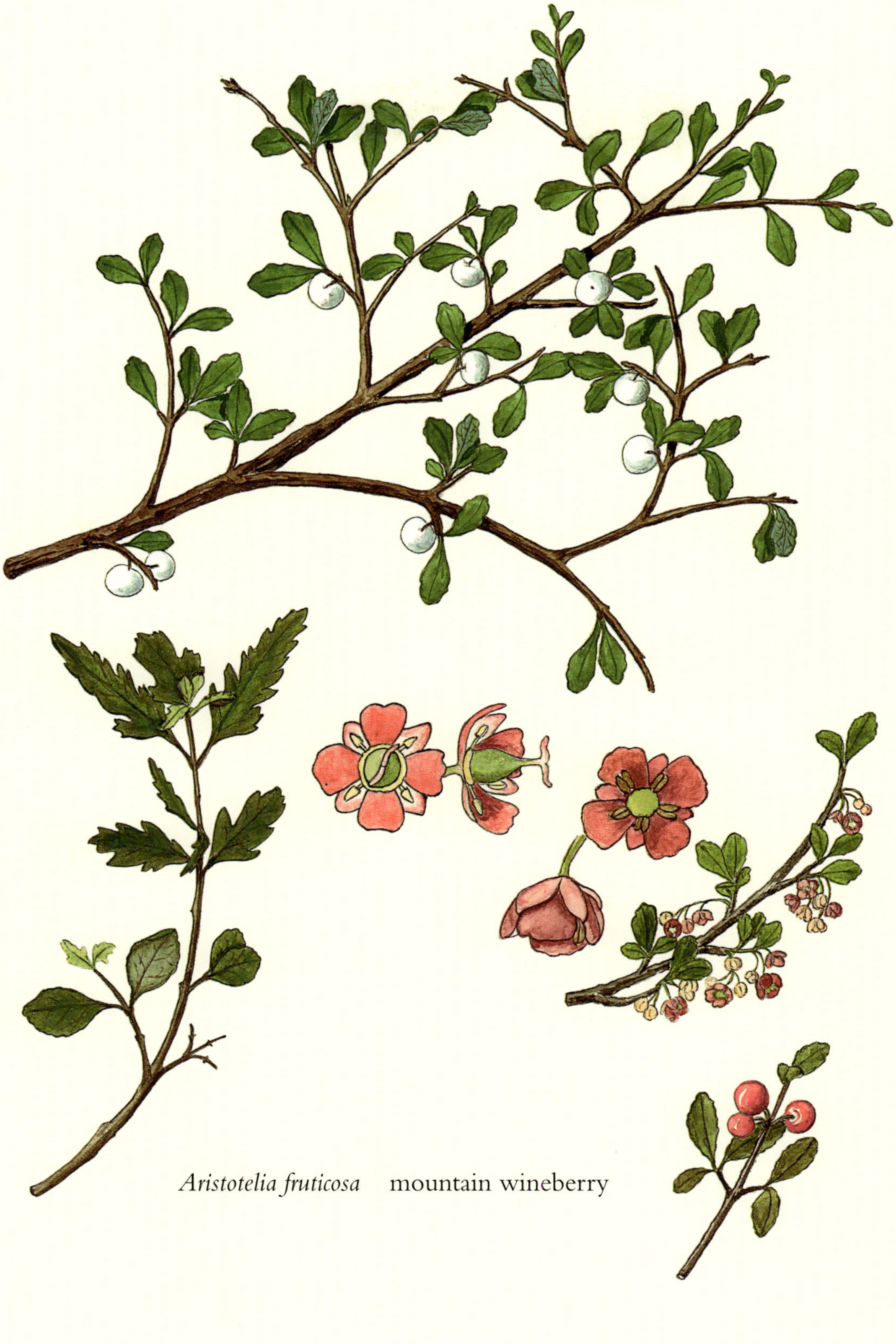

Aristotelia fruticosa mountain wineberry

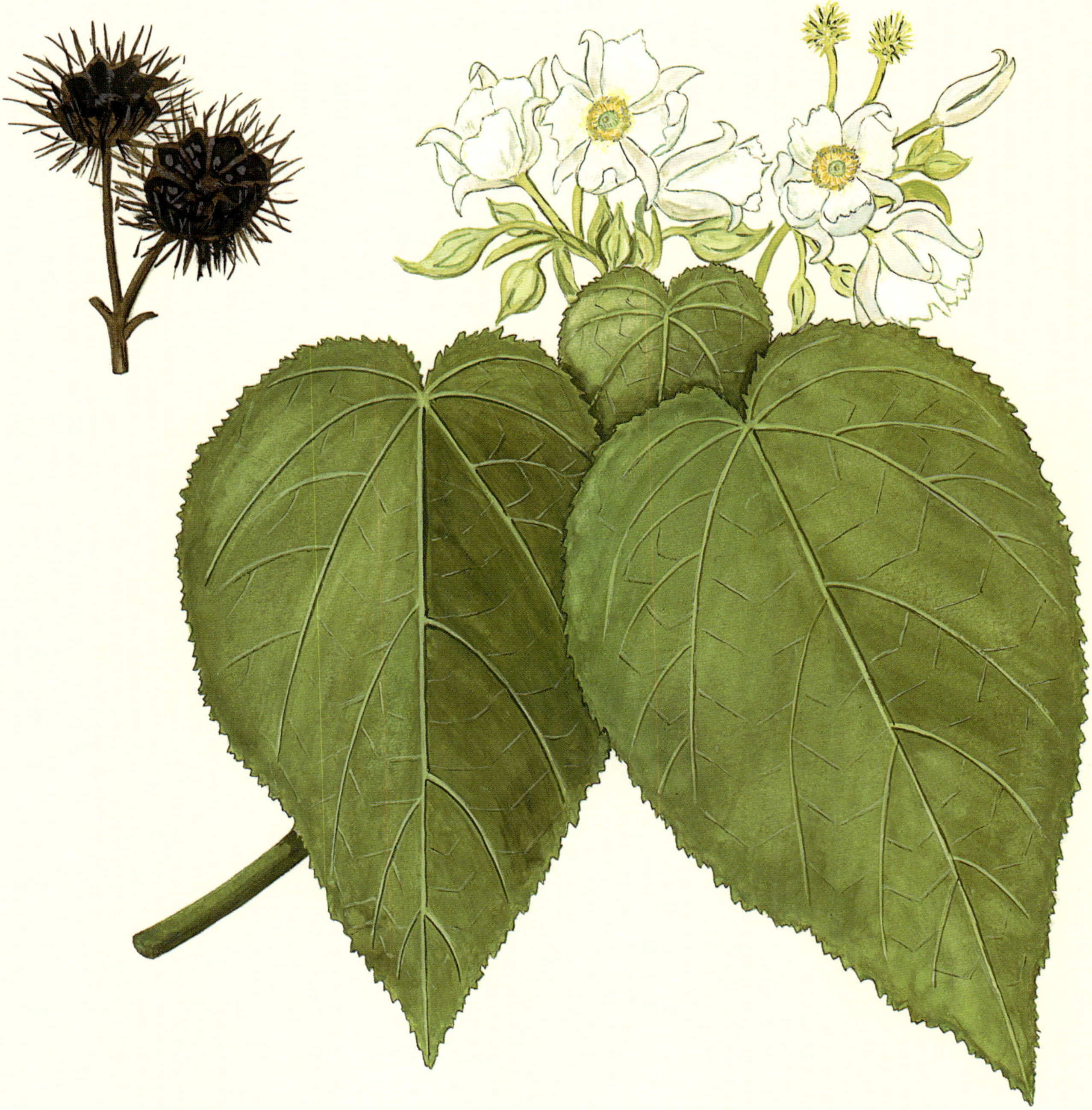

Entelea arborescens whau, cork tree

PLATE 50

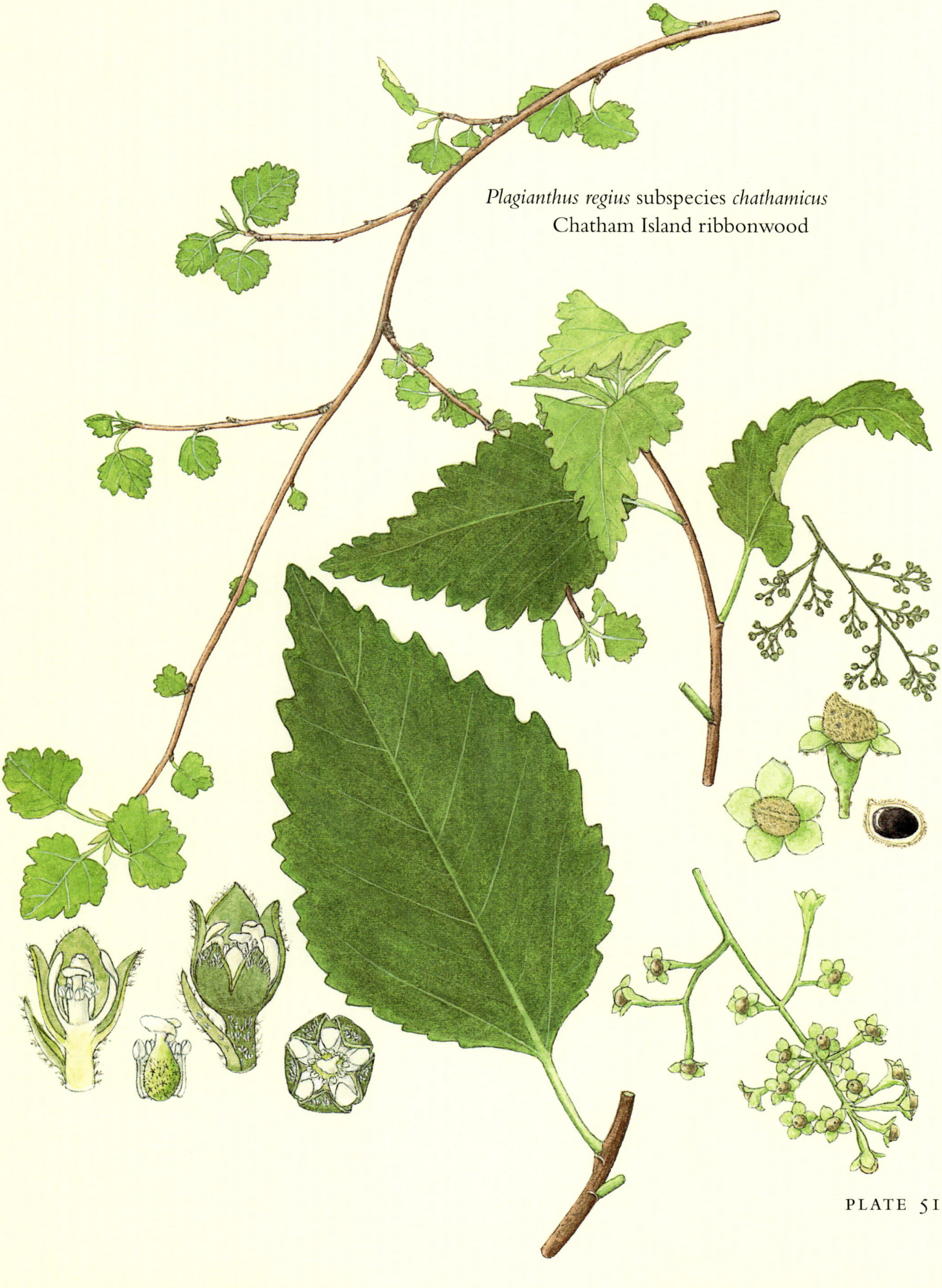

Plagianthus regius subspecies *chathamicus*
Chatham Island ribbonwood

PLATE 51

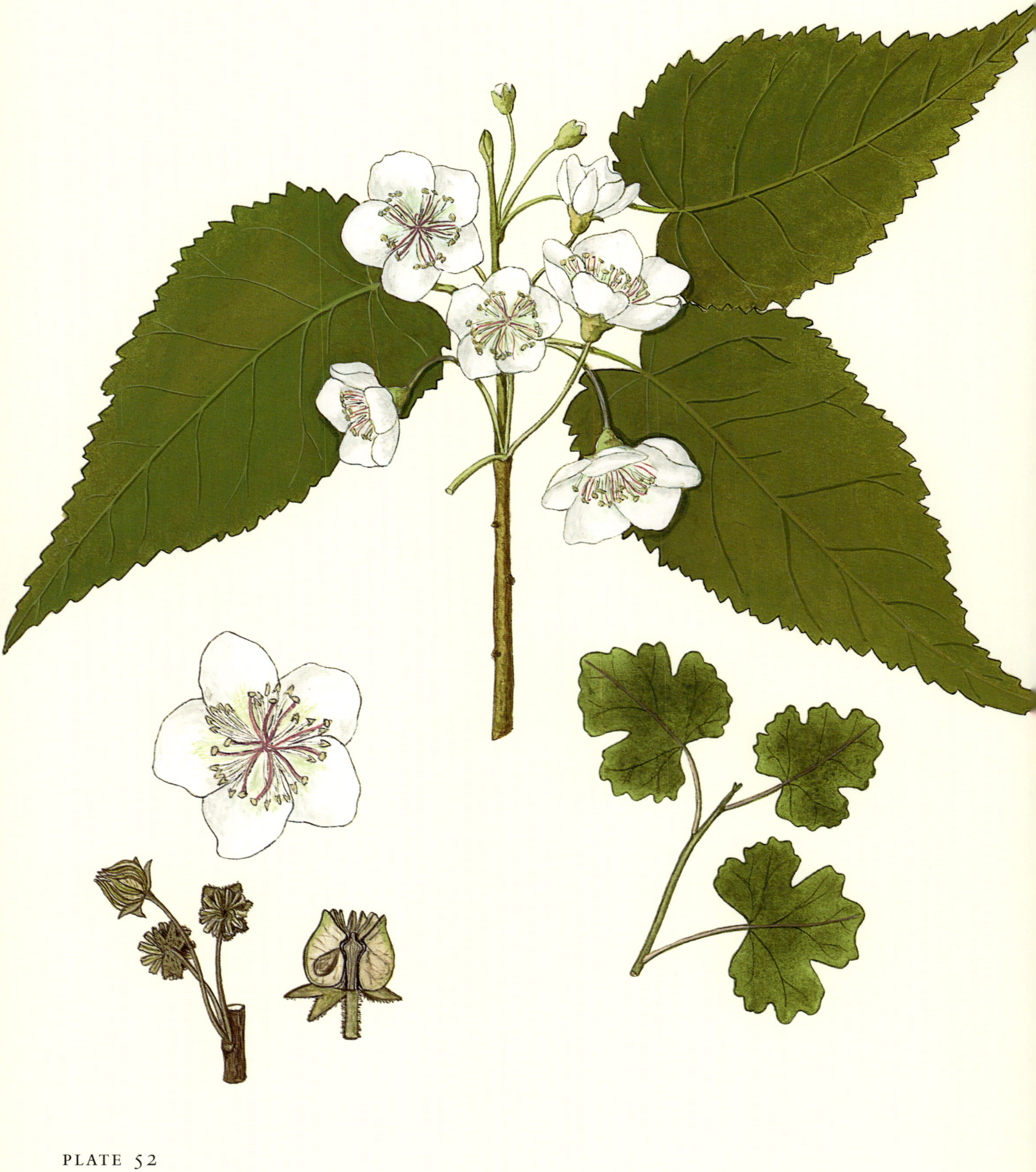

PLATE 52

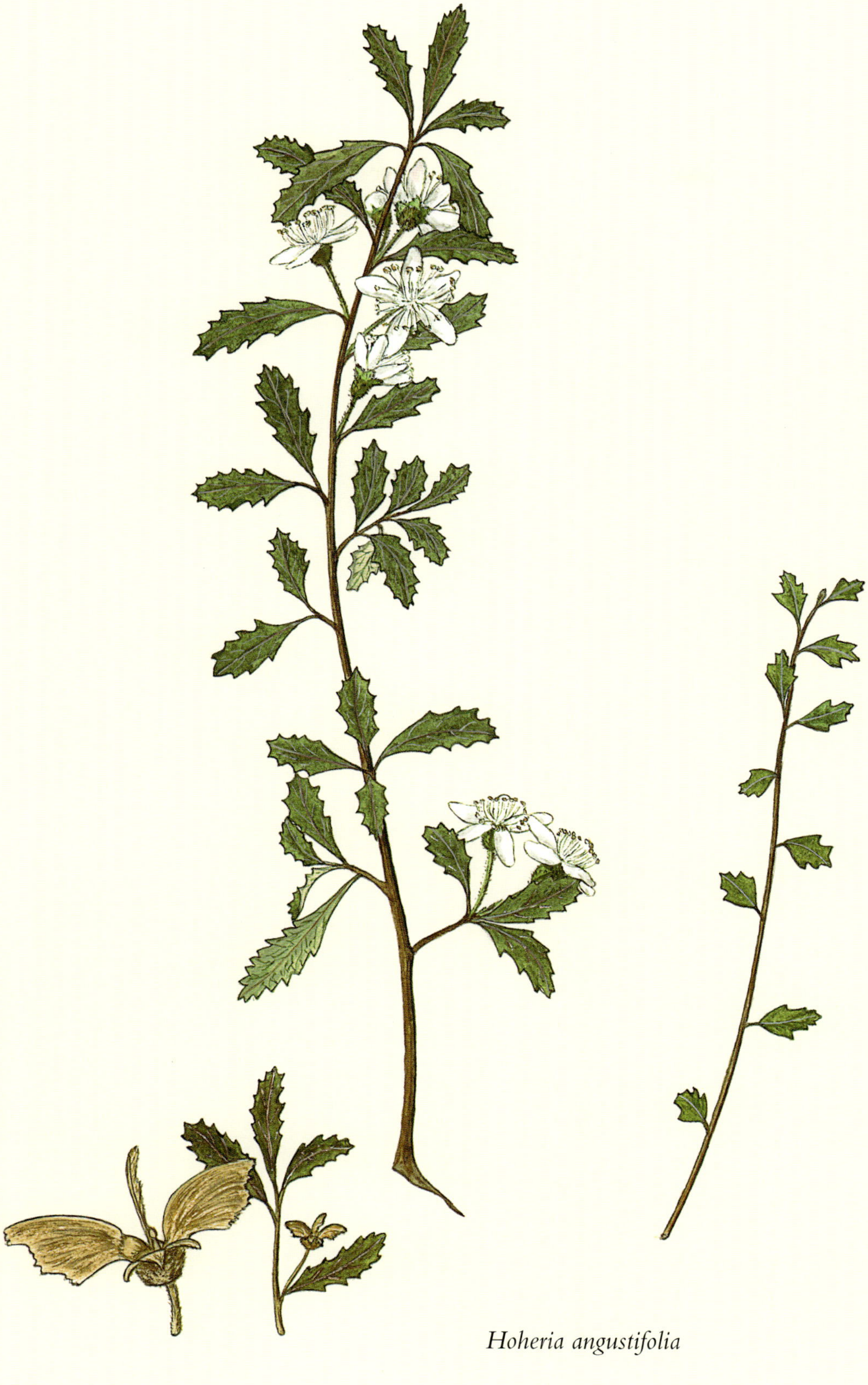

Hoheria angustifolia

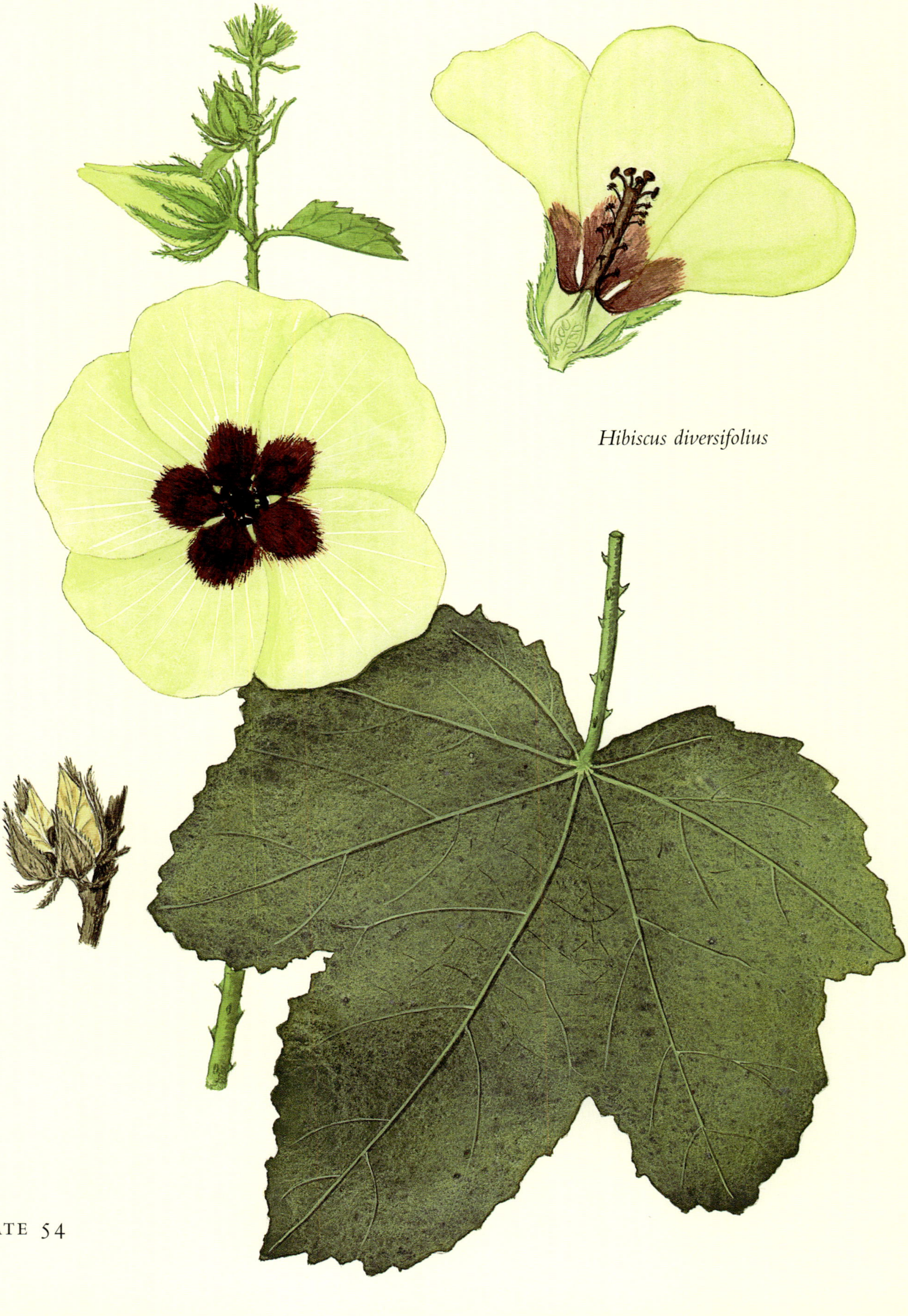

Hibiscus diversifolius

PLATE 54

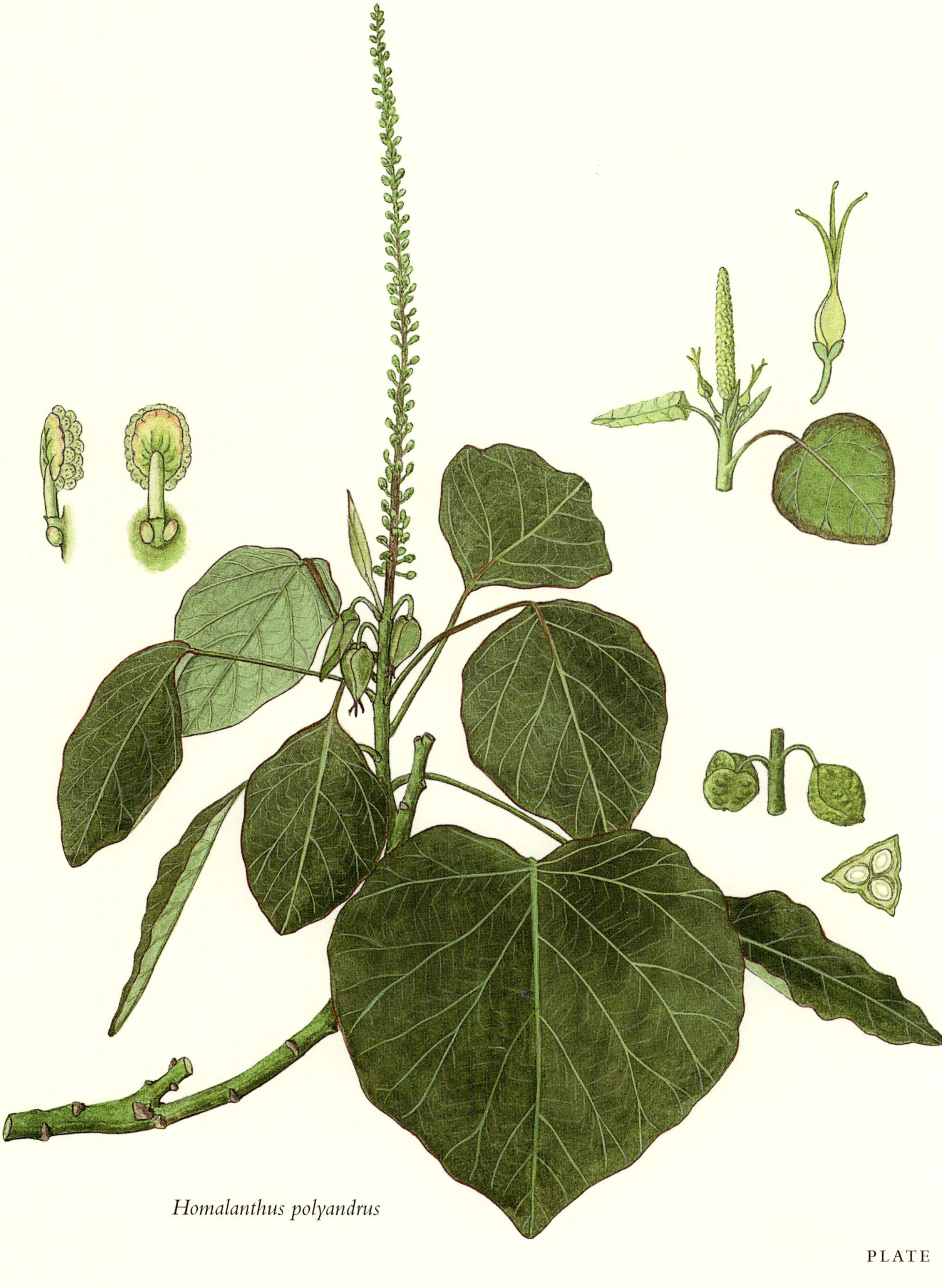

Homalanthus polyandrus

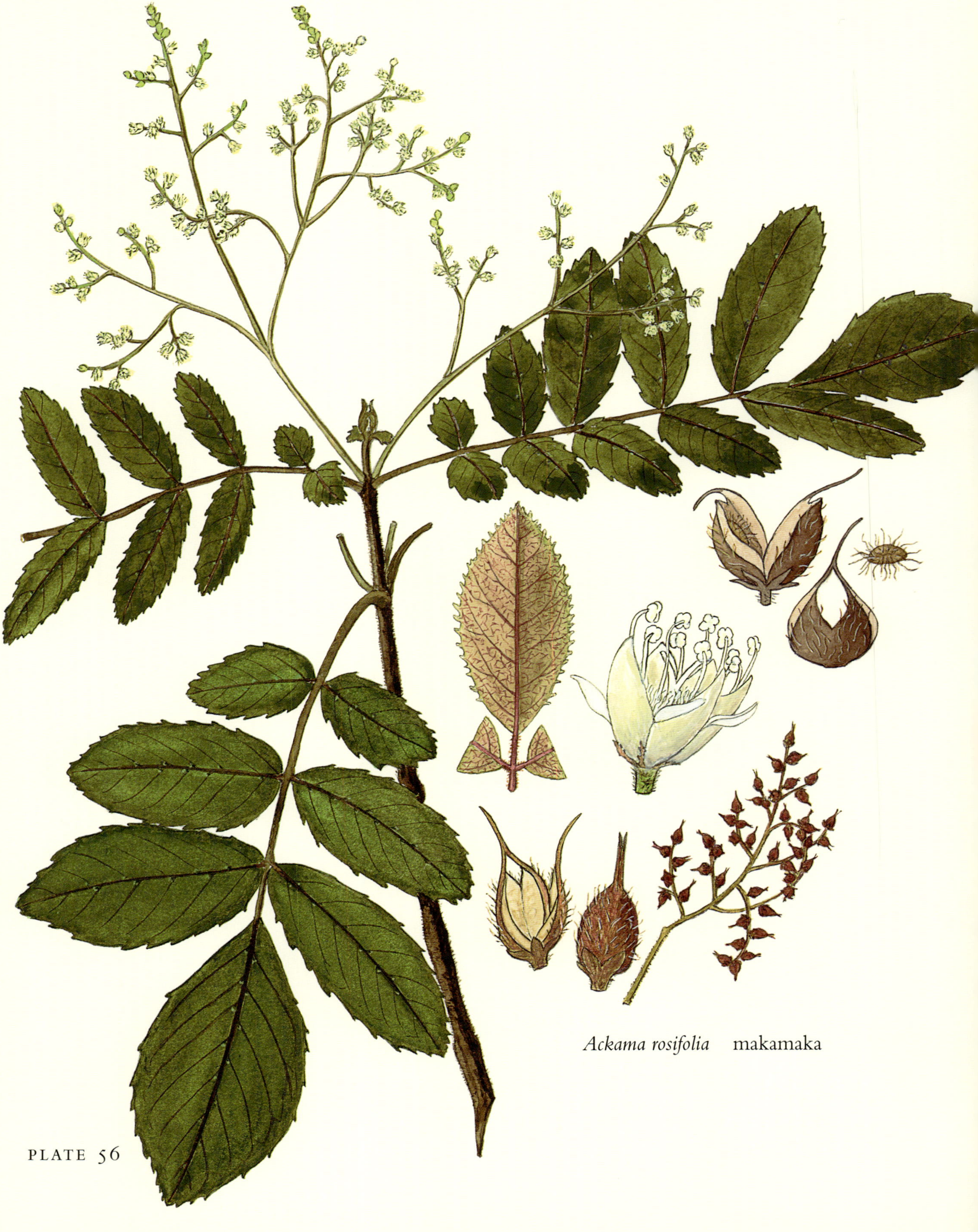

Ackama rosifolia makamaka

PLATE 56

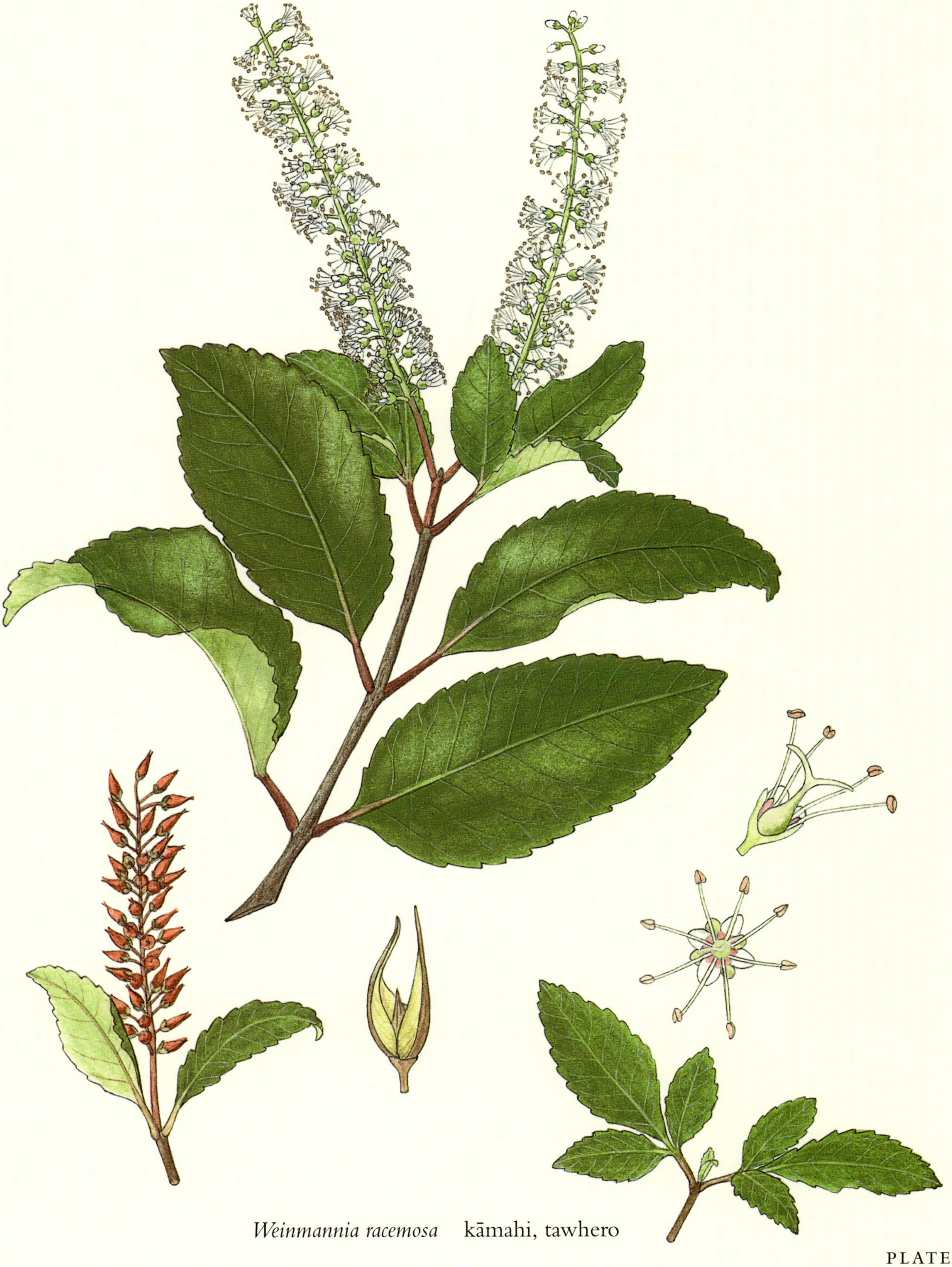

Weinmannia racemosa kāmahi, tawhero

PLATE 57

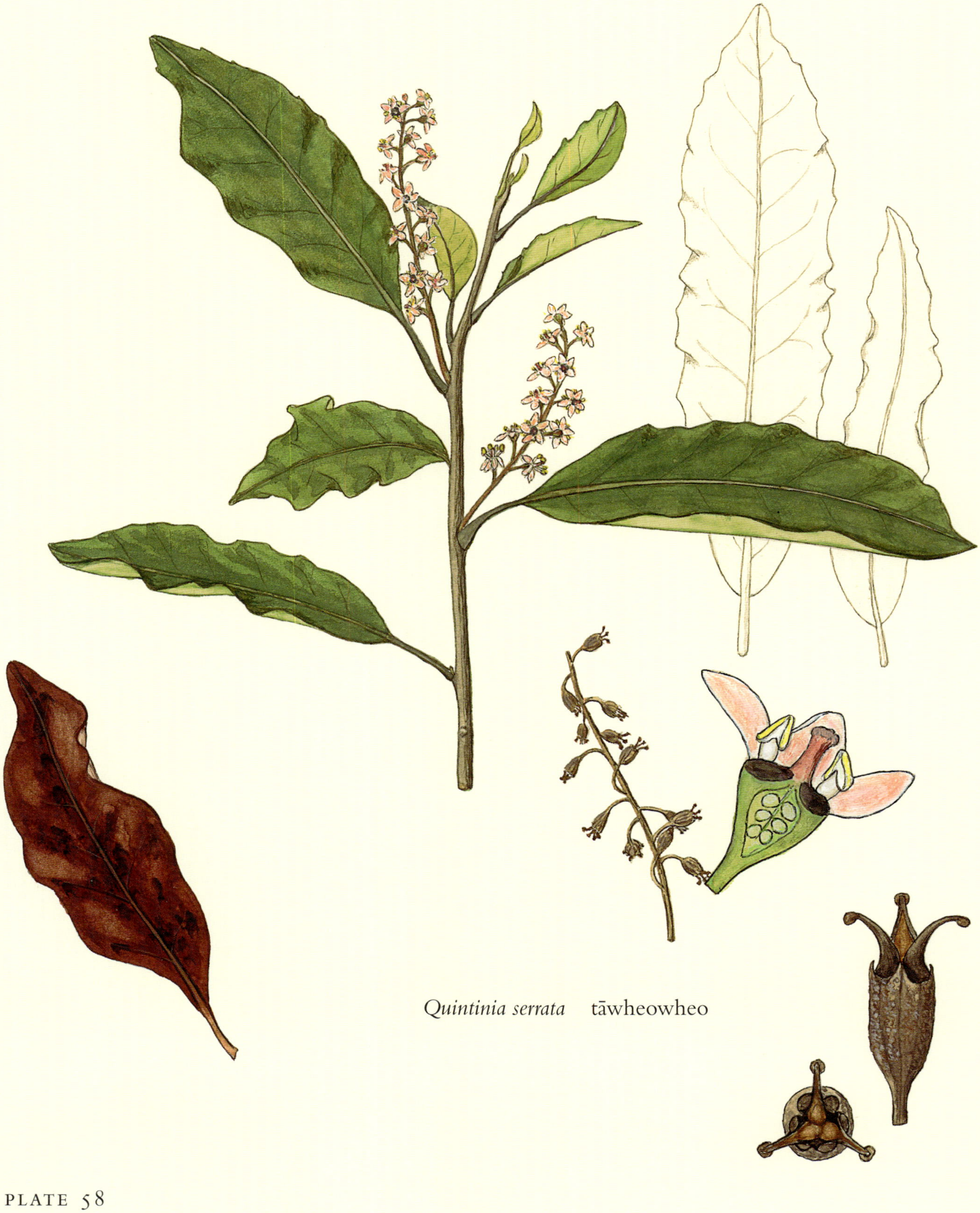

Quintinia serrata tāwheowheo

PLATE 58

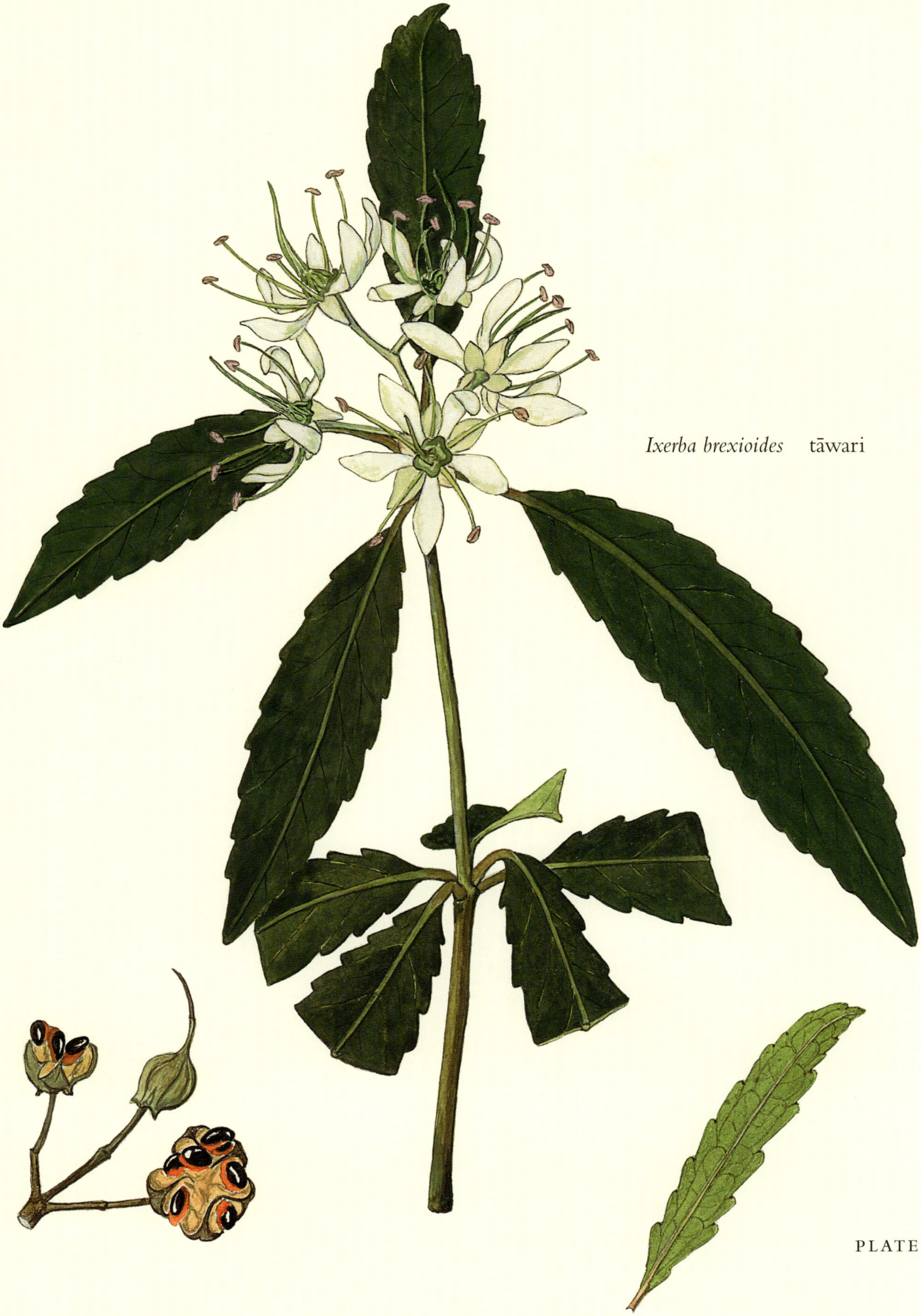

Ixerba brexioides tāwari

PLATE 59

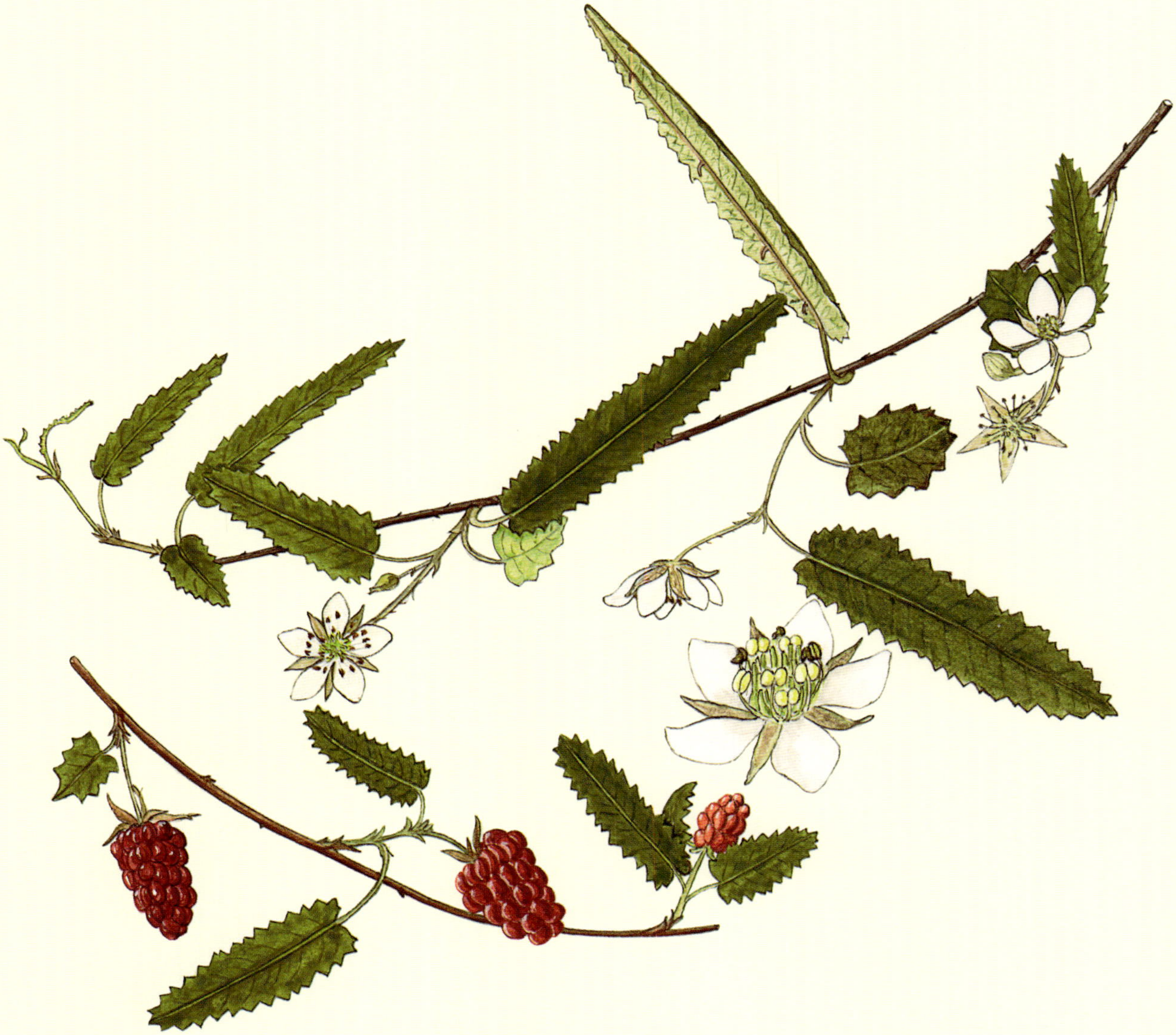

Rubus parvus creeping lawyer

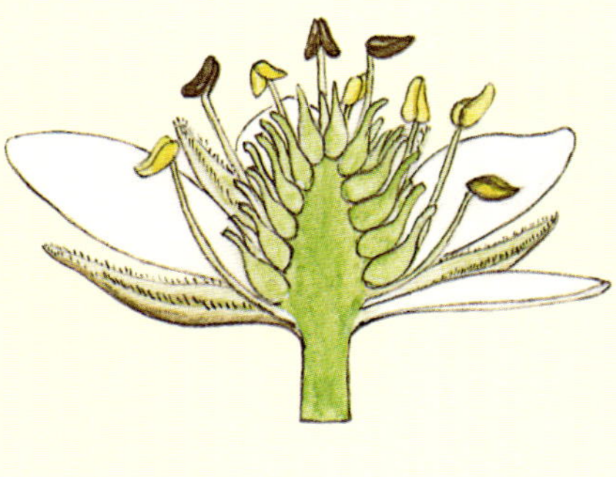

PLATE 60

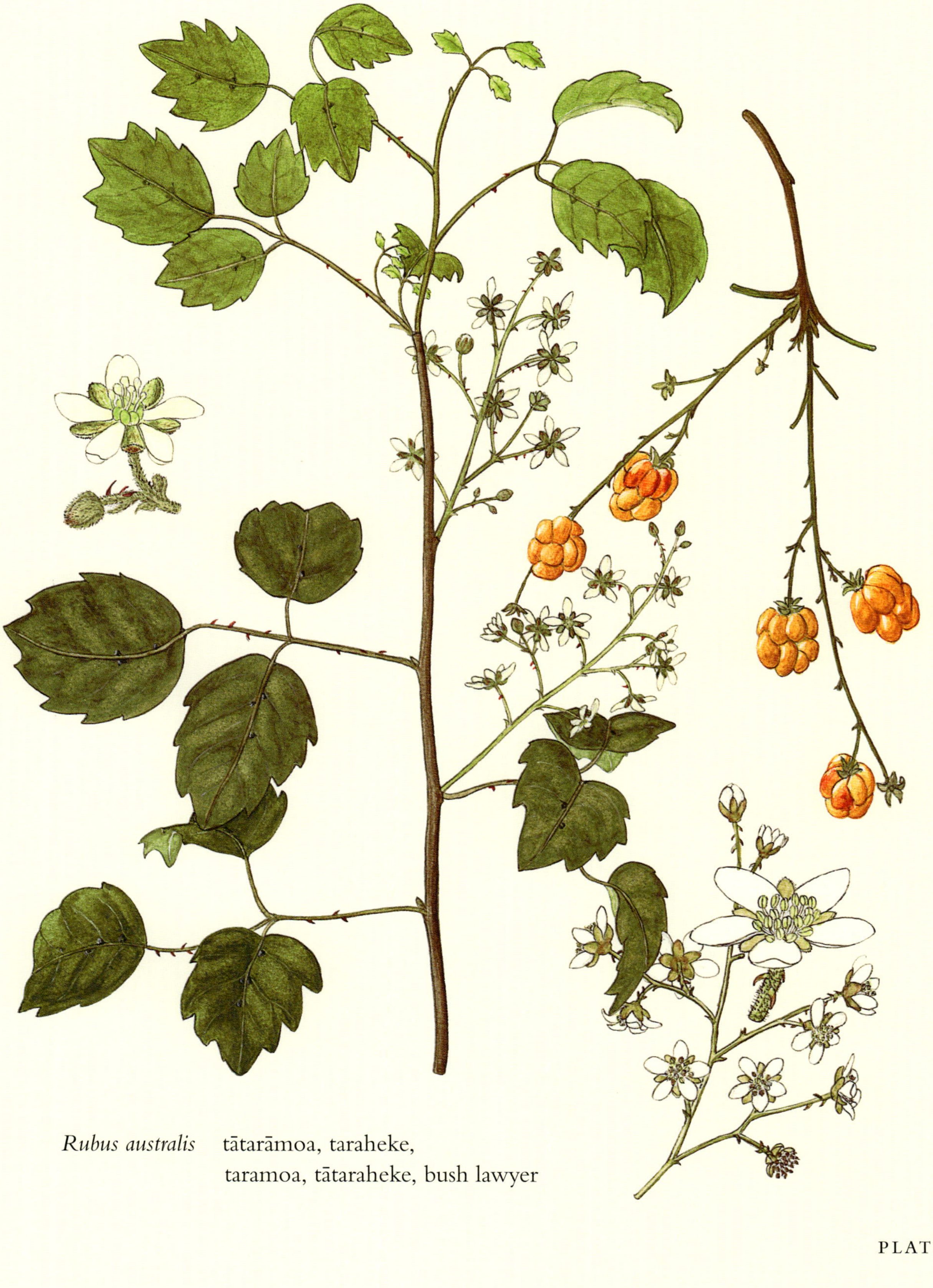

Rubus australis tātarāmoa, taraheke,
taramoa, tātaraheke, bush lawyer

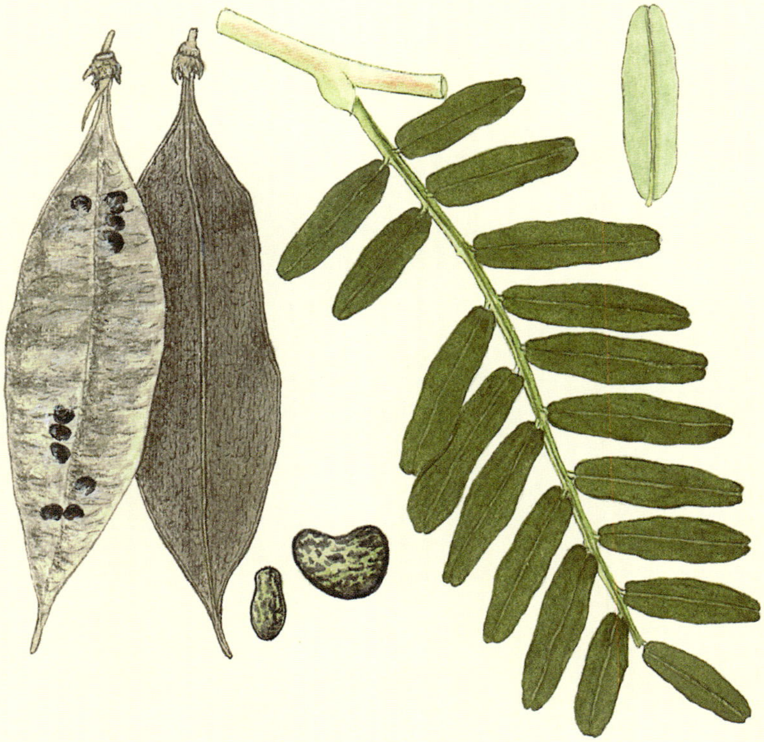

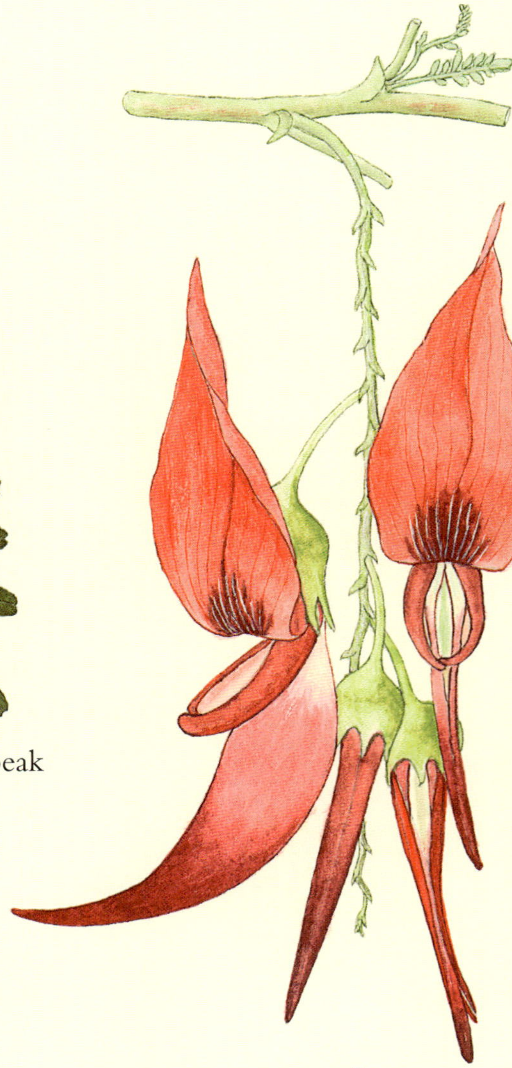

Clianthus puniceus   kōwhai-ngutu-kākā, kakabeak

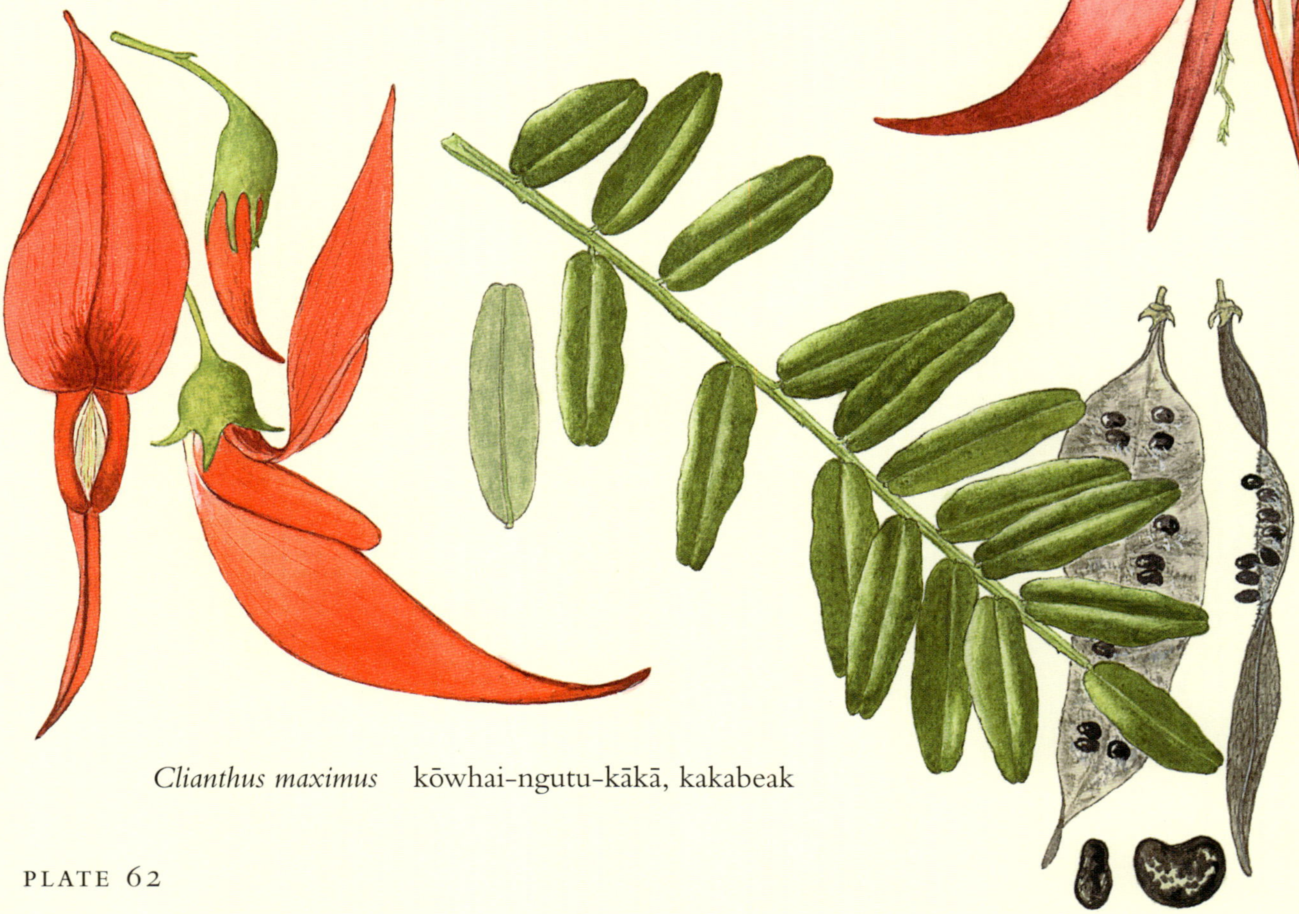

Clianthus maximus   kōwhai-ngutu-kākā, kakabeak

PLATE 62

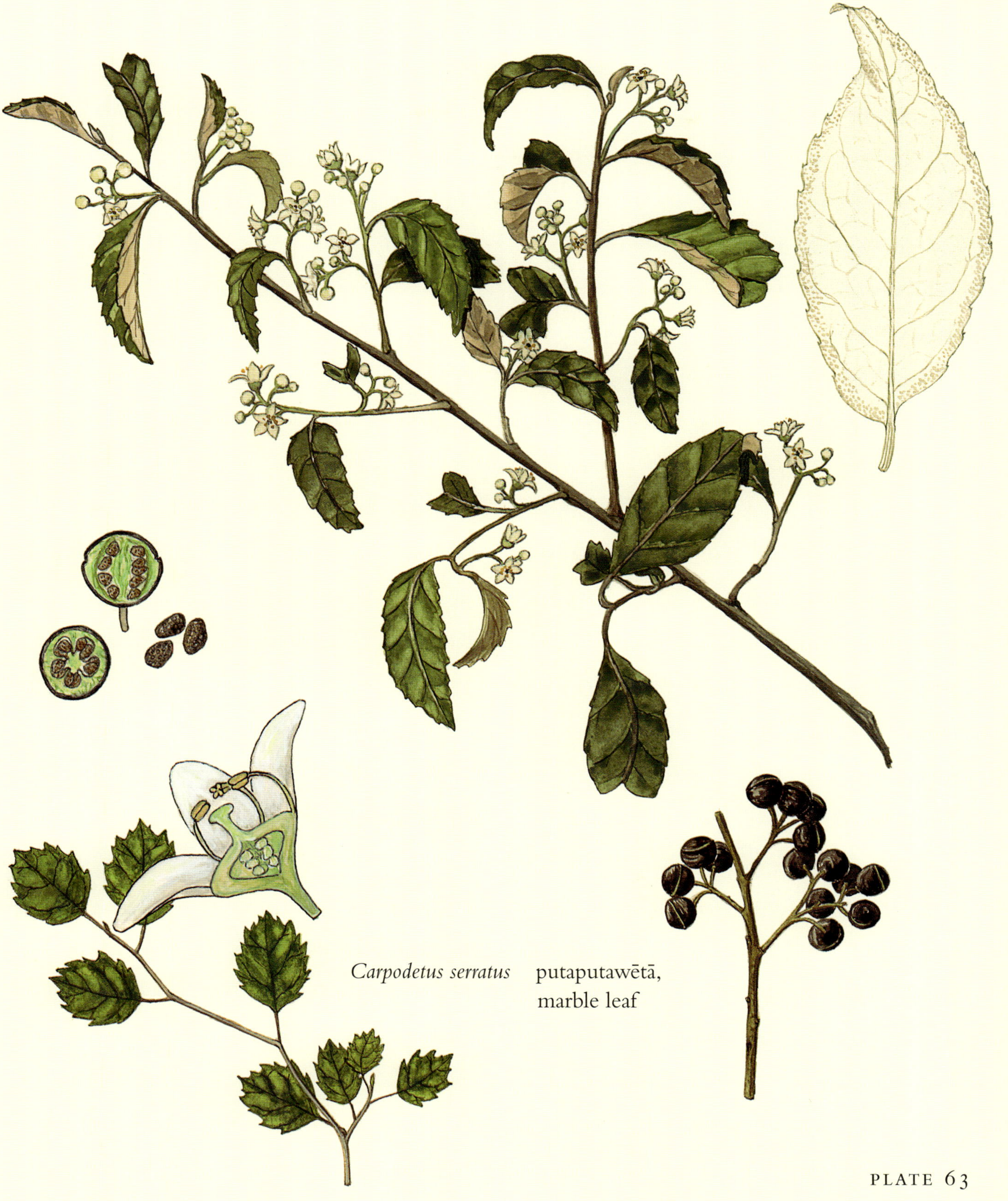

Carpodetus serratus putaputawētā, marble leaf

Sophora godleyi

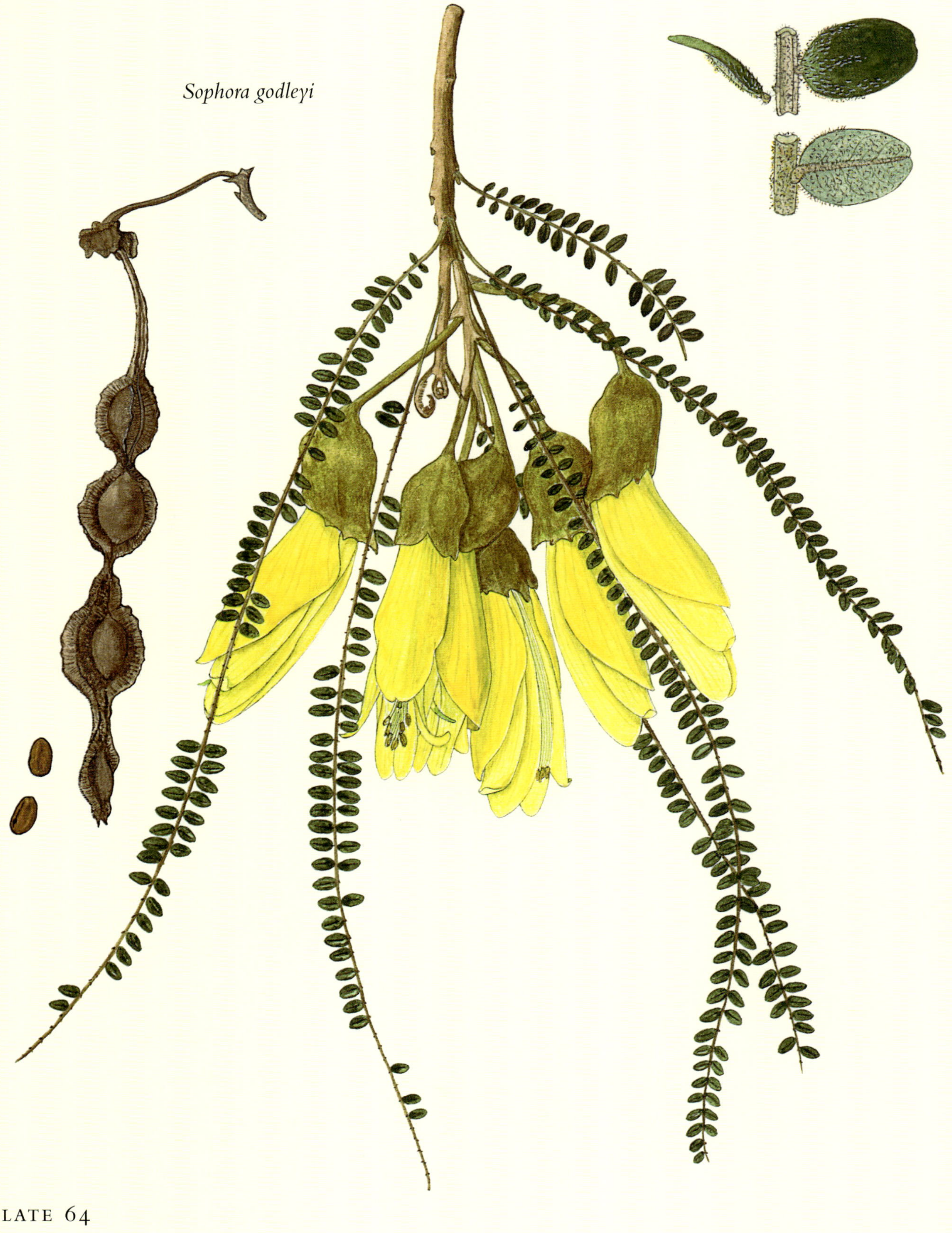

PLATE 64

Sophora microphylla kōwhai

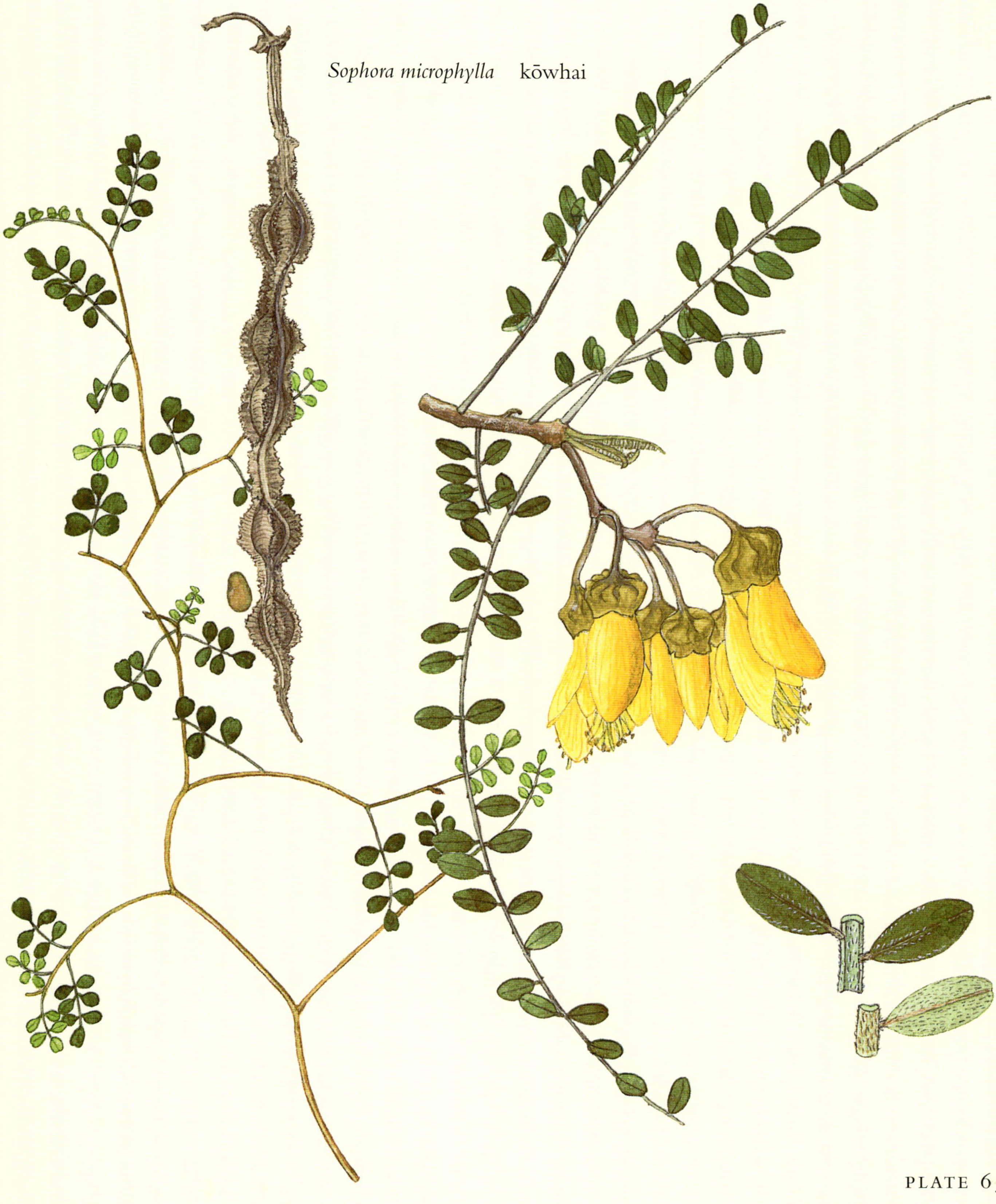

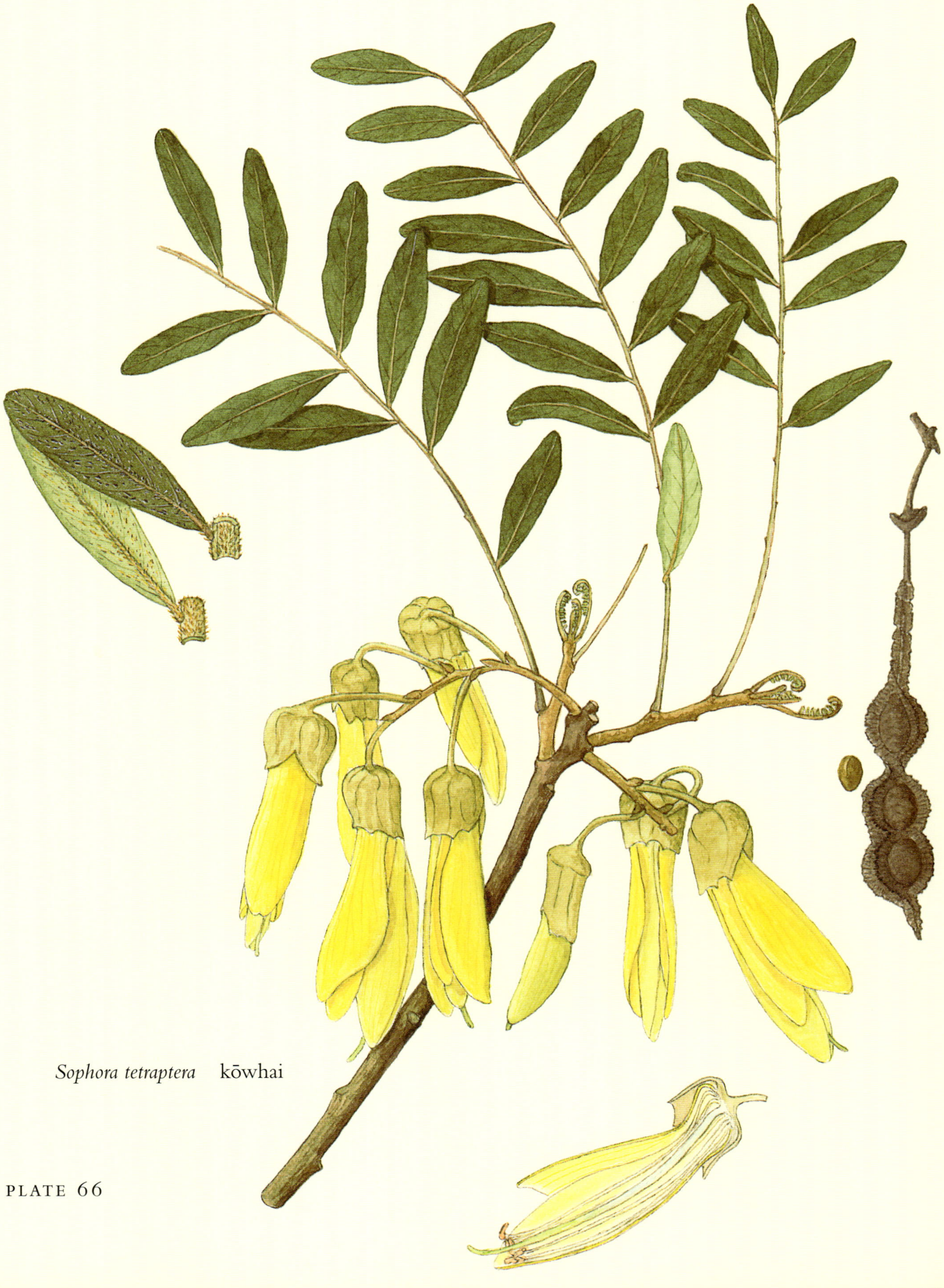

Sophora tetraptera kōwhai

PLATE 66

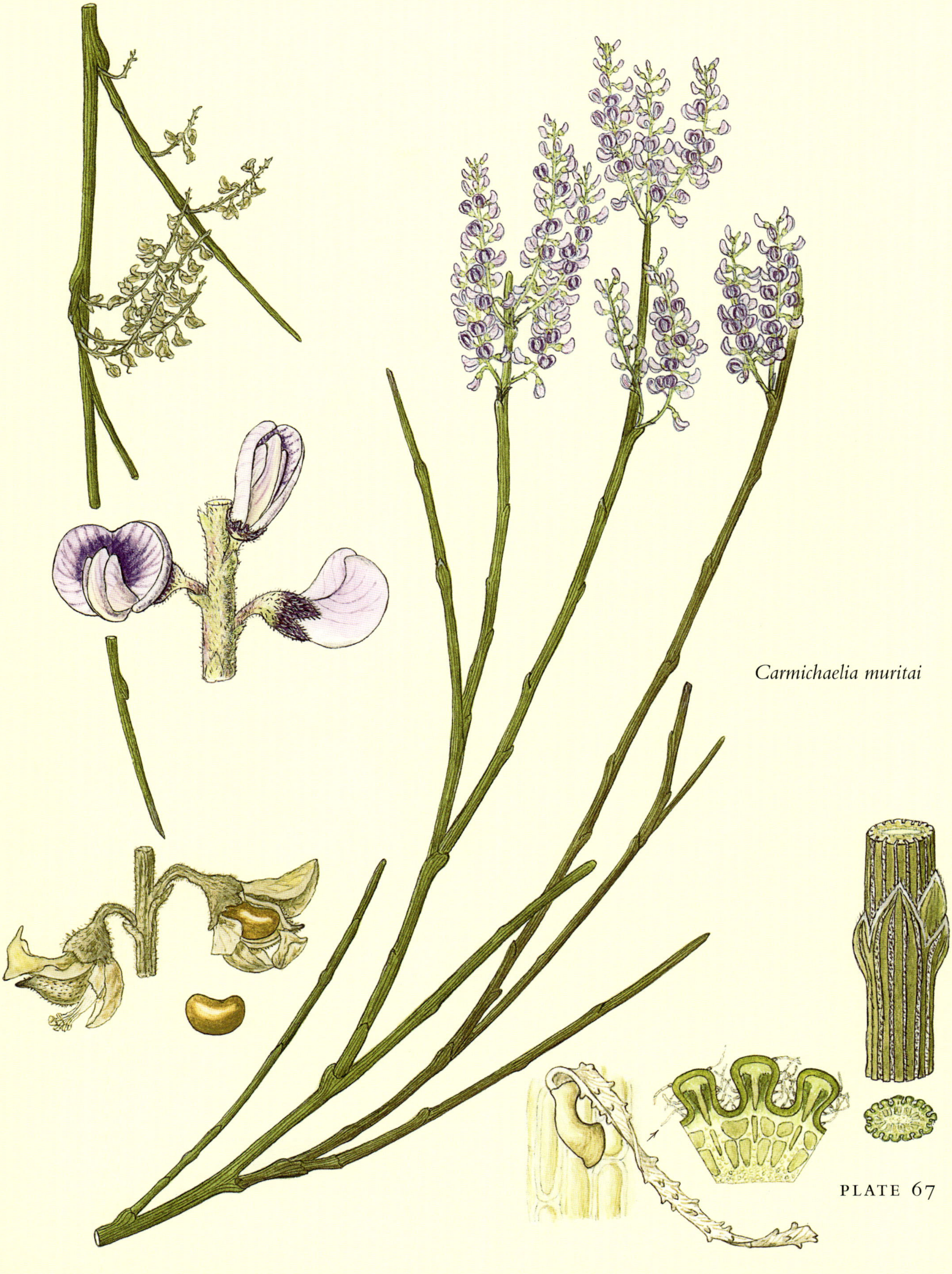

Carmichaelia muritai

PLATE 67

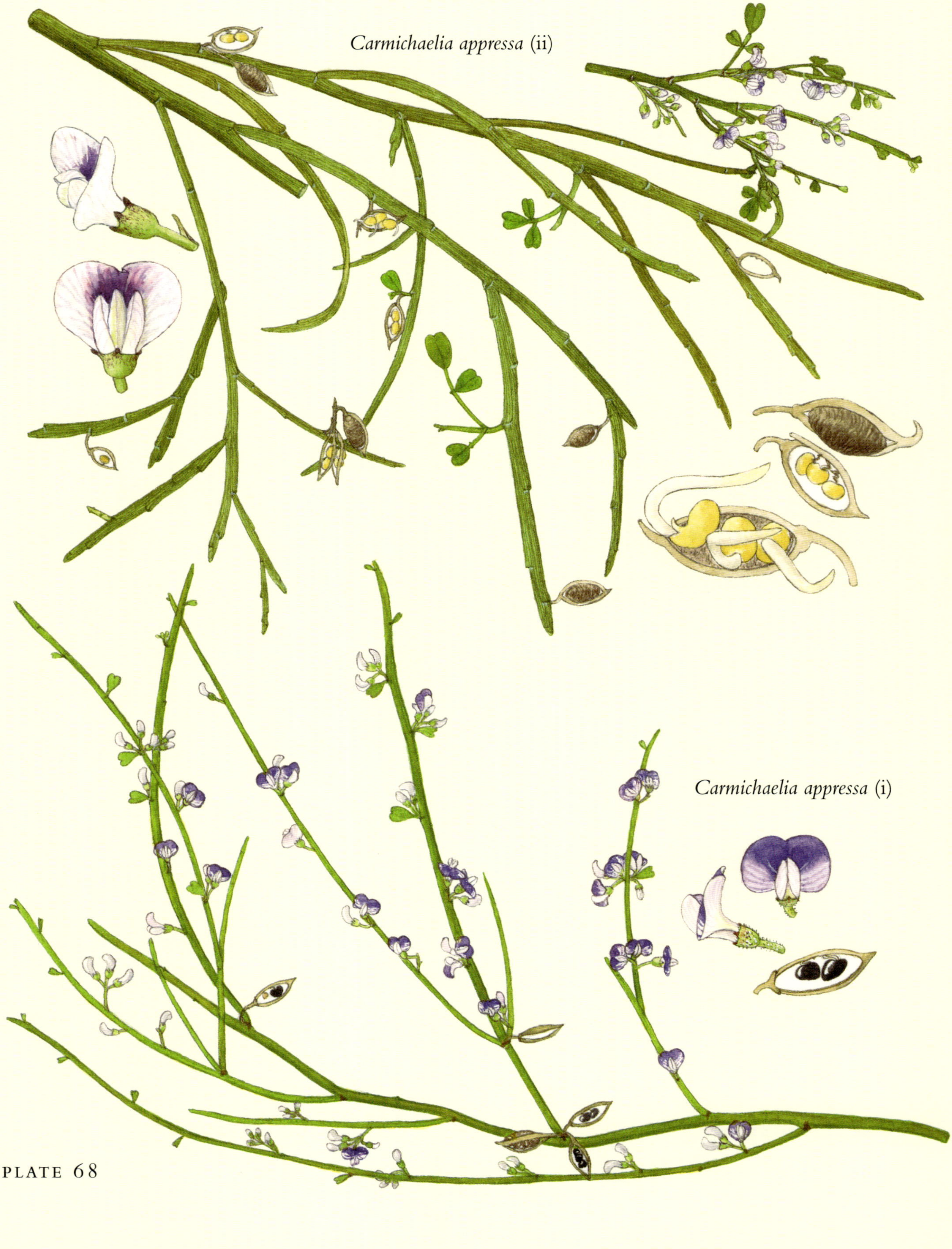

PLATE 68

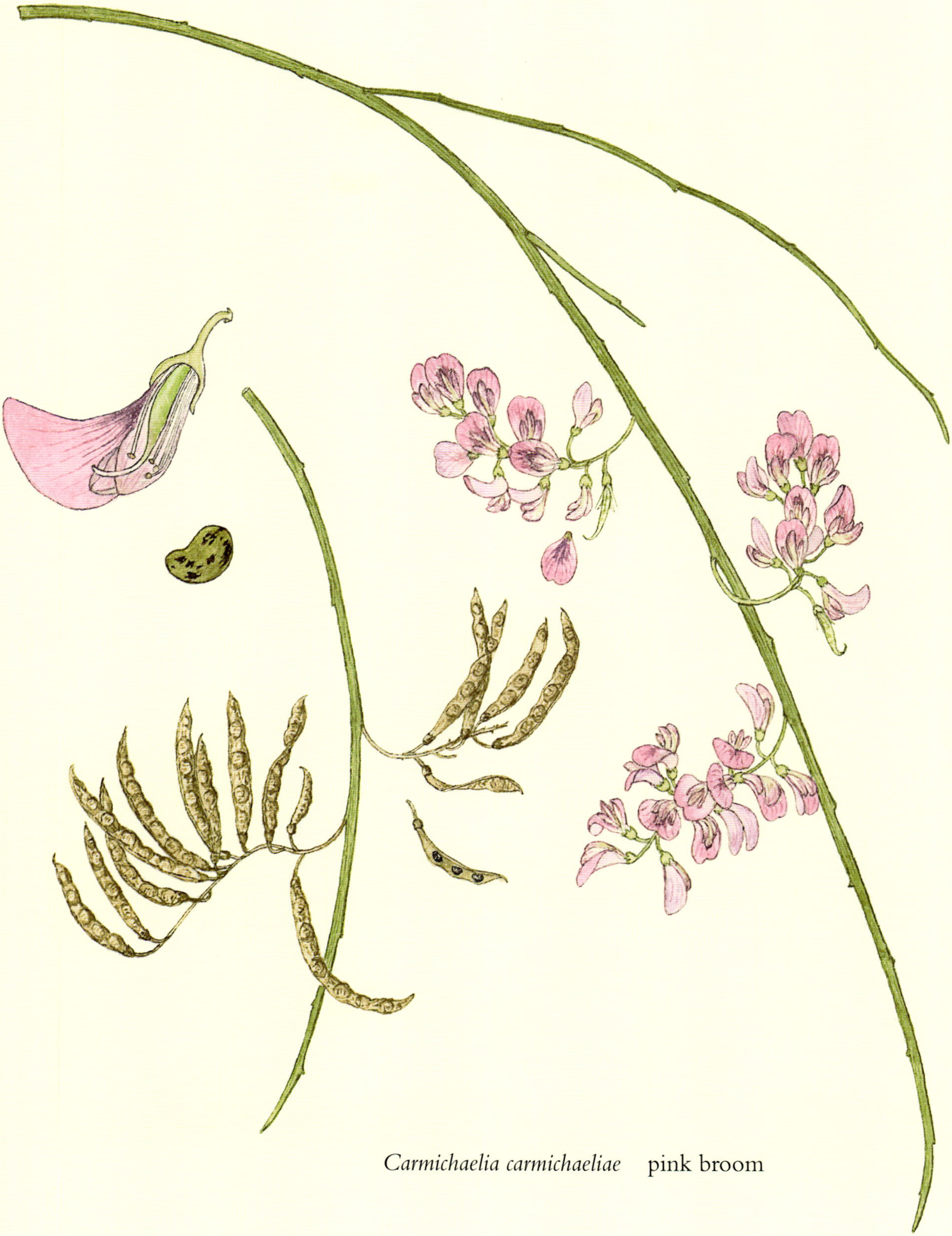

Carmichaelia carmichaeliae pink broom

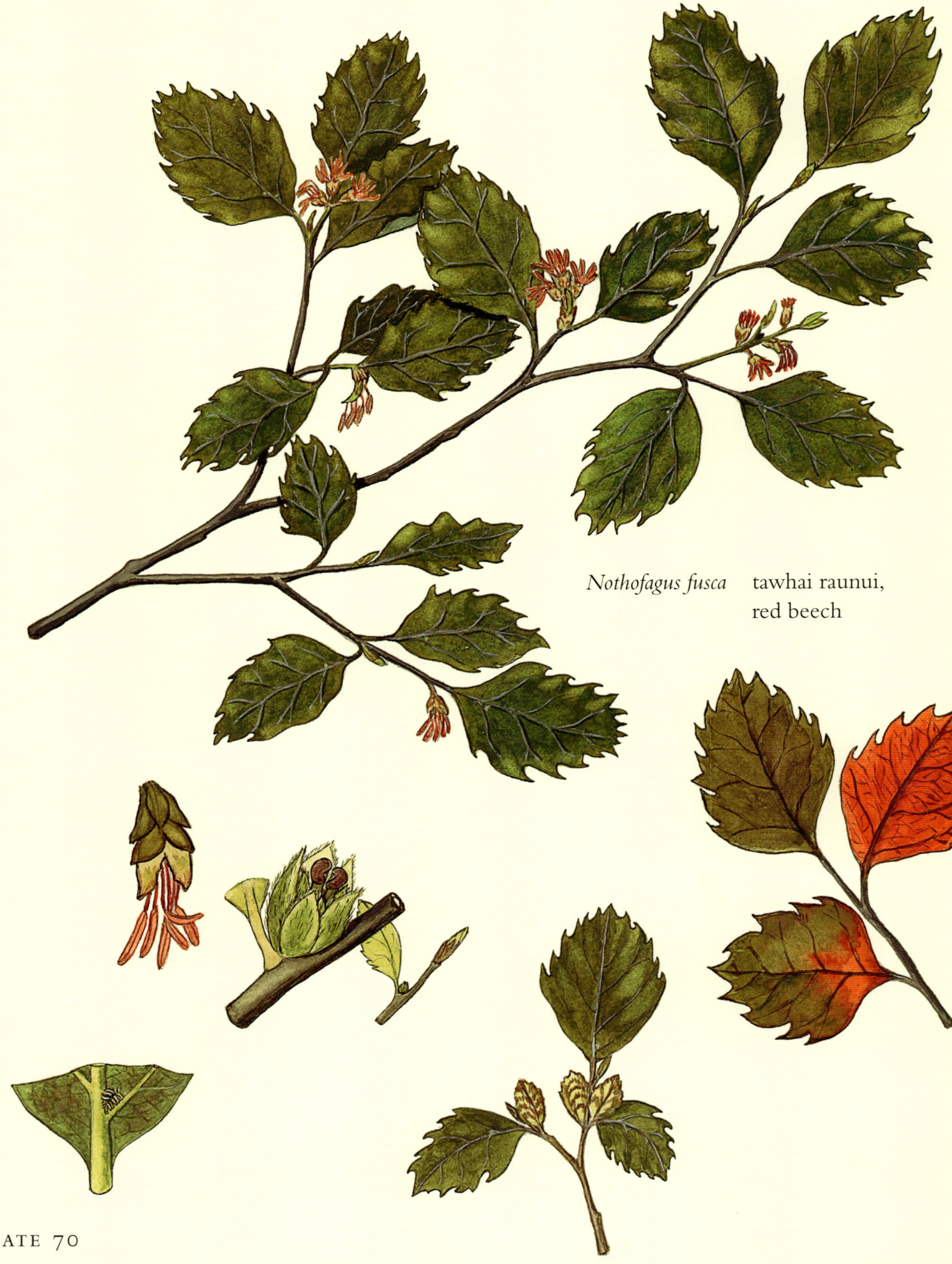

Nothofagus fusca tawhai raunui, red beech

PLATE 70

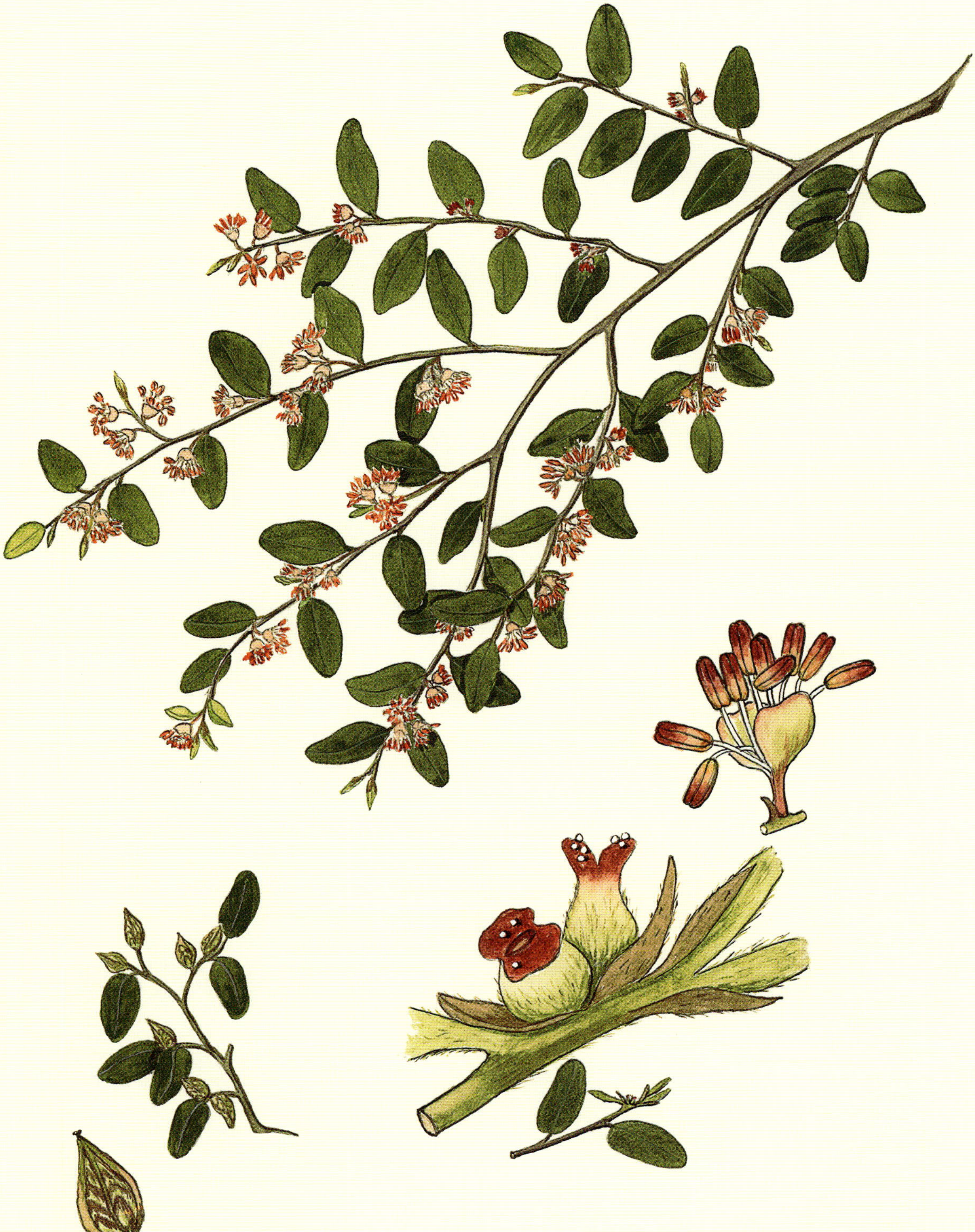

Nothofagus solandri var. *solandri* tawhai rauriki, black beech

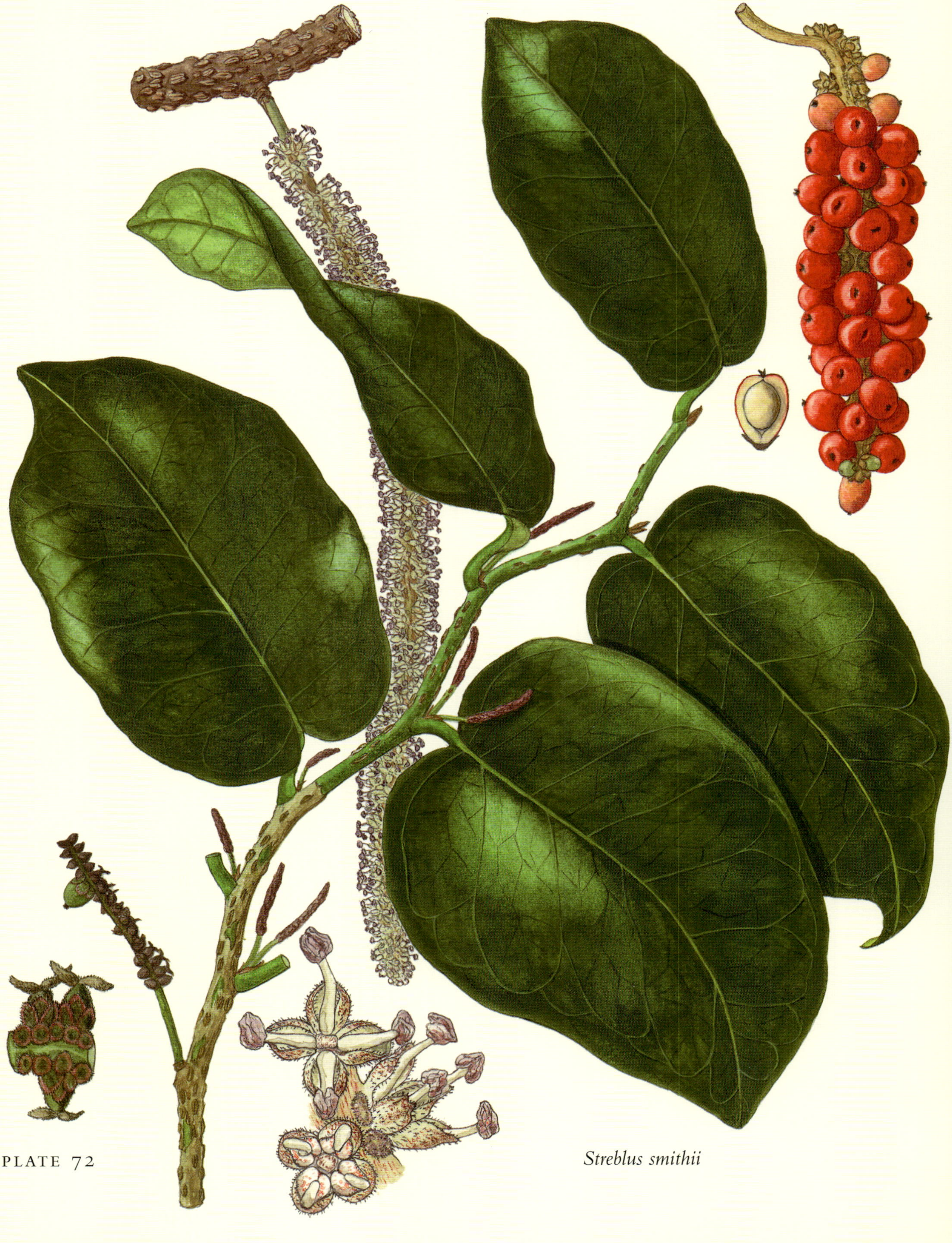

PLATE 72

Streblus smithii

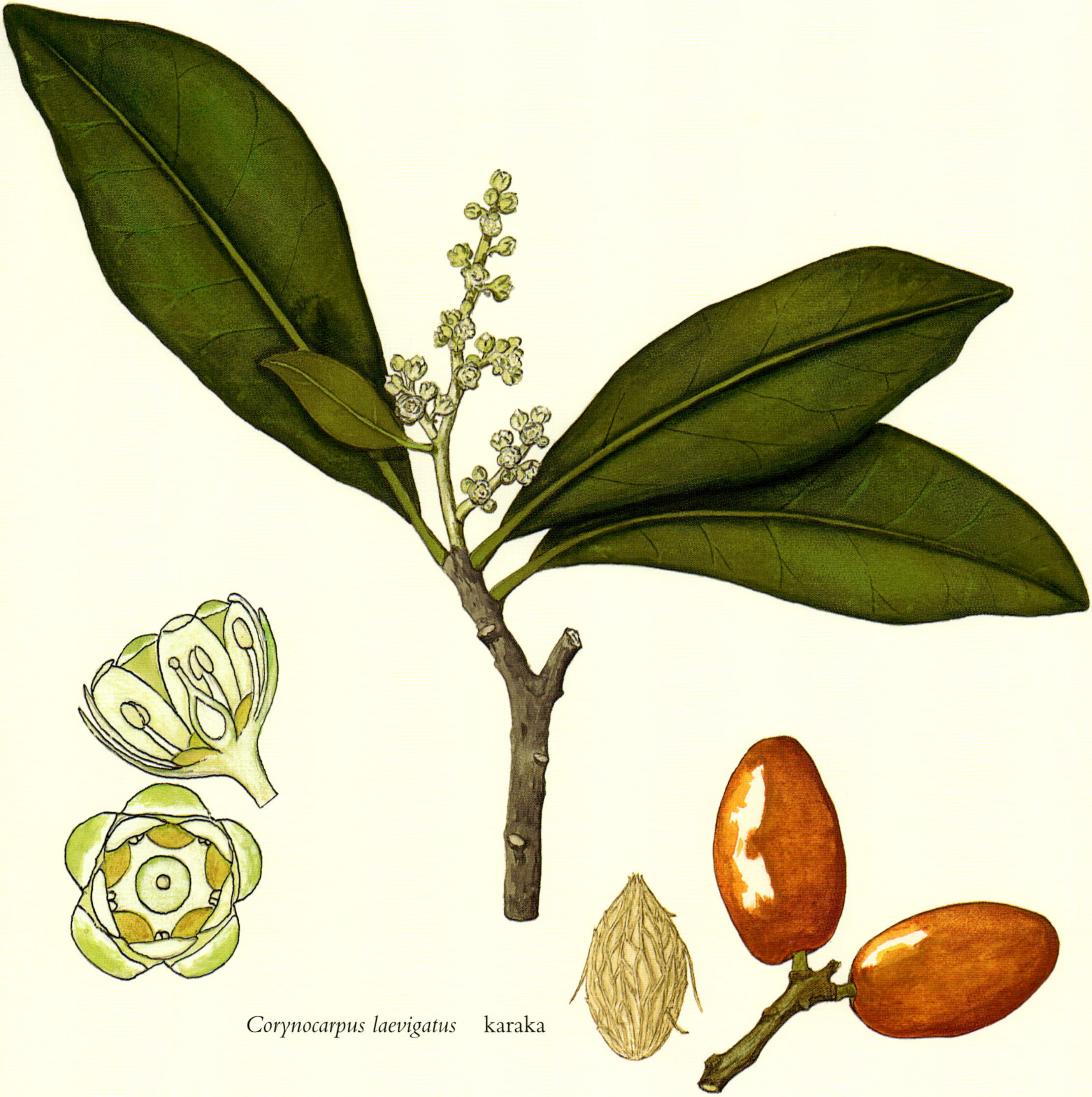

Corynocarpus laevigatus karaka

PLATE 73

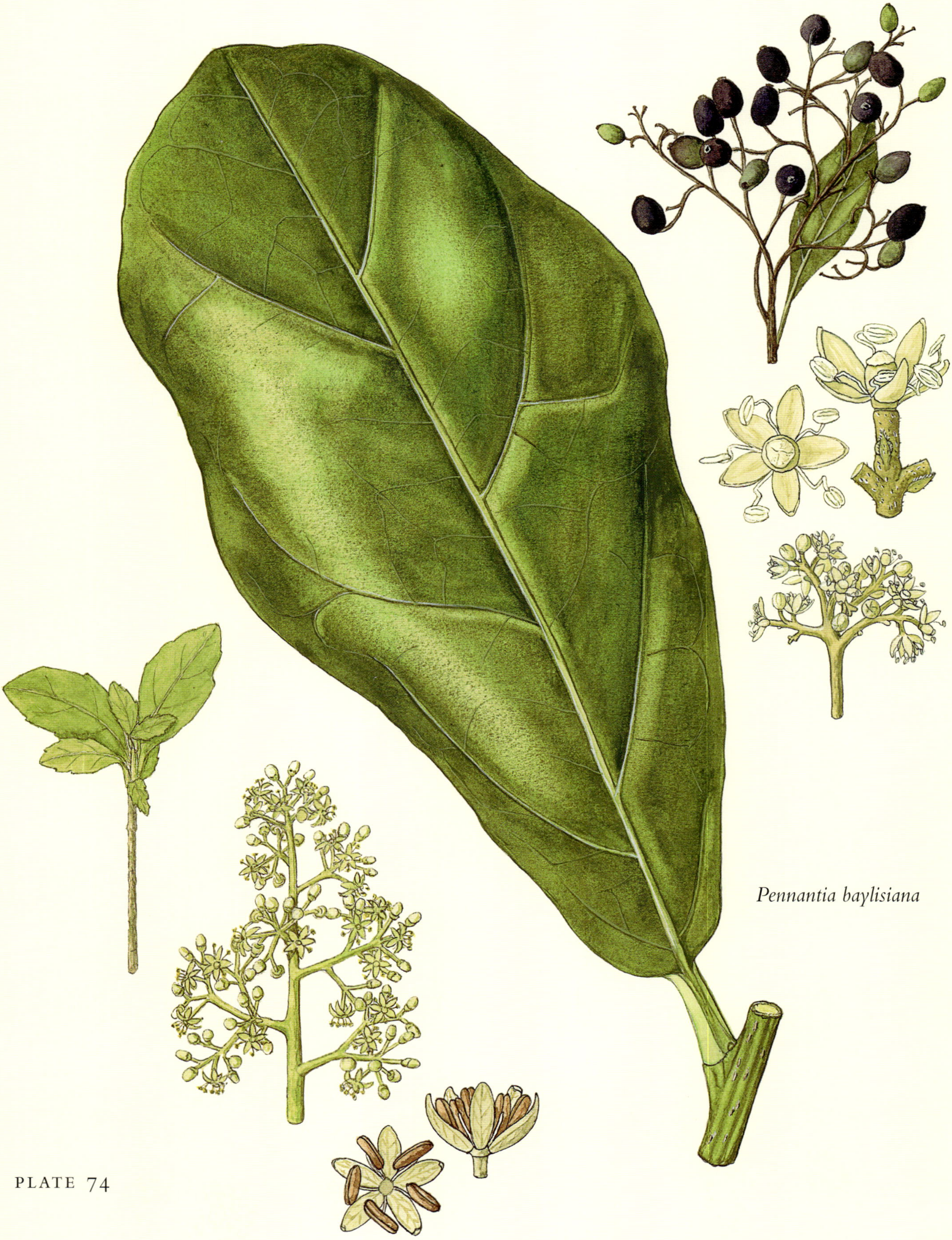

Pennantia baylisiana

PLATE 74

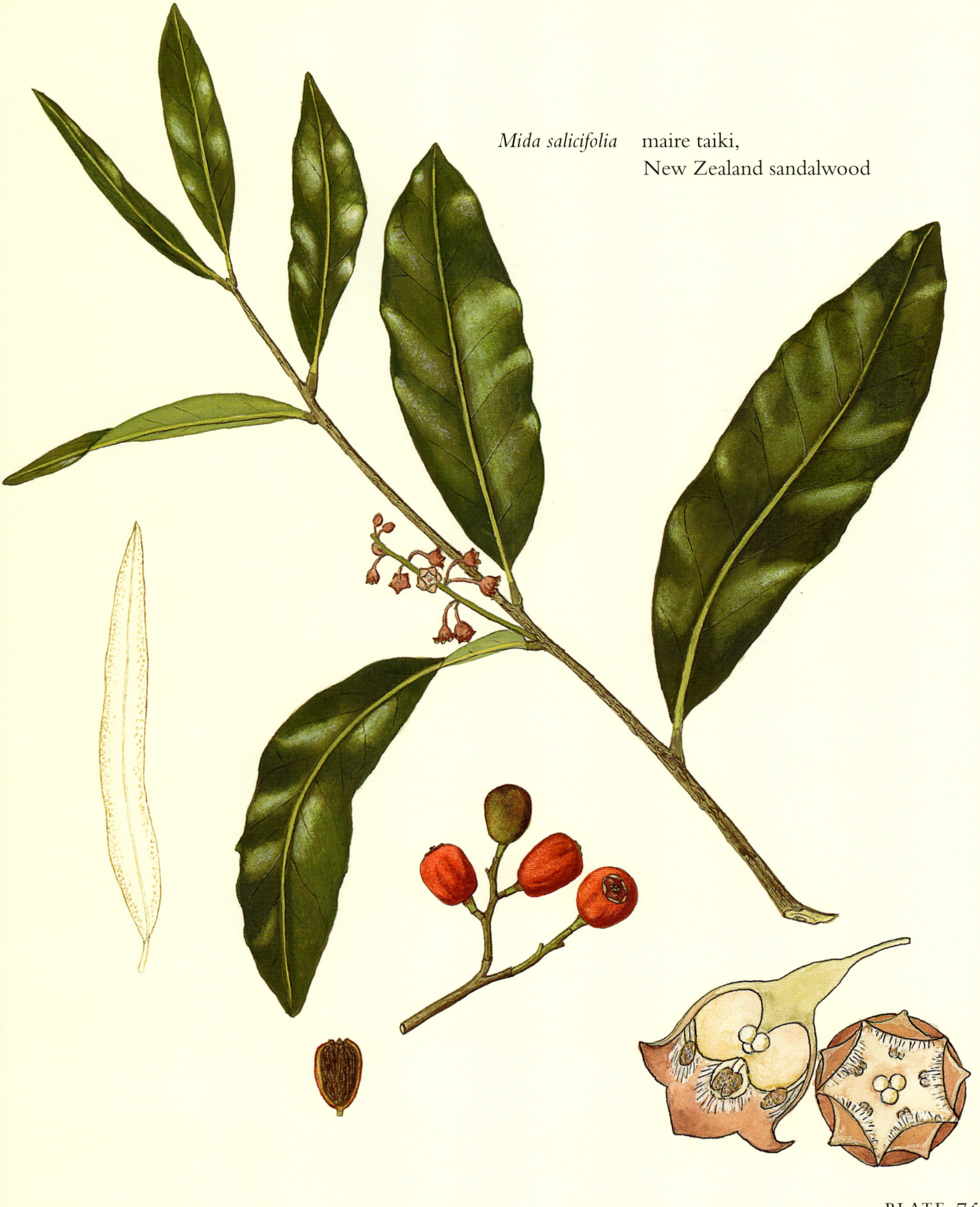

Mida salicifolia maire taiki,
New Zealand sandalwood

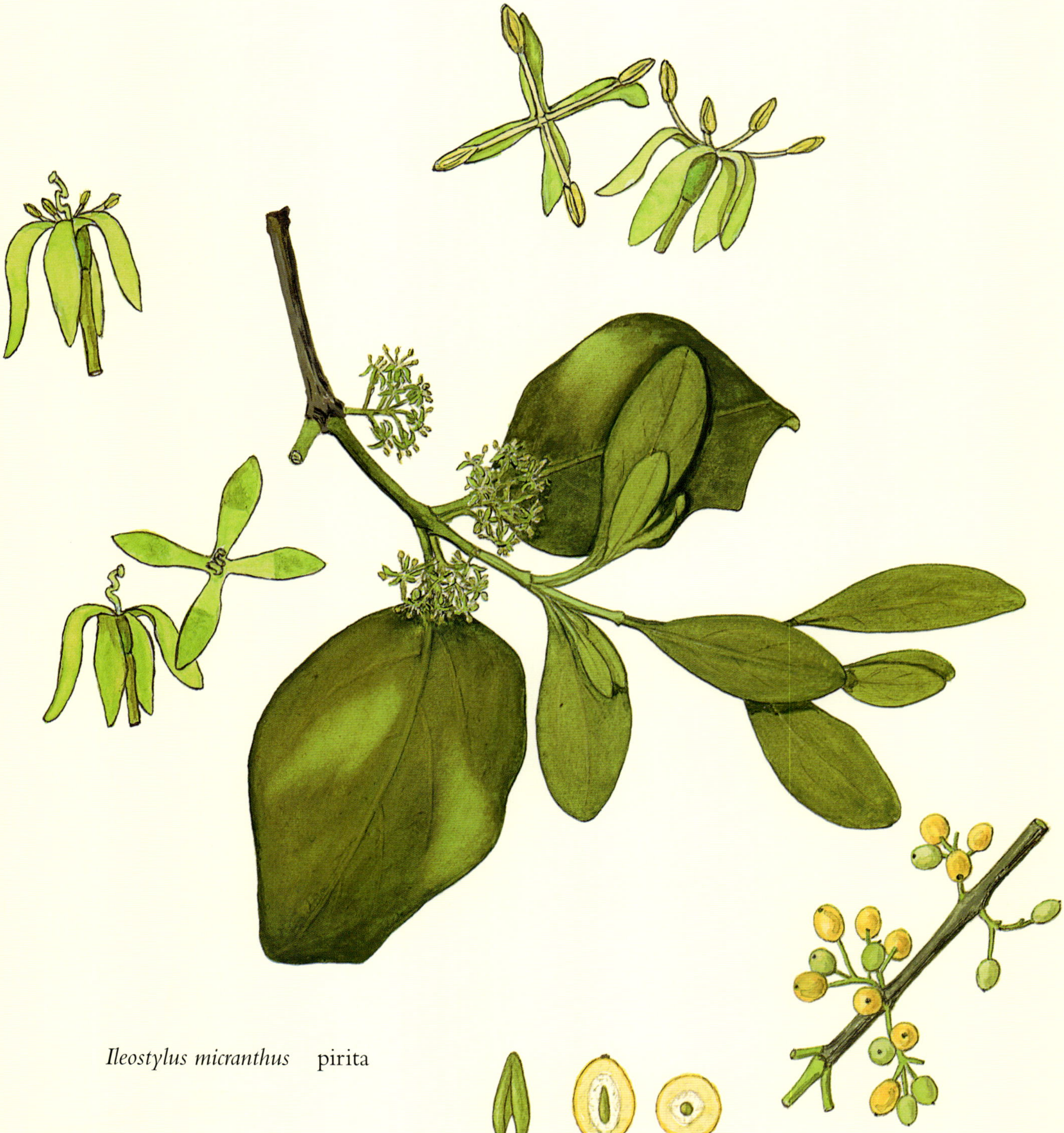

Ileostylus micranthus pirita

PLATE 76

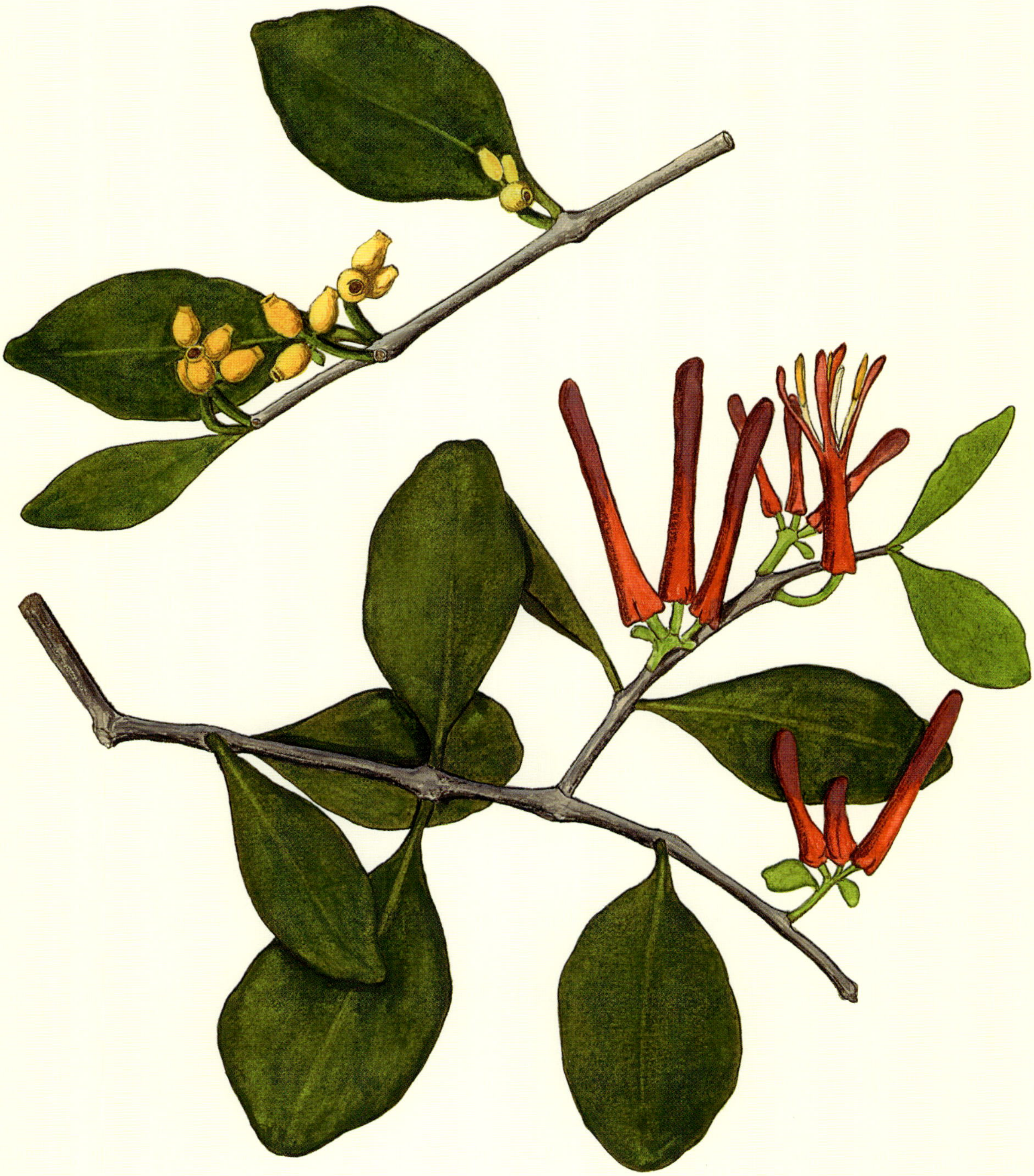

Peraxilla colensoi pirita, korukoru (when in flower),
scarlet-flowered beech mistletoe

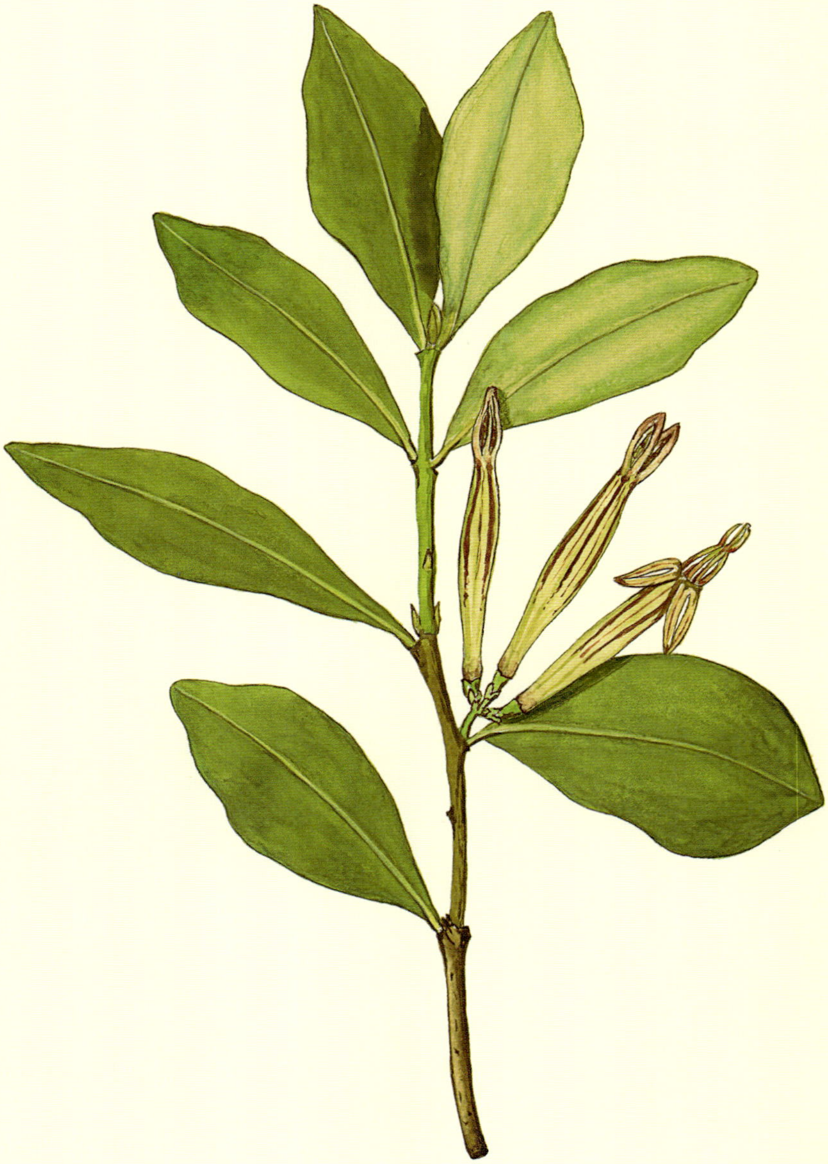

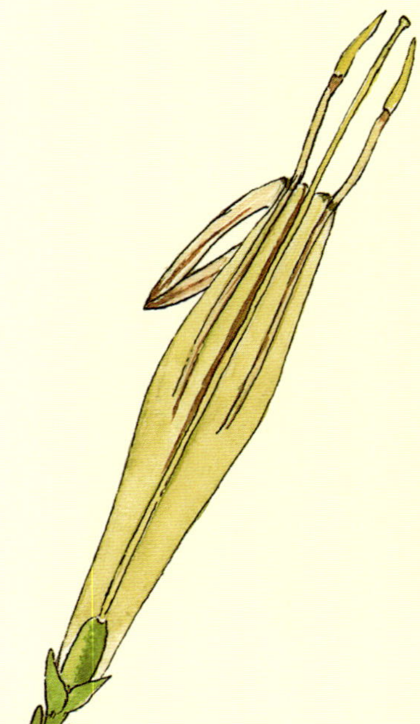

Trilepidea adamsii Adams' mistletoe

PLATE 78

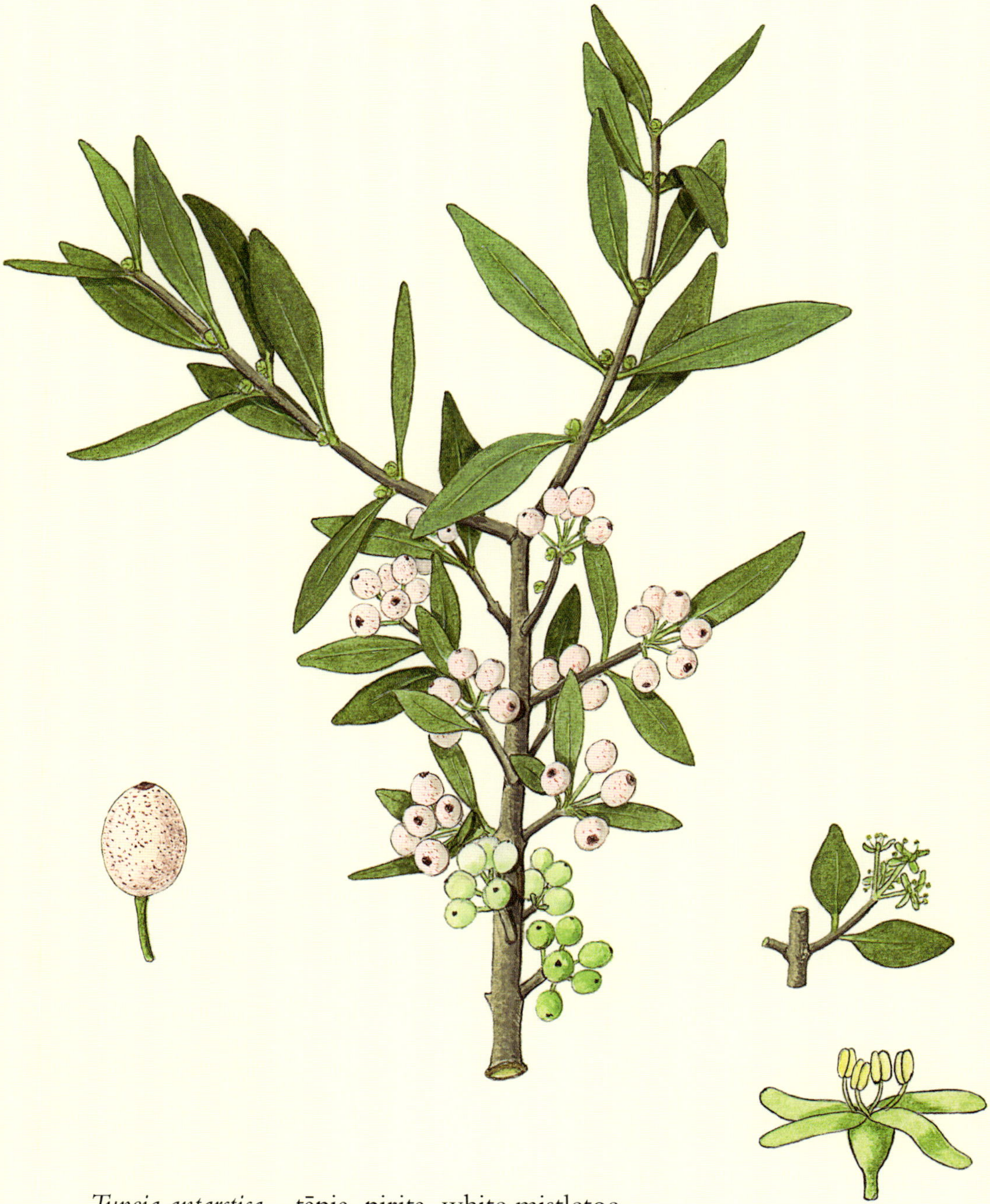

Tupeia antarctica tāpia, pirita, white mistletoe

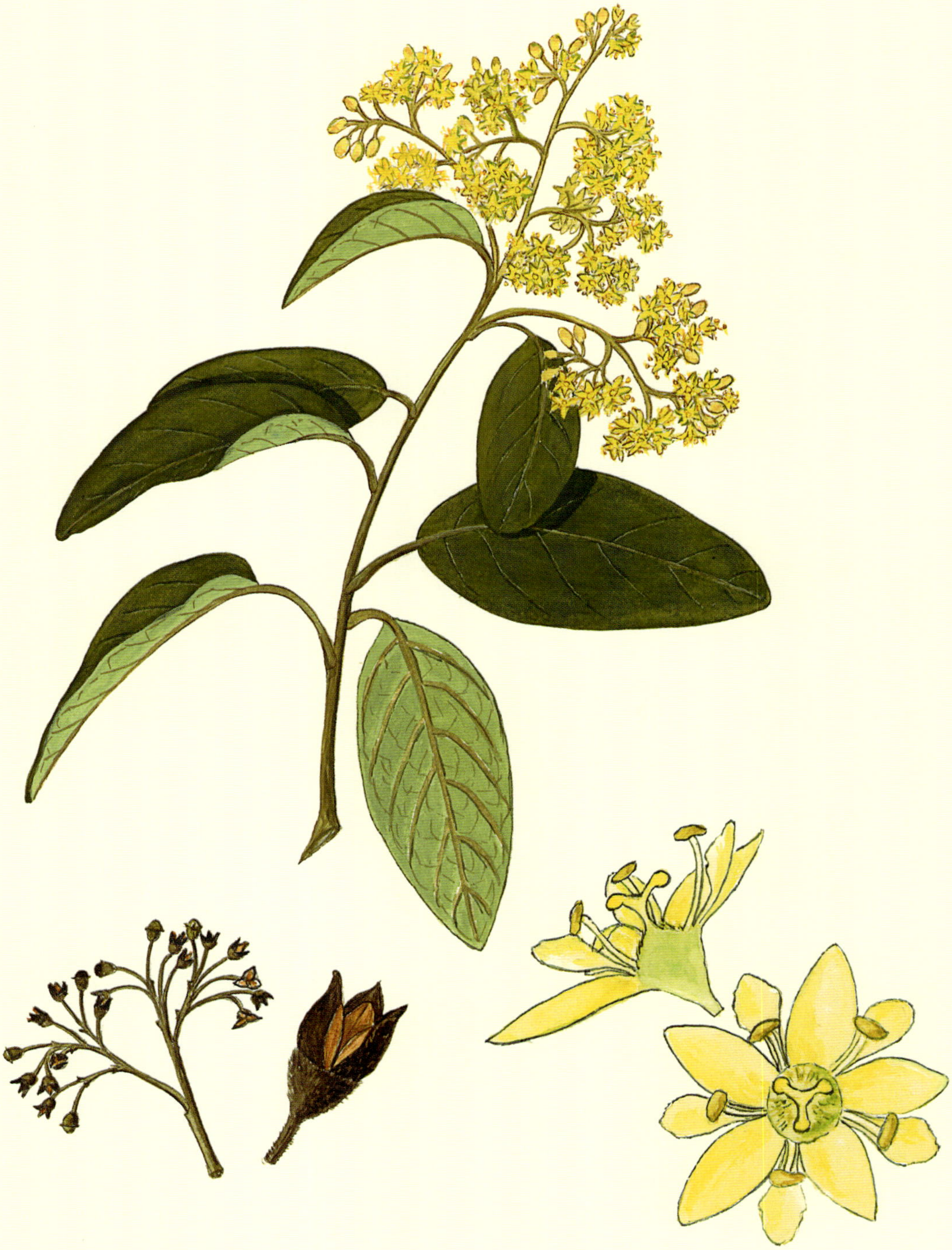

Pomaderris kumeraho kūmarahou

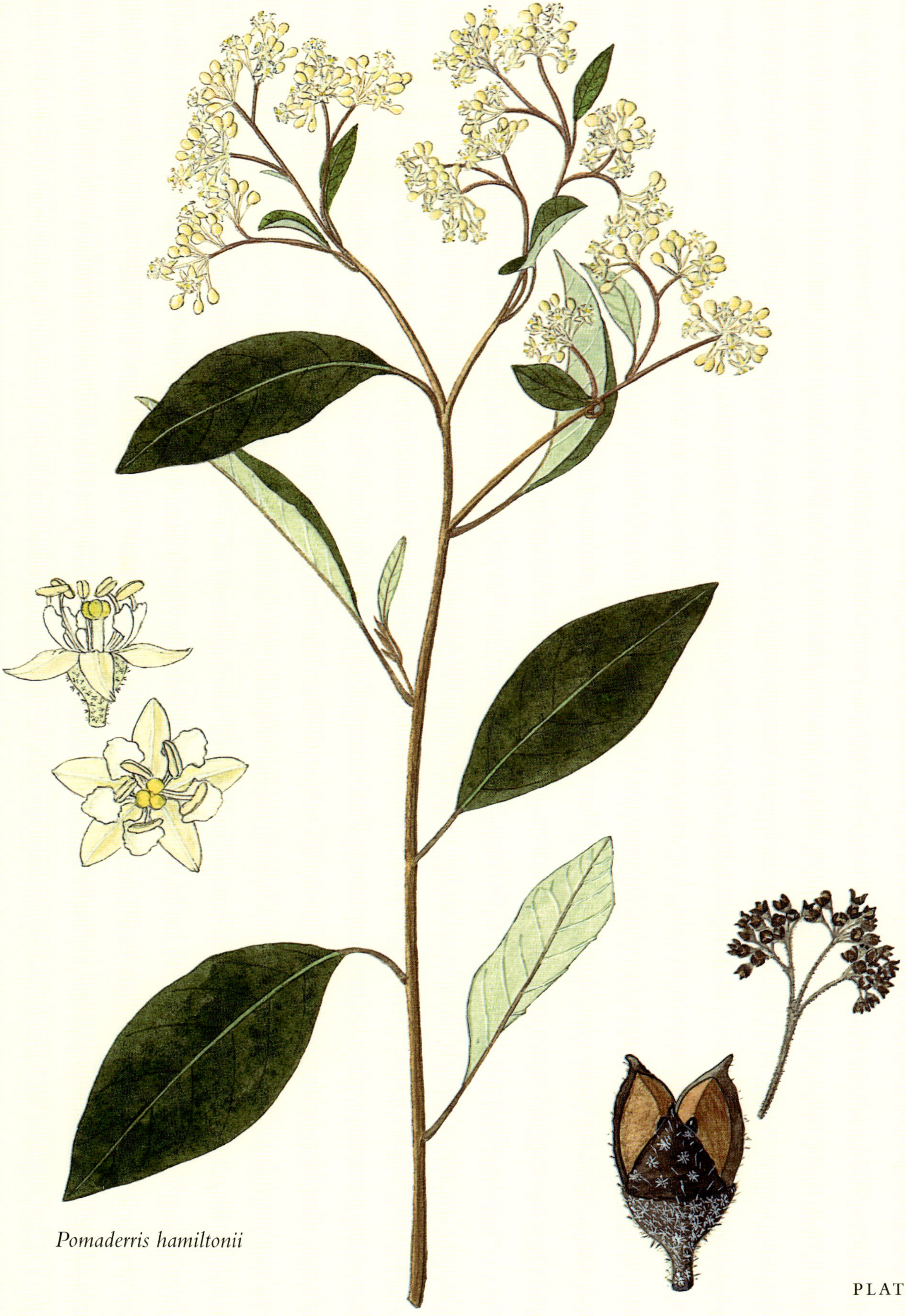

Pomaderris hamiltonii

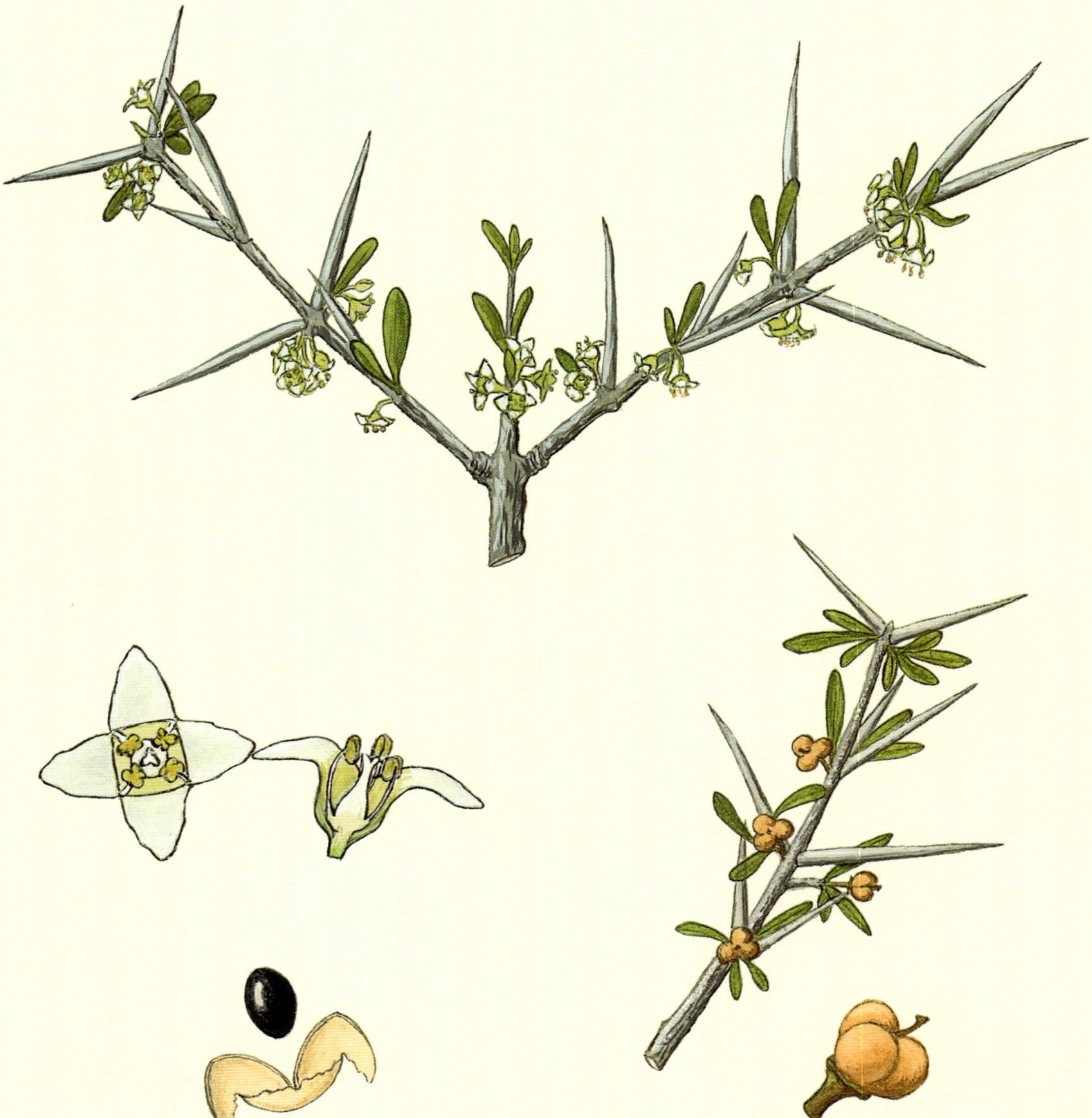

Discaria toumatou tūmatakuru, wild Irishman, matagouri

PLATE 82

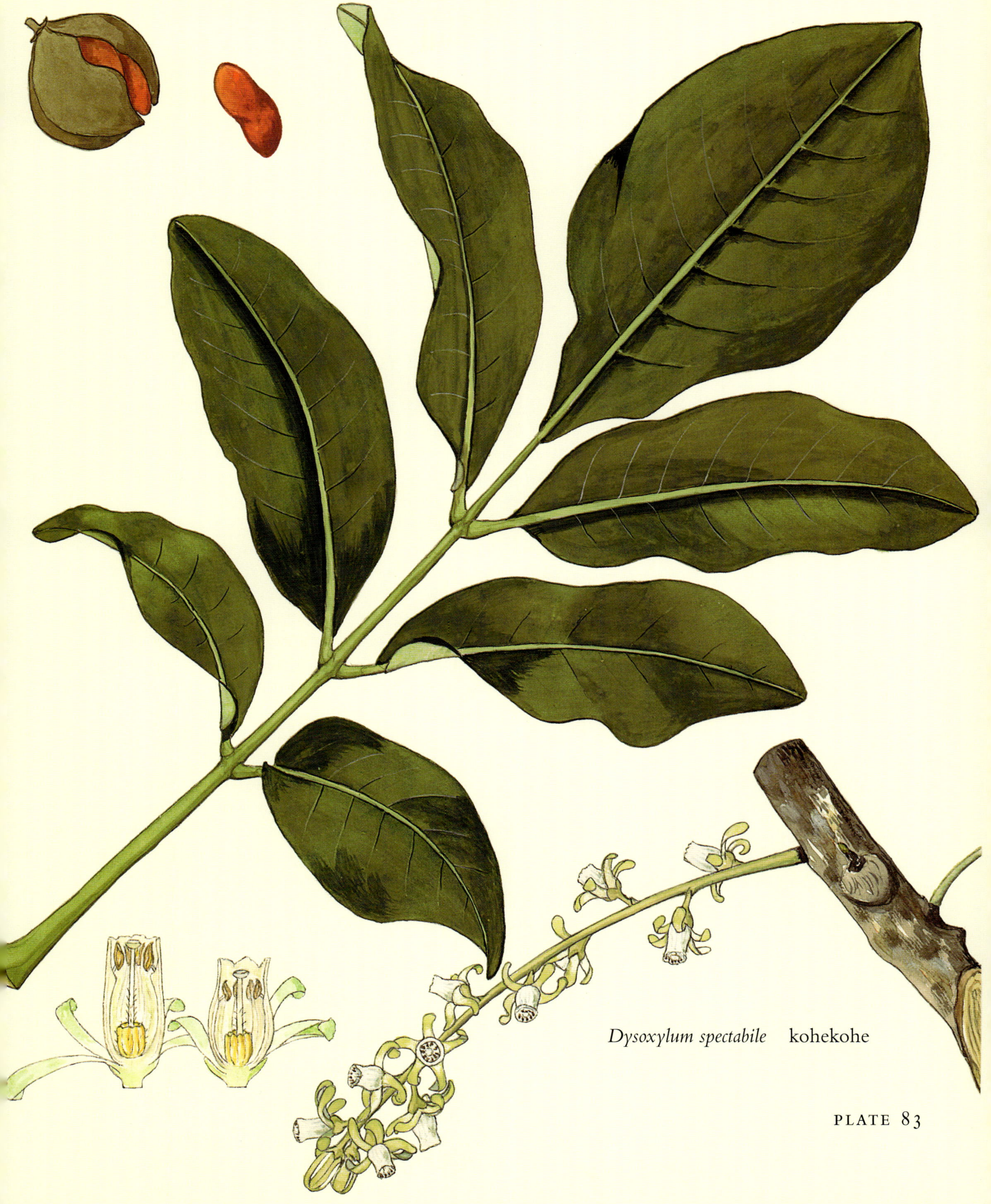

Dysoxylum spectabile kohekohe

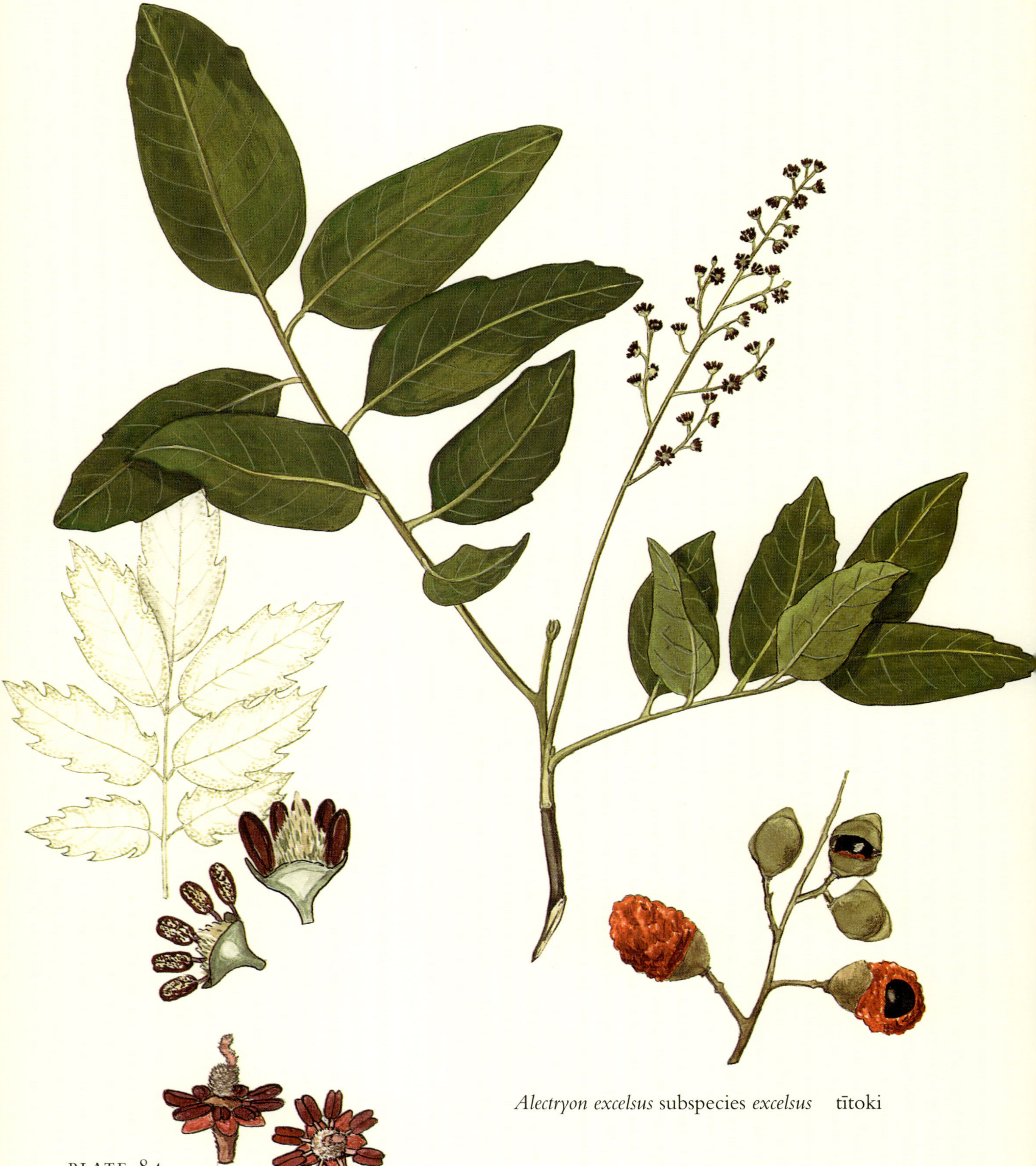

Alectryon excelsus subspecies *excelsus* tītoki

PLATE 84

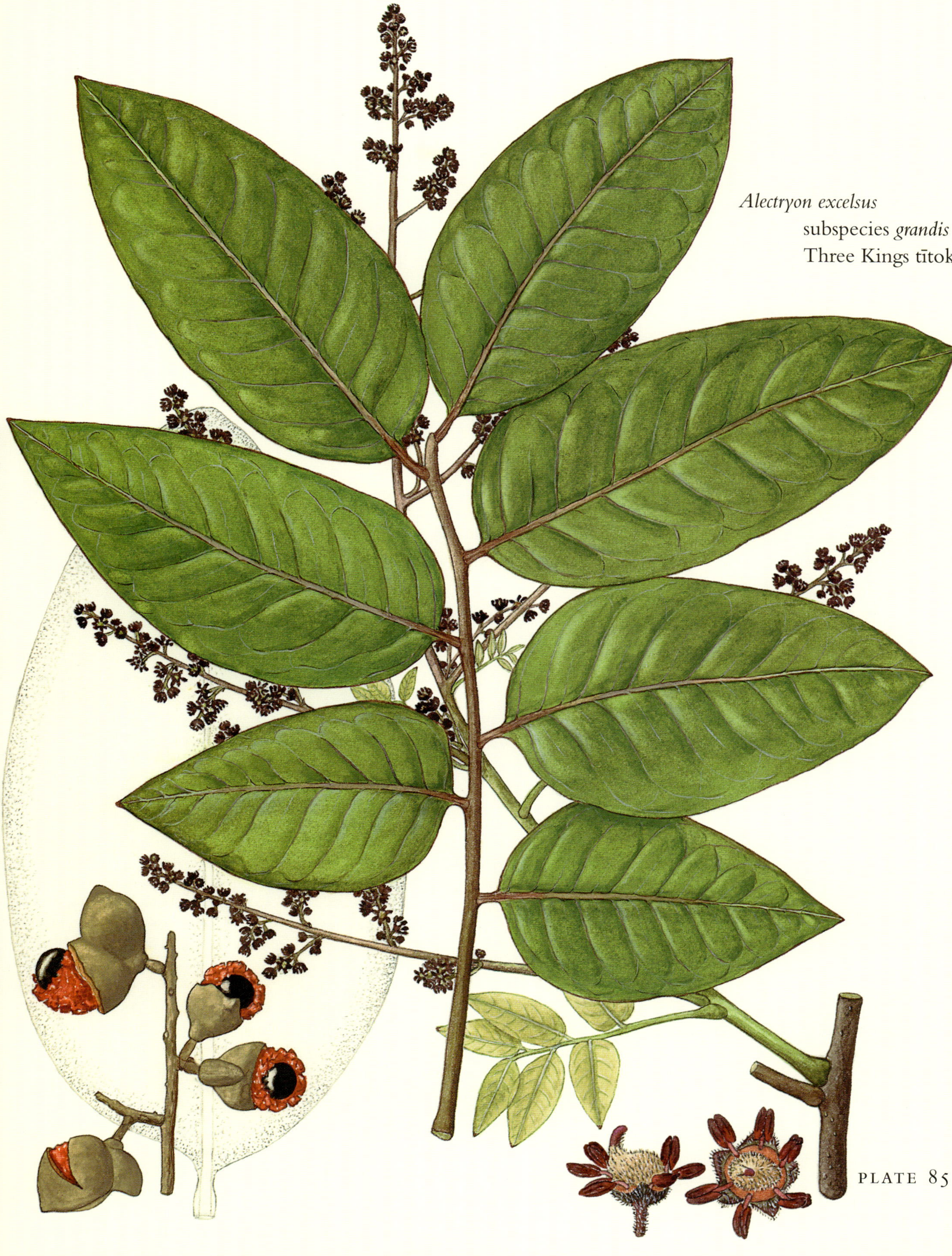

Alectryon excelsus
subspecies *grandis*
Three Kings tītoki

PLATE 85

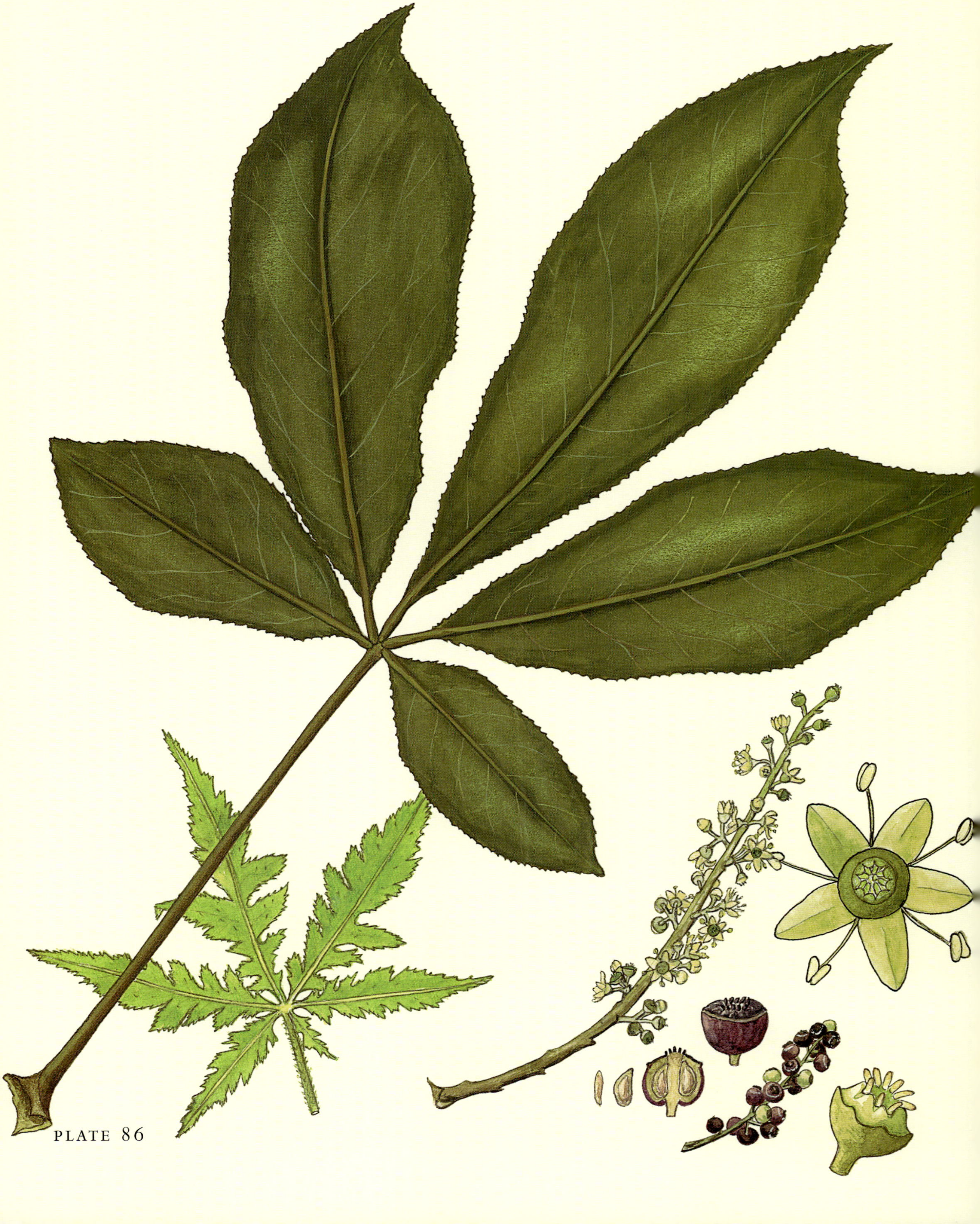

PLATE 86

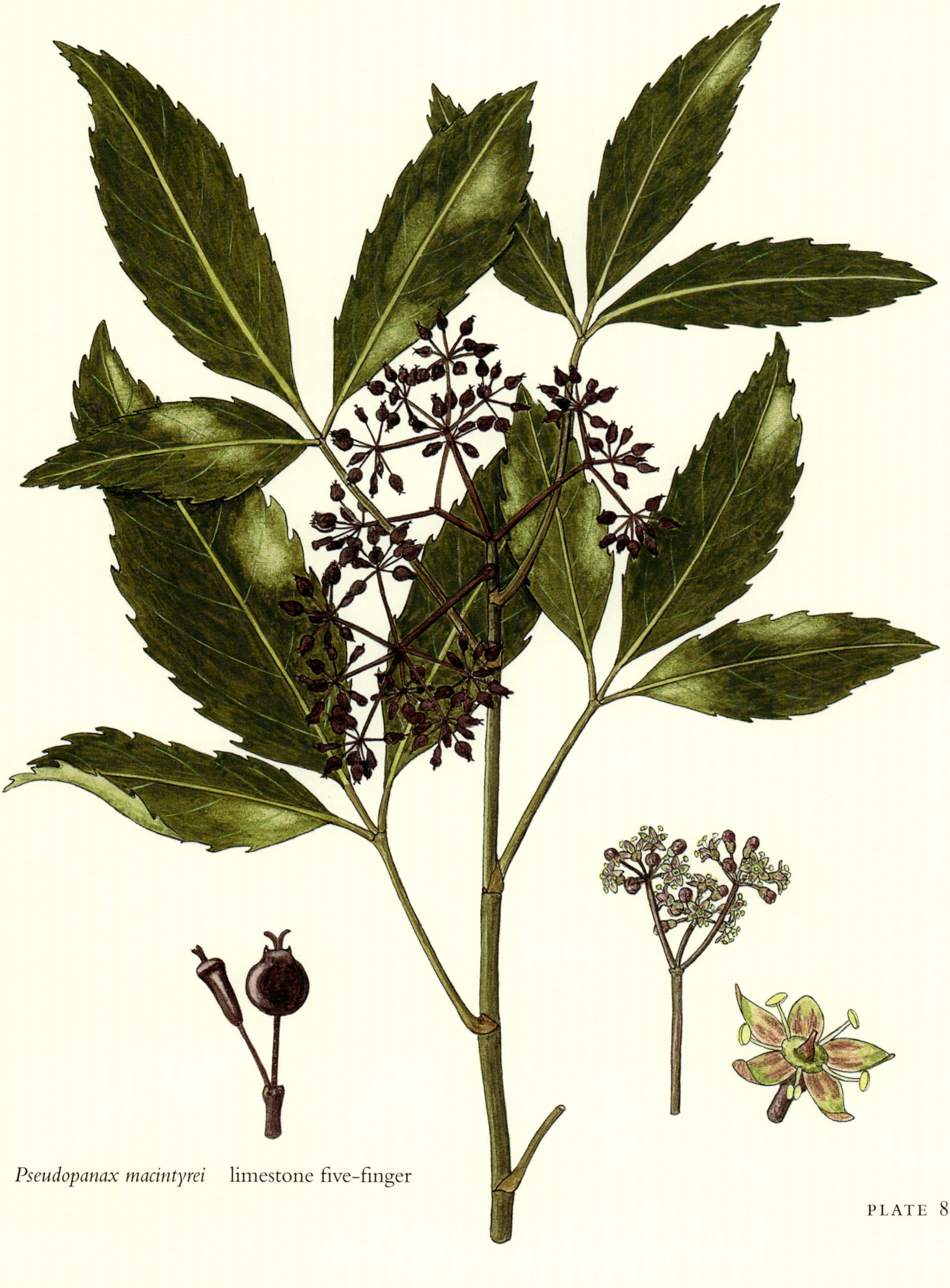

Pseudopanax macintyrei limestone five-finger

PLATE 87

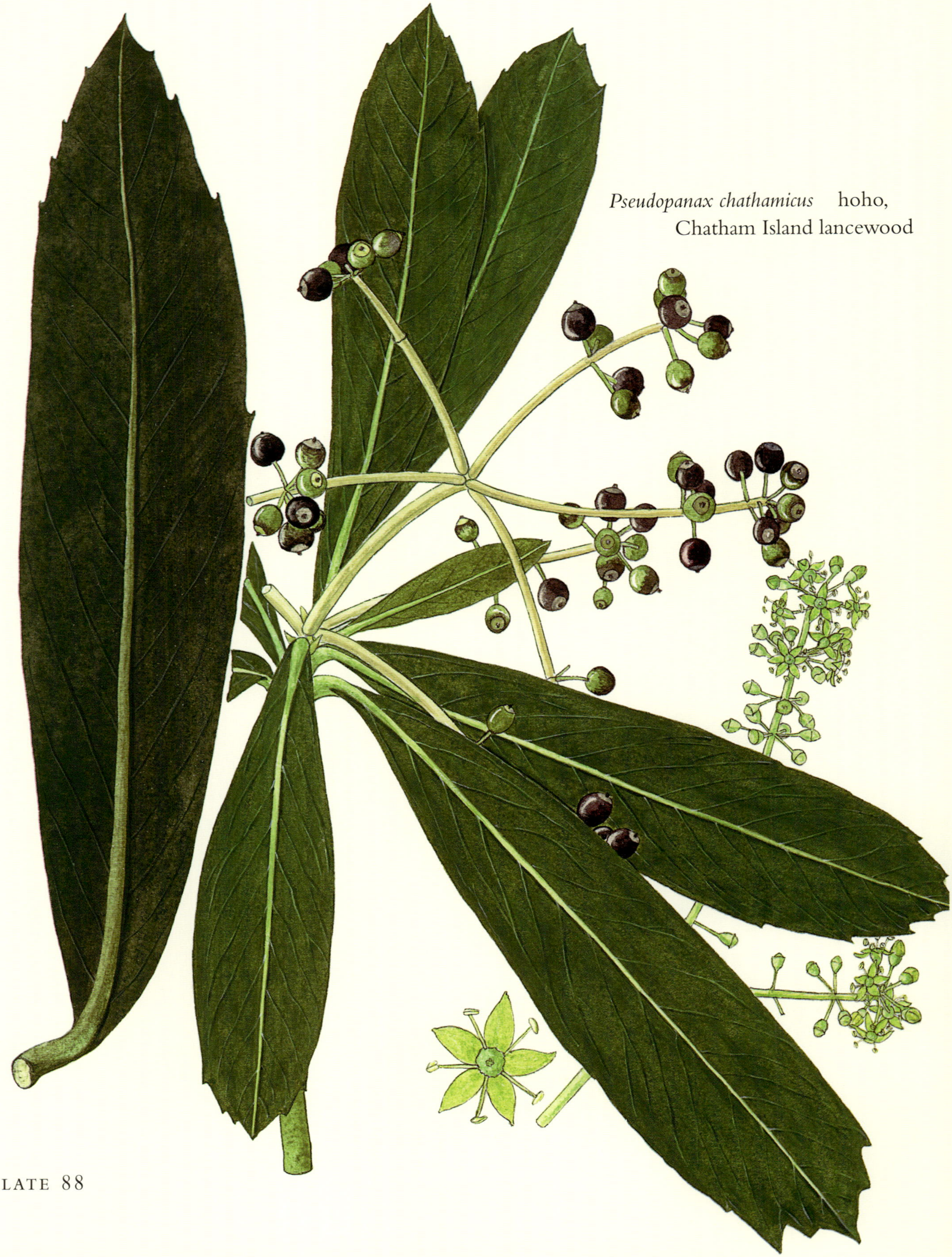

Pseudopanax chathamicus hoho,
Chatham Island lancewood

PLATE 88

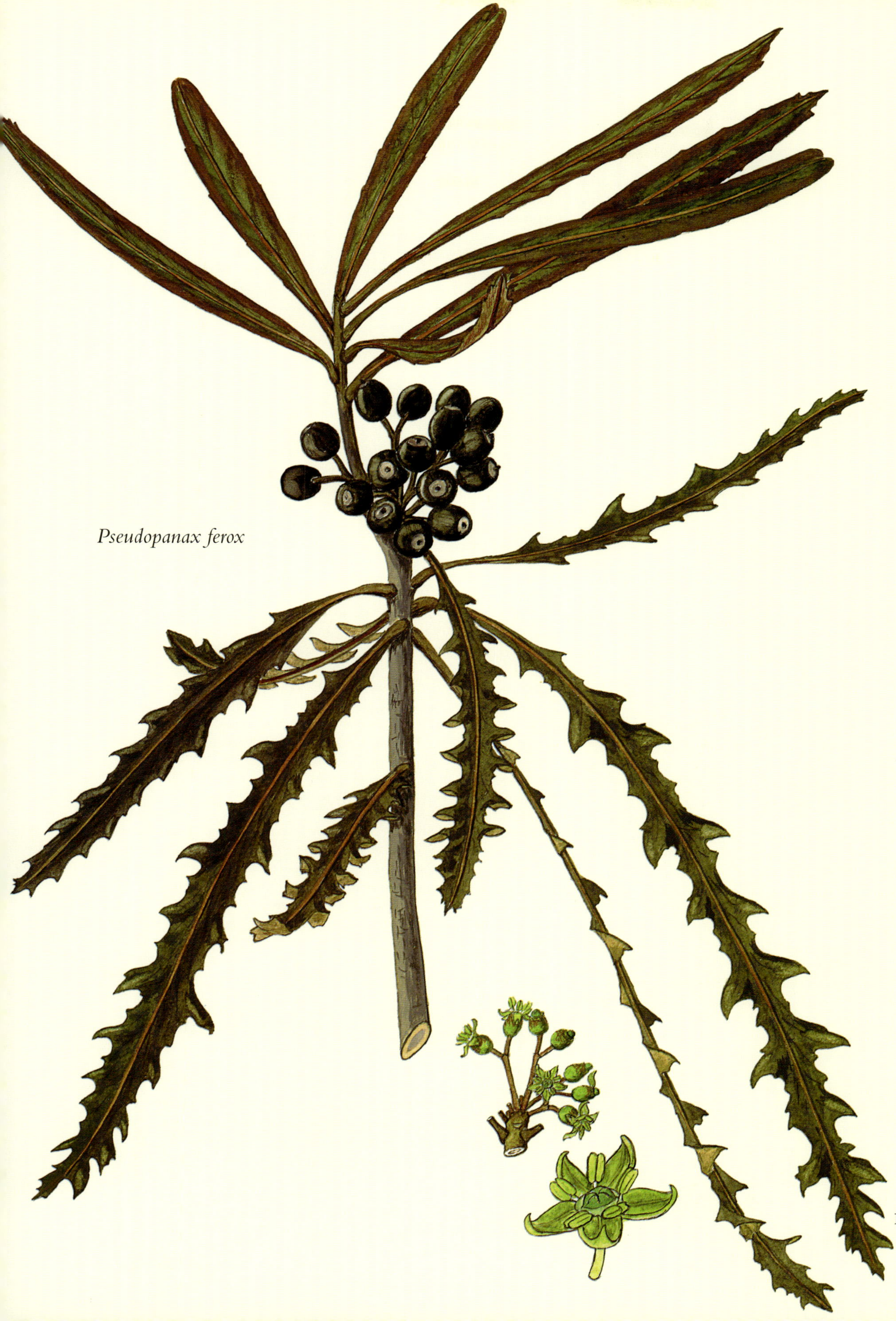

Pseudopanax ferox

PLATE 89

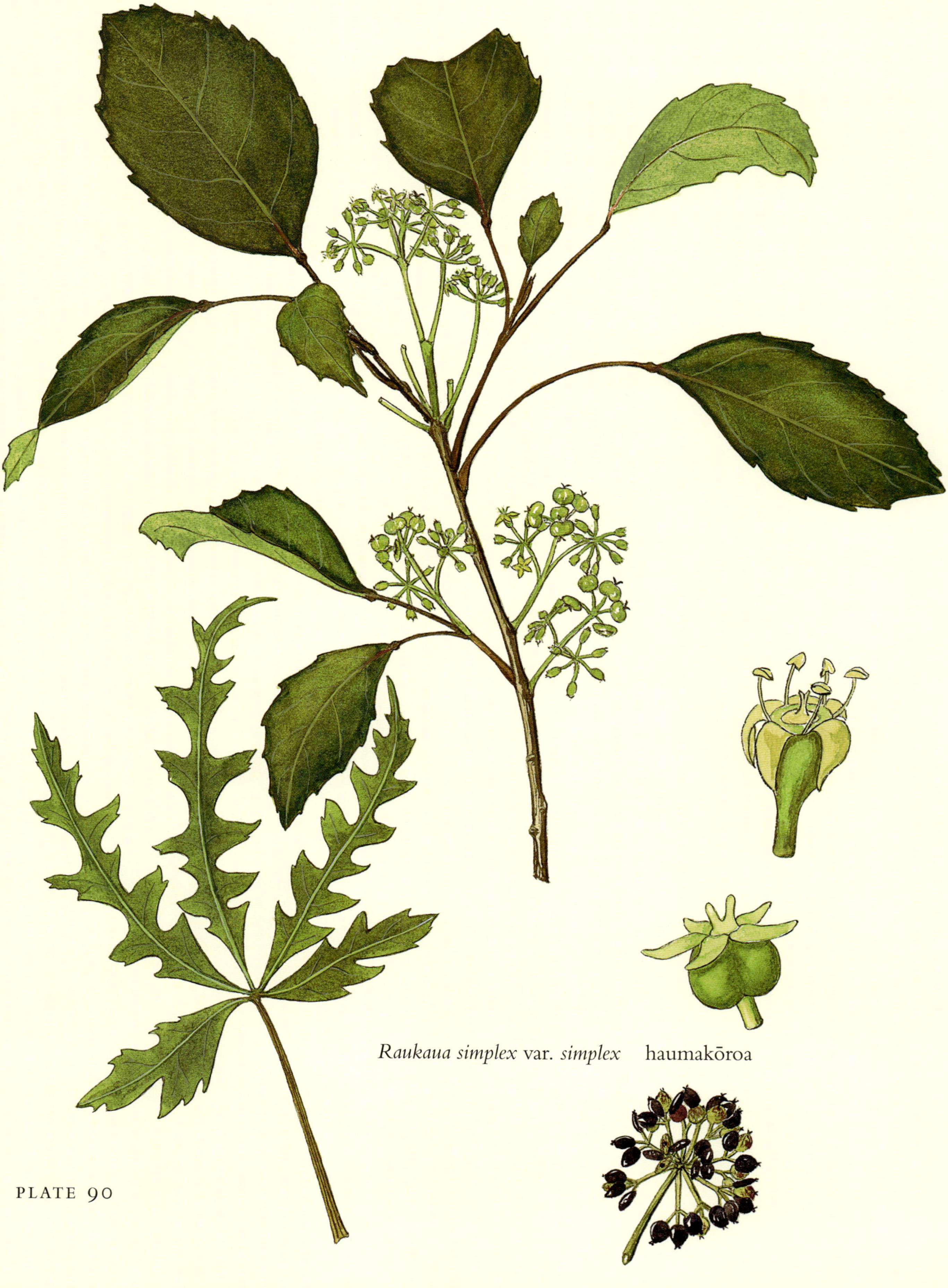

Raukaua simplex var. *simplex* haumakōroa

PLATE 90

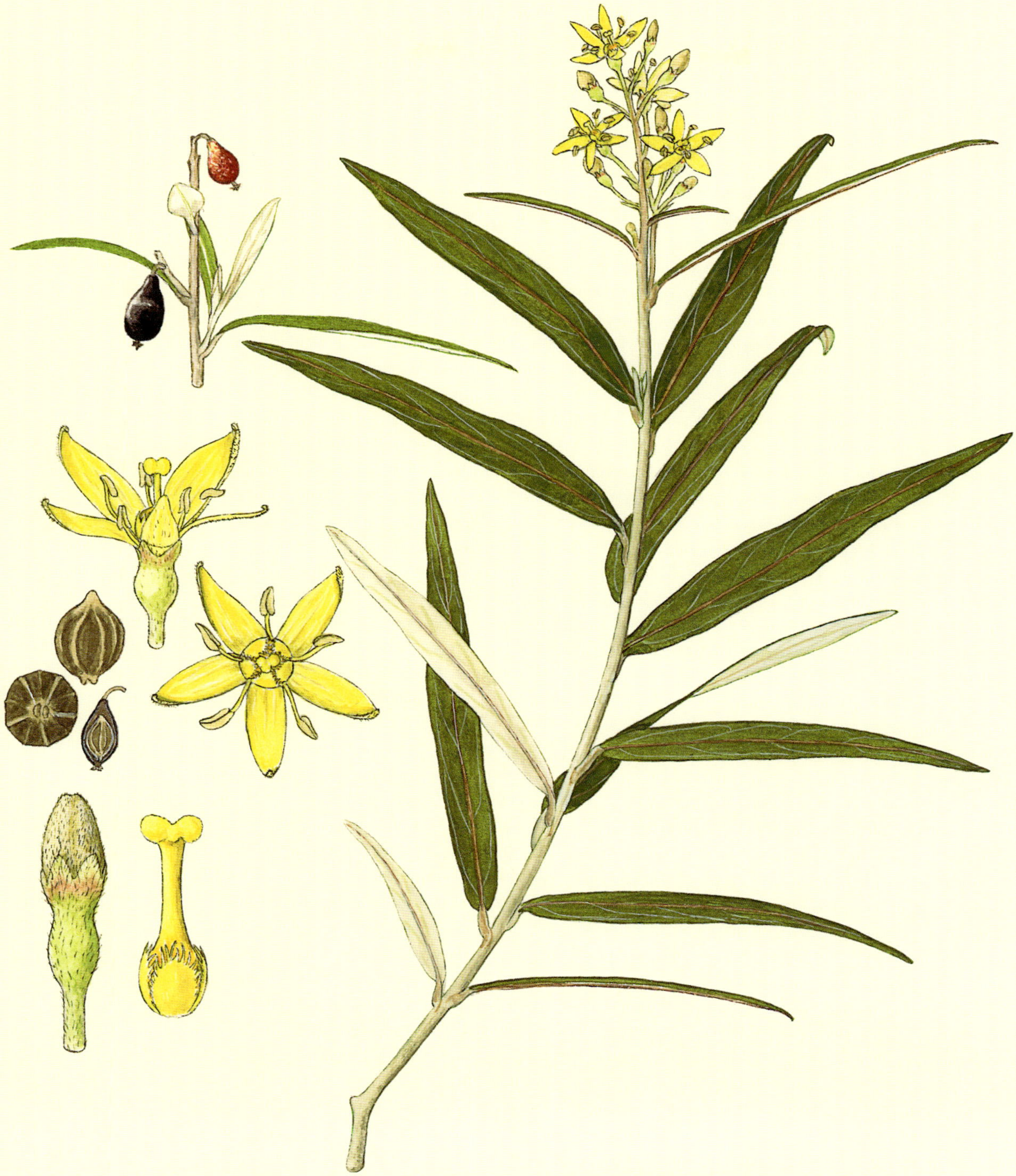

Corokia buddleioides var. *linearis* korokio

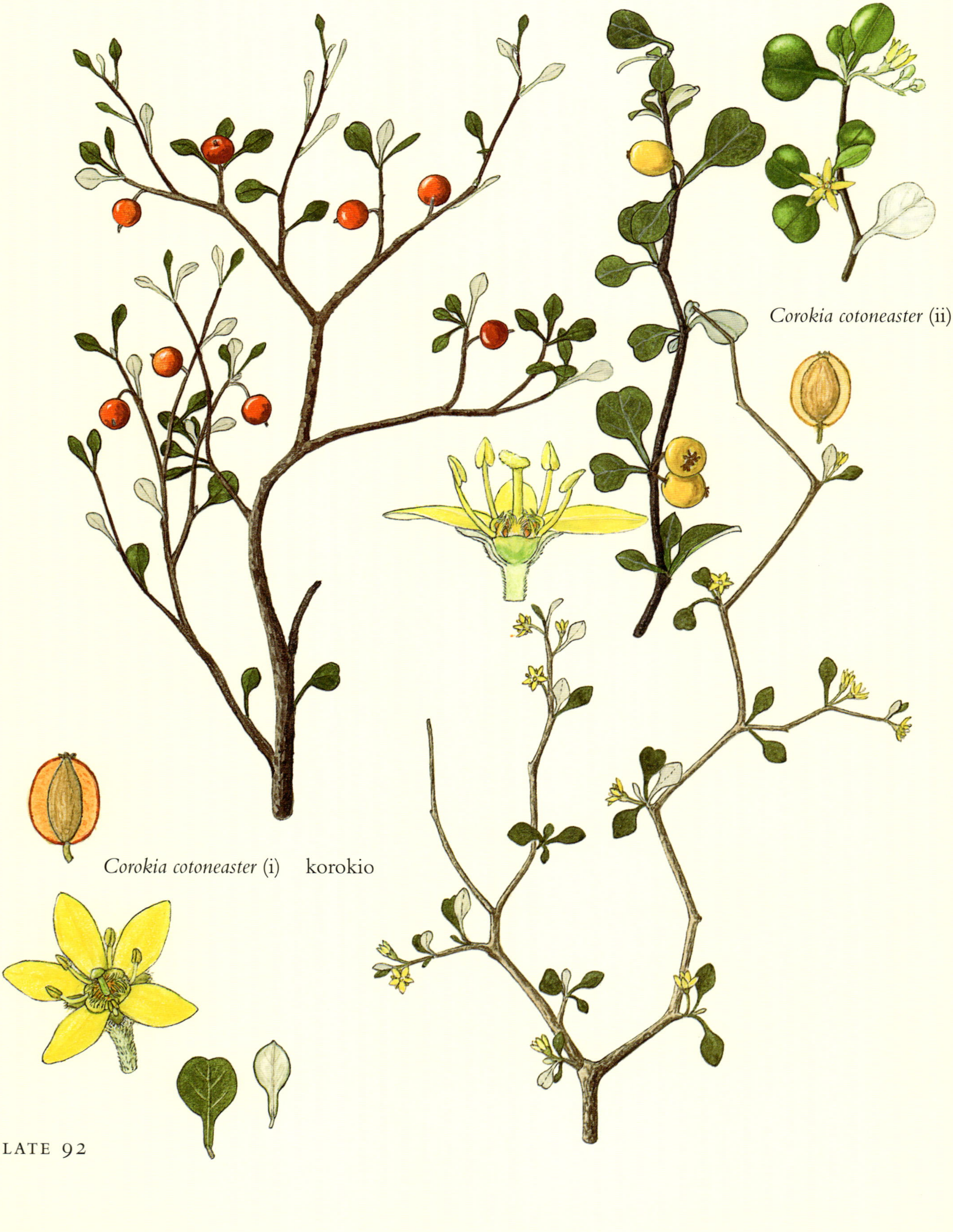

Corokia cotoneaster (ii)

Corokia cotoneaster (i) korokio

PLATE 92

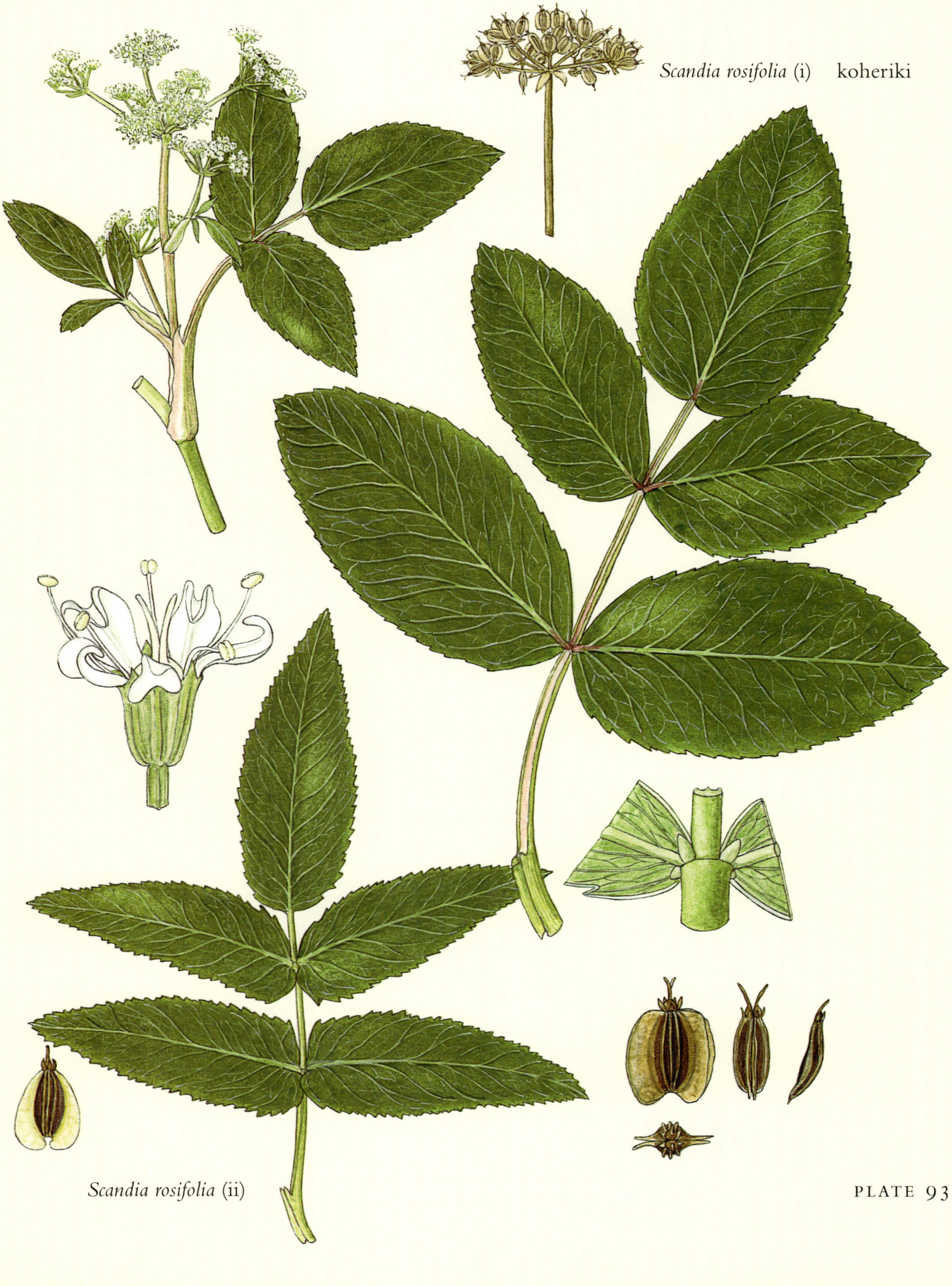

Scandia rosifolia (i) koheriki
Scandia rosifolia (ii)

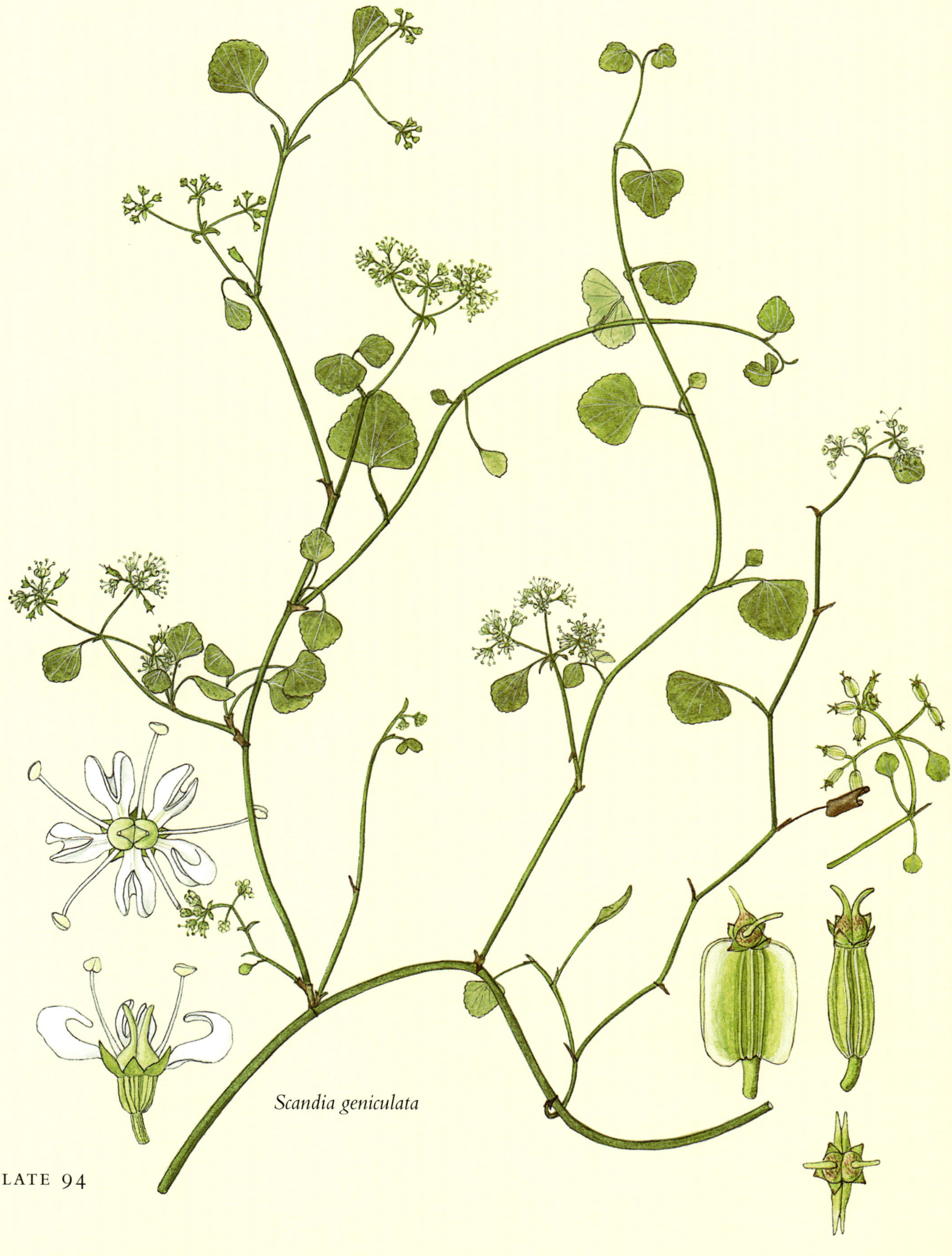

Scandia geniculata

PLATE 94

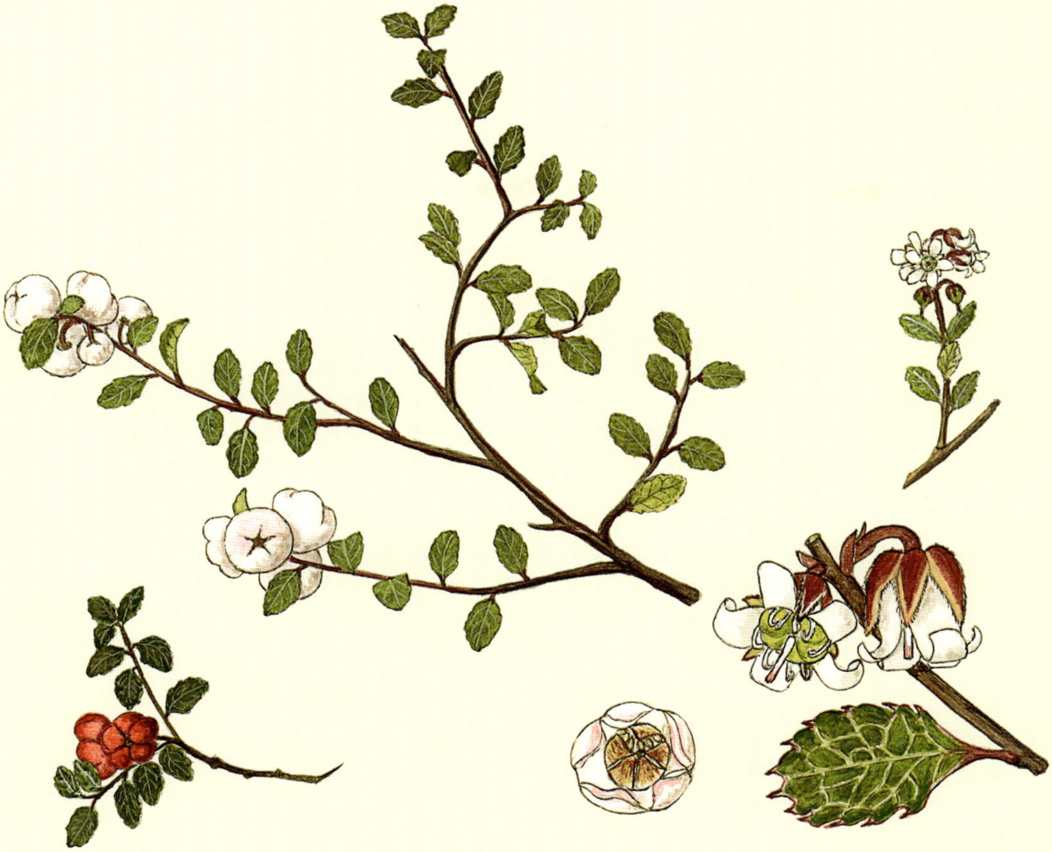

Gaultheria depressa var. *novae-zealandiae* mountain snowberry

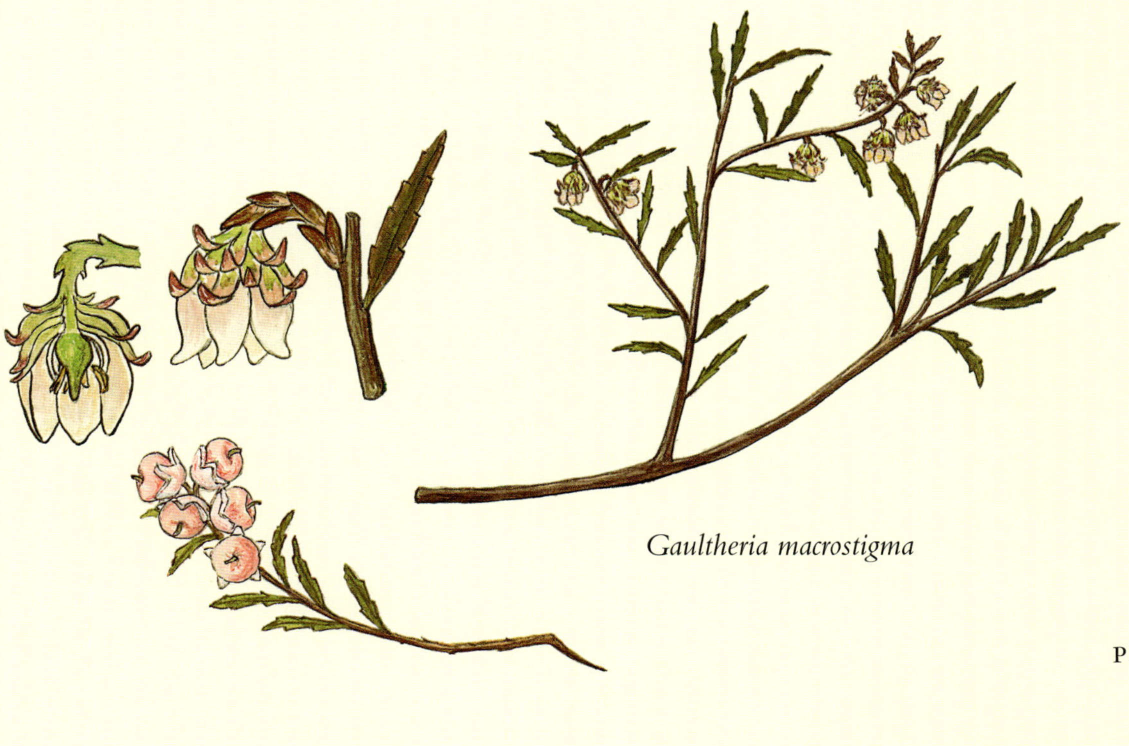

Gaultheria macrostigma

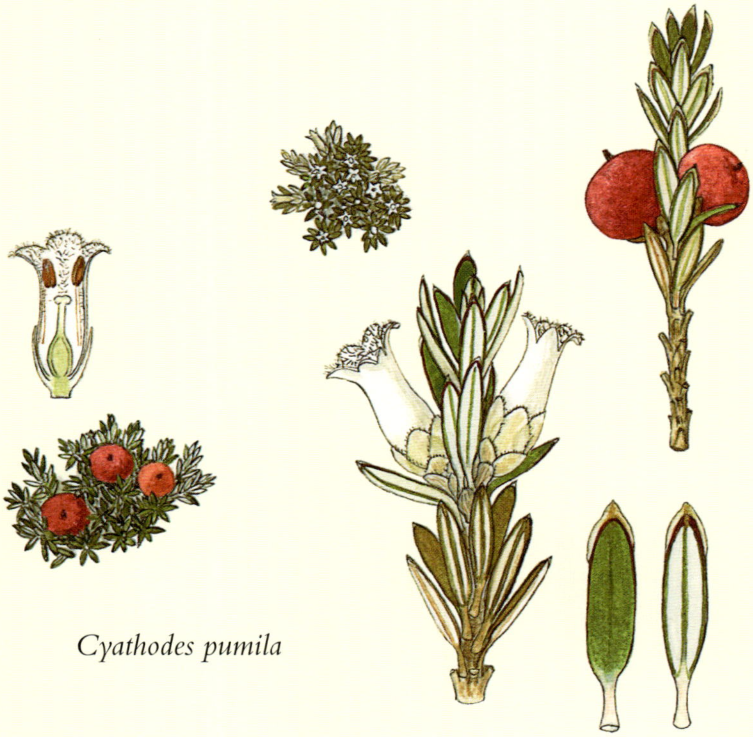

Cyathodes pumila

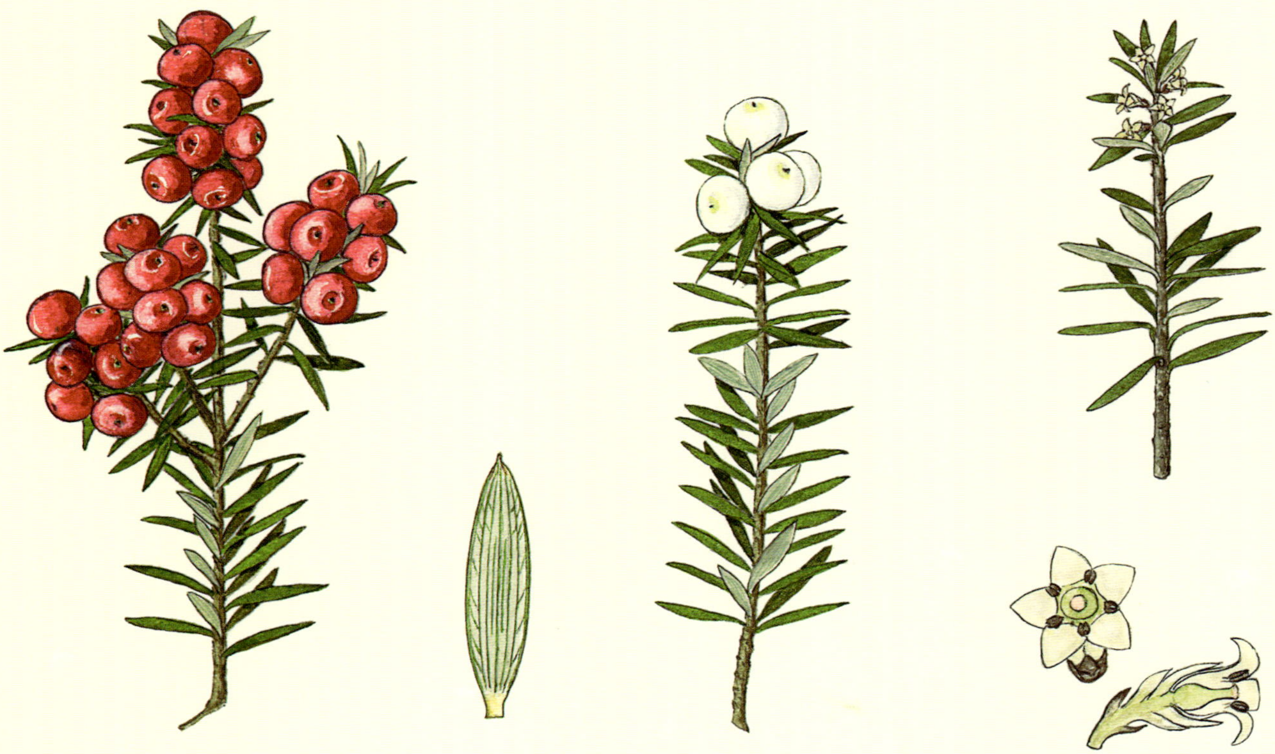

Leptecophylla robusta poteriteri,
Chatham Island mingimingi

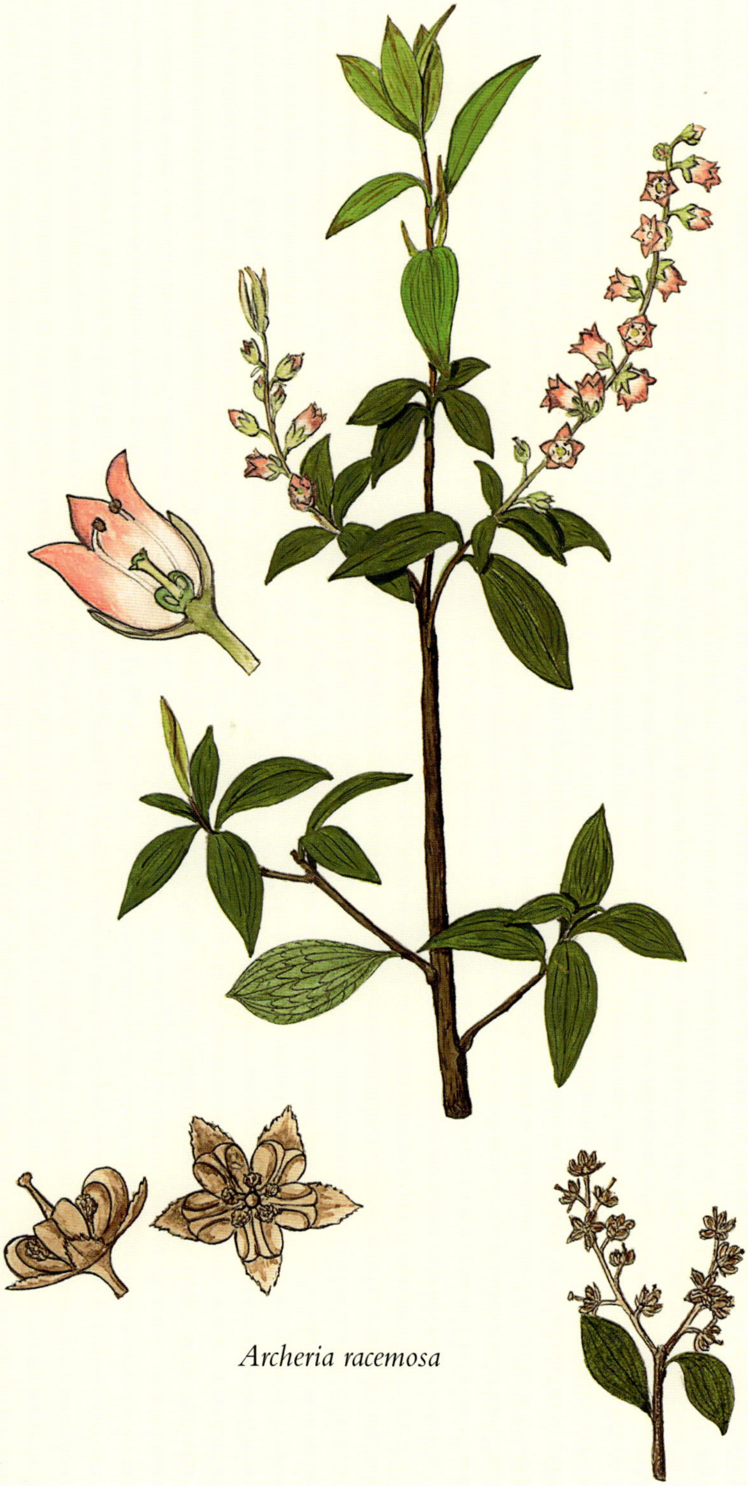

Archeria racemosa

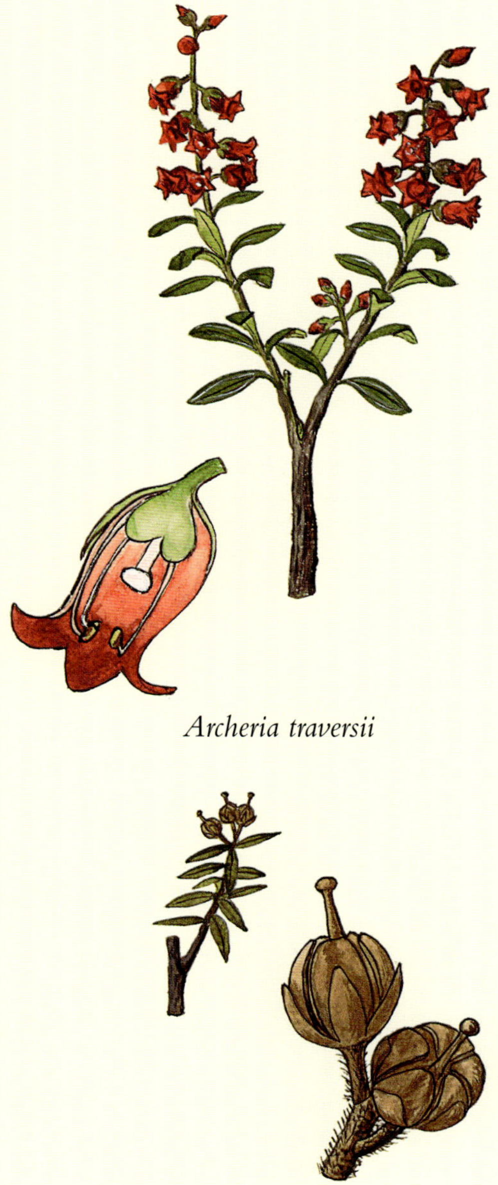

Archeria traversii

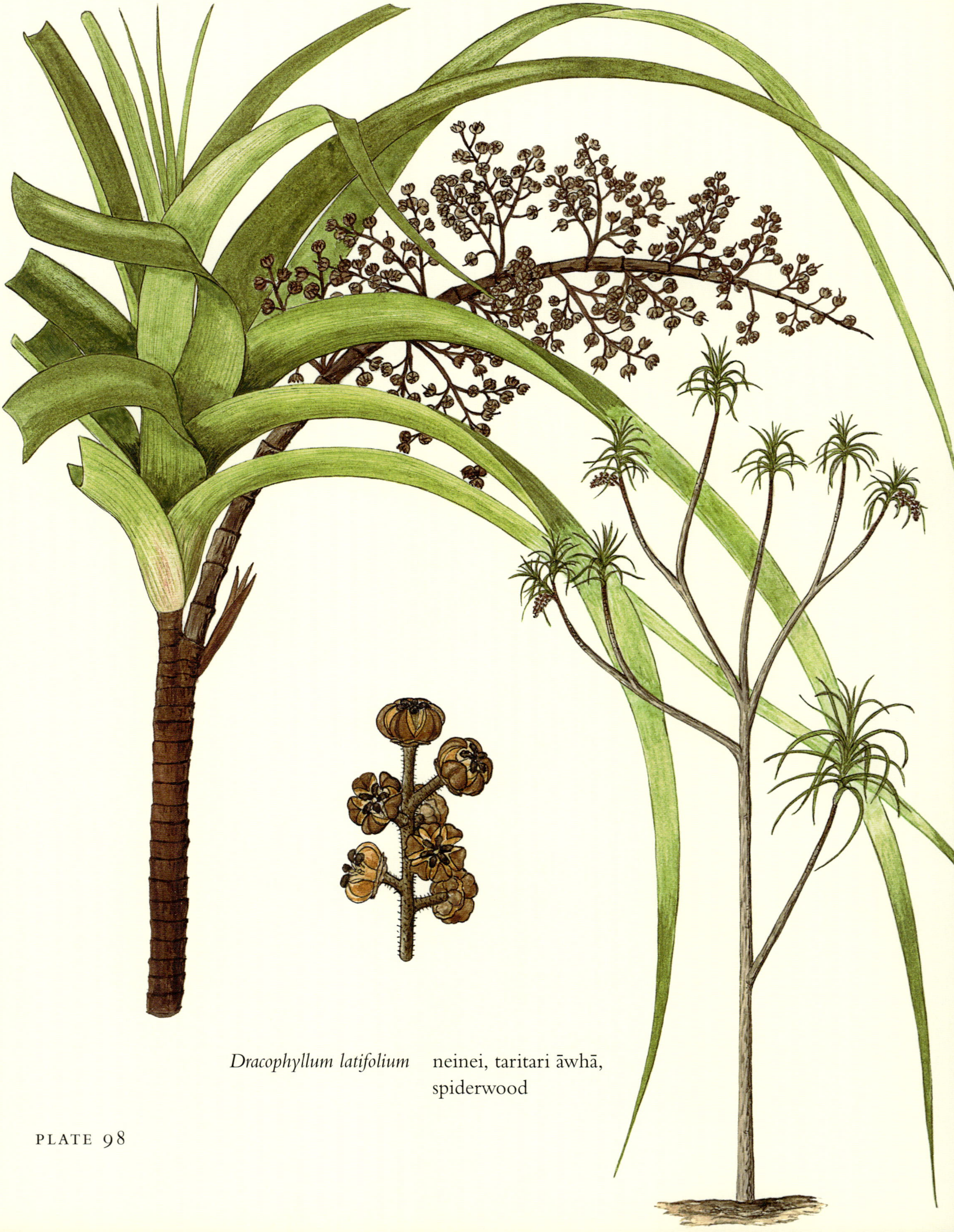

Dracophyllum latifolium neinei, taritari āwhā, spiderwood

PLATE 98

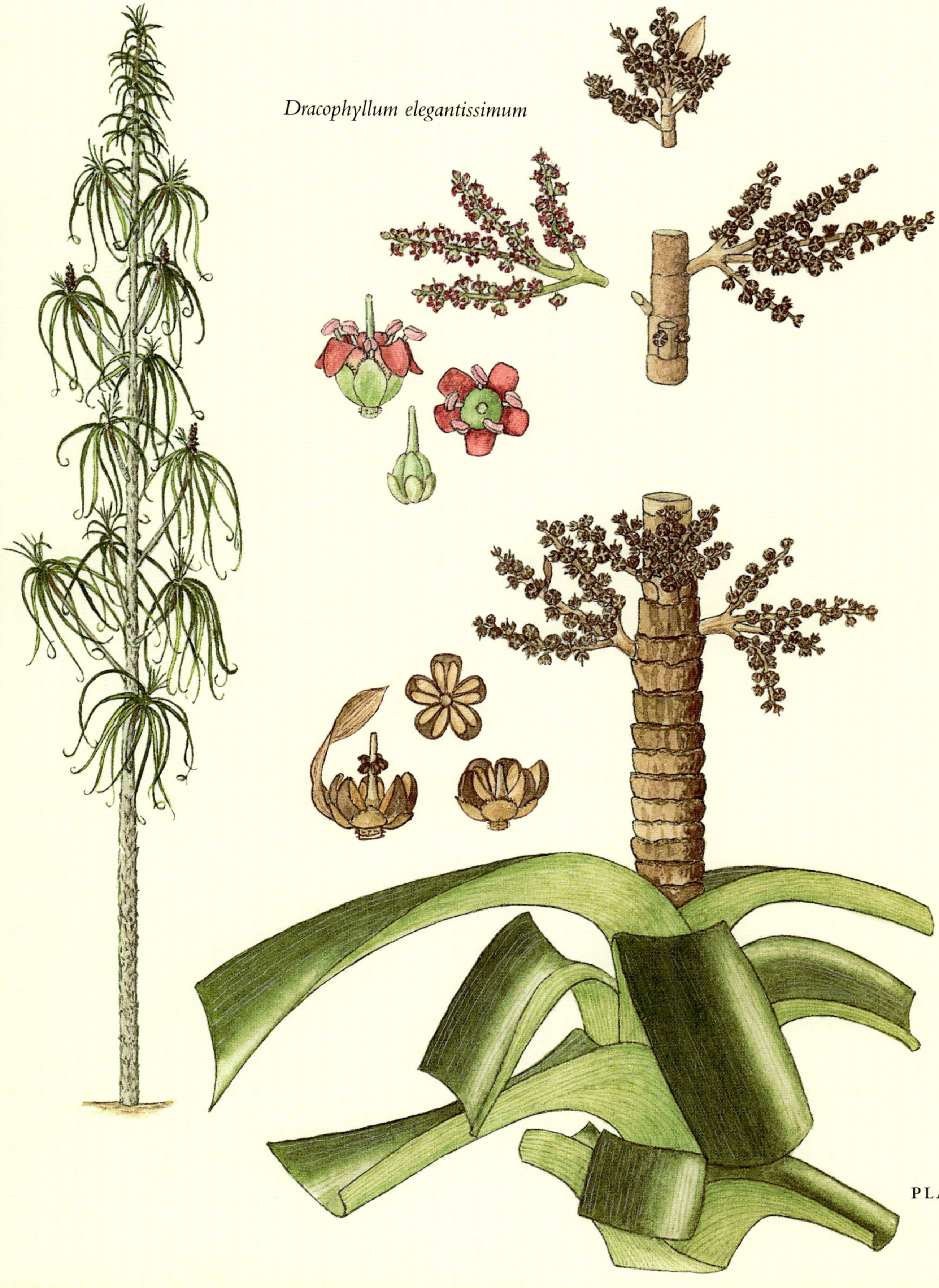

Dracophyllum elegantissimum

PLATE 99

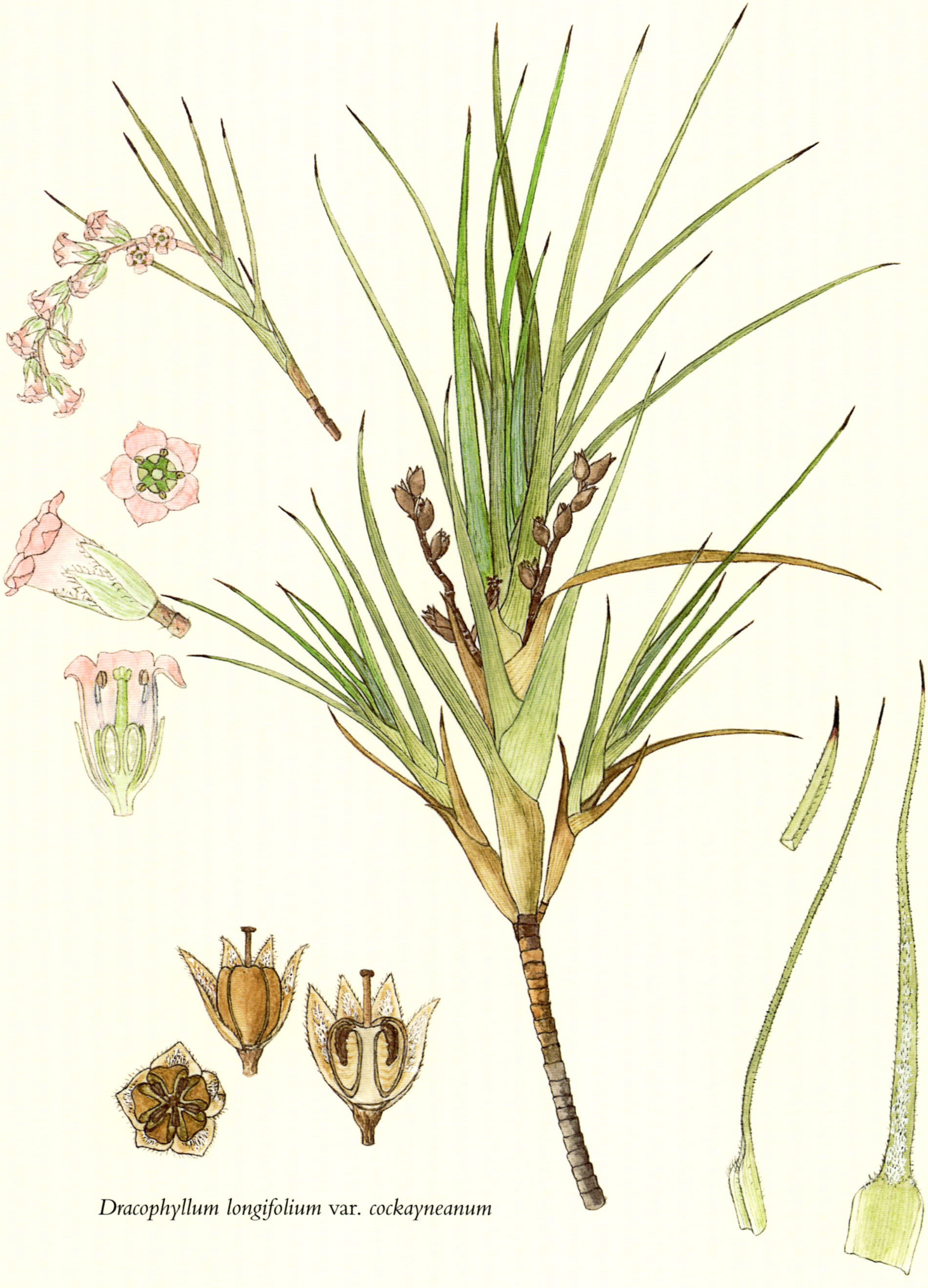

Dracophyllum longifolium var. *cockayneanum*

PLATE 100

Dracophyllum sinclairii

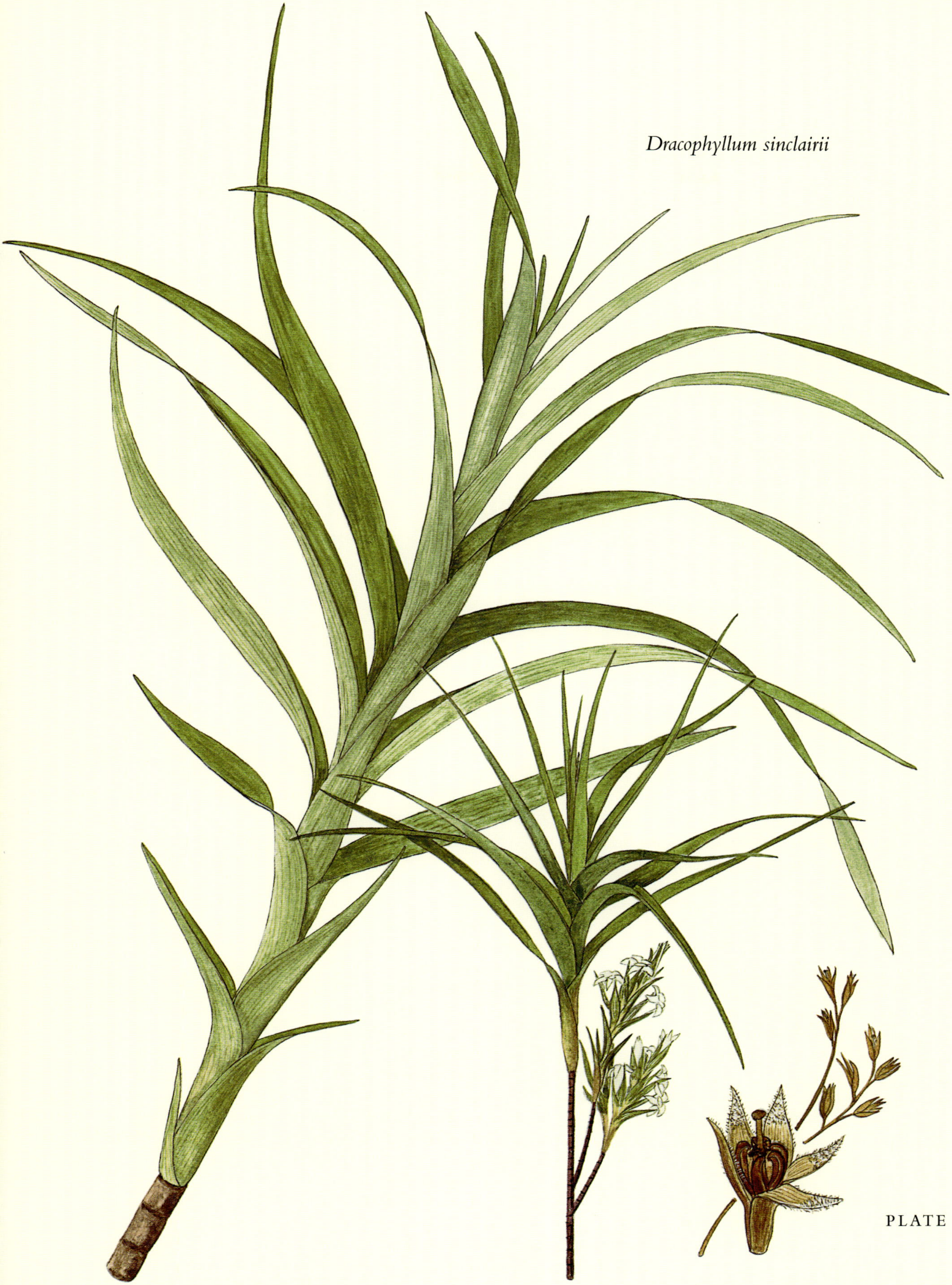

PLATE 101

Dracophyllum menziesii

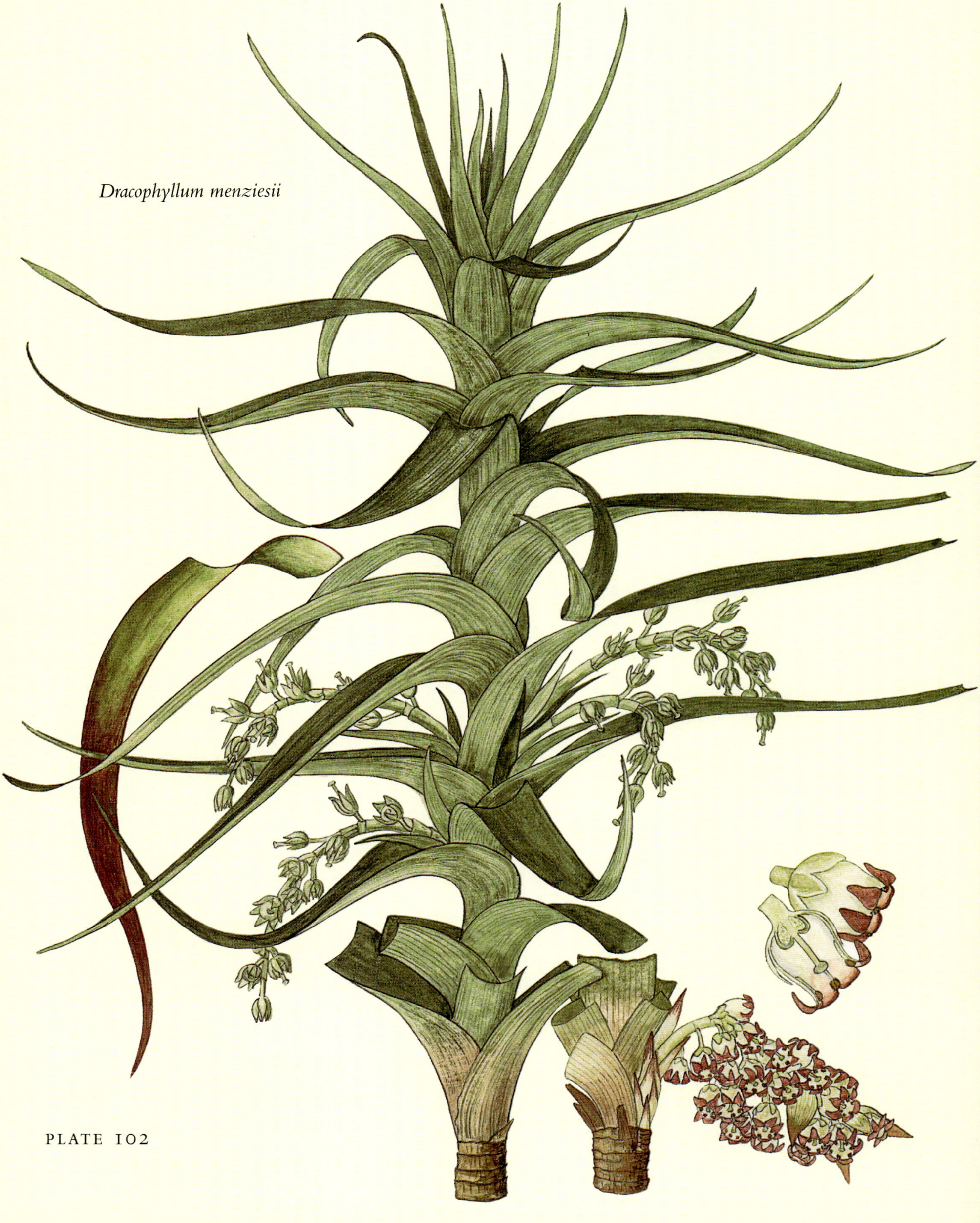

PLATE 102

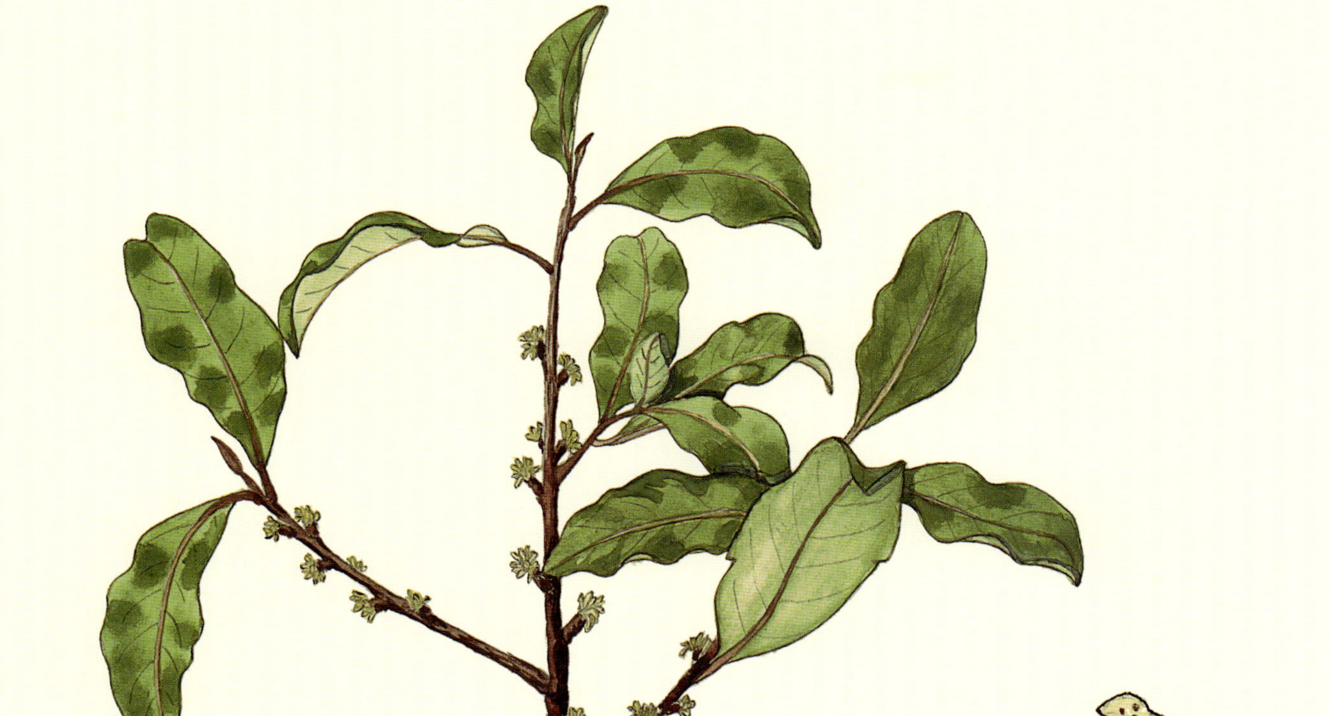

Myrsine australis māpou, māpau, red matipo

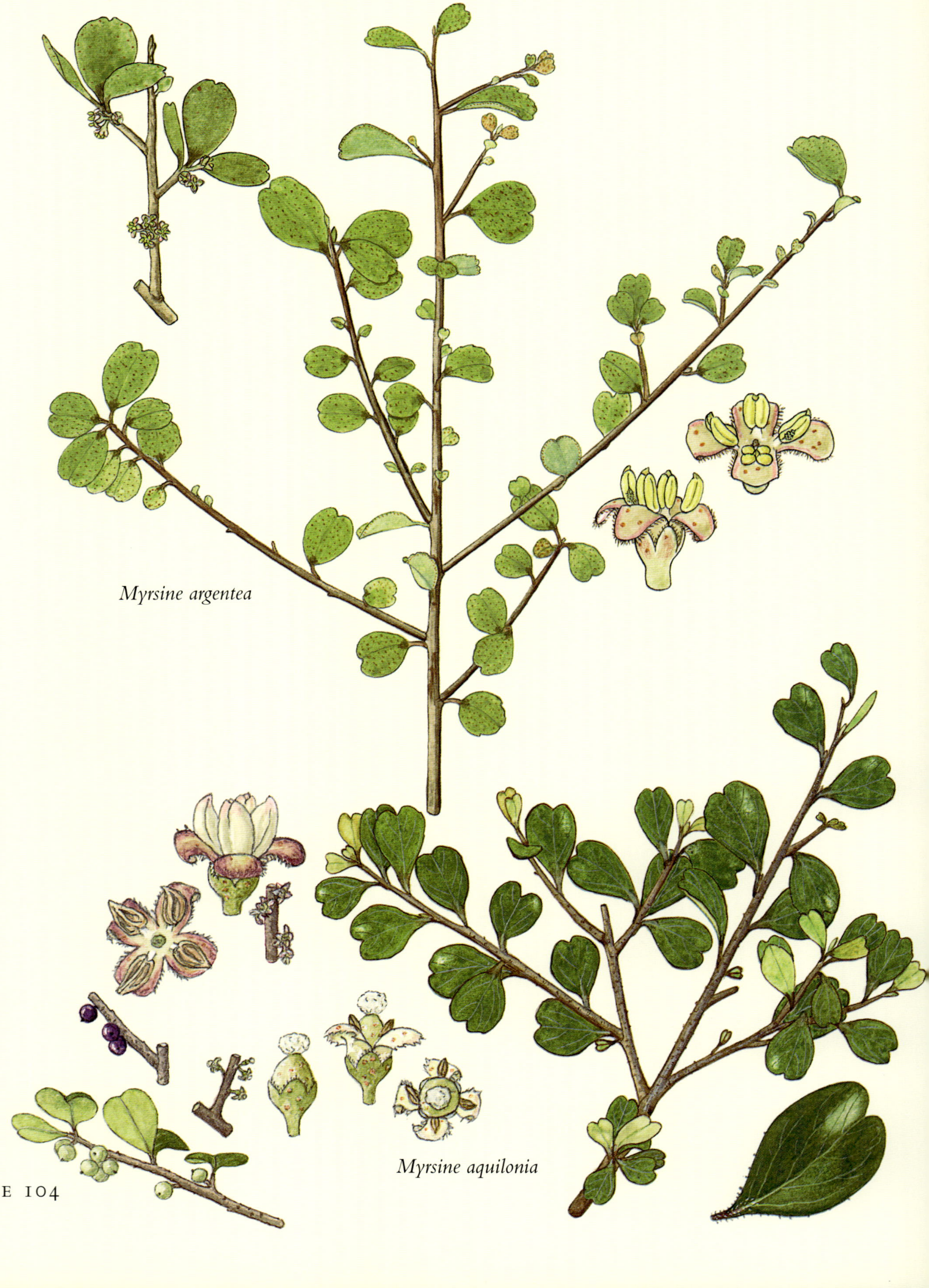

PLATE 104

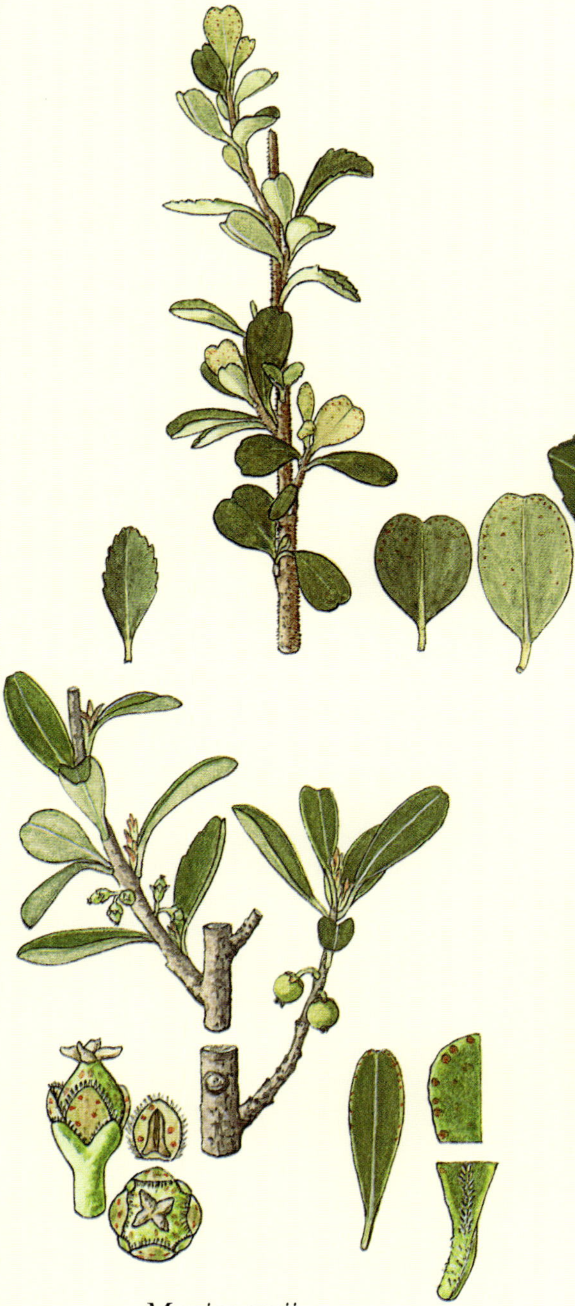

Myrsine coxii

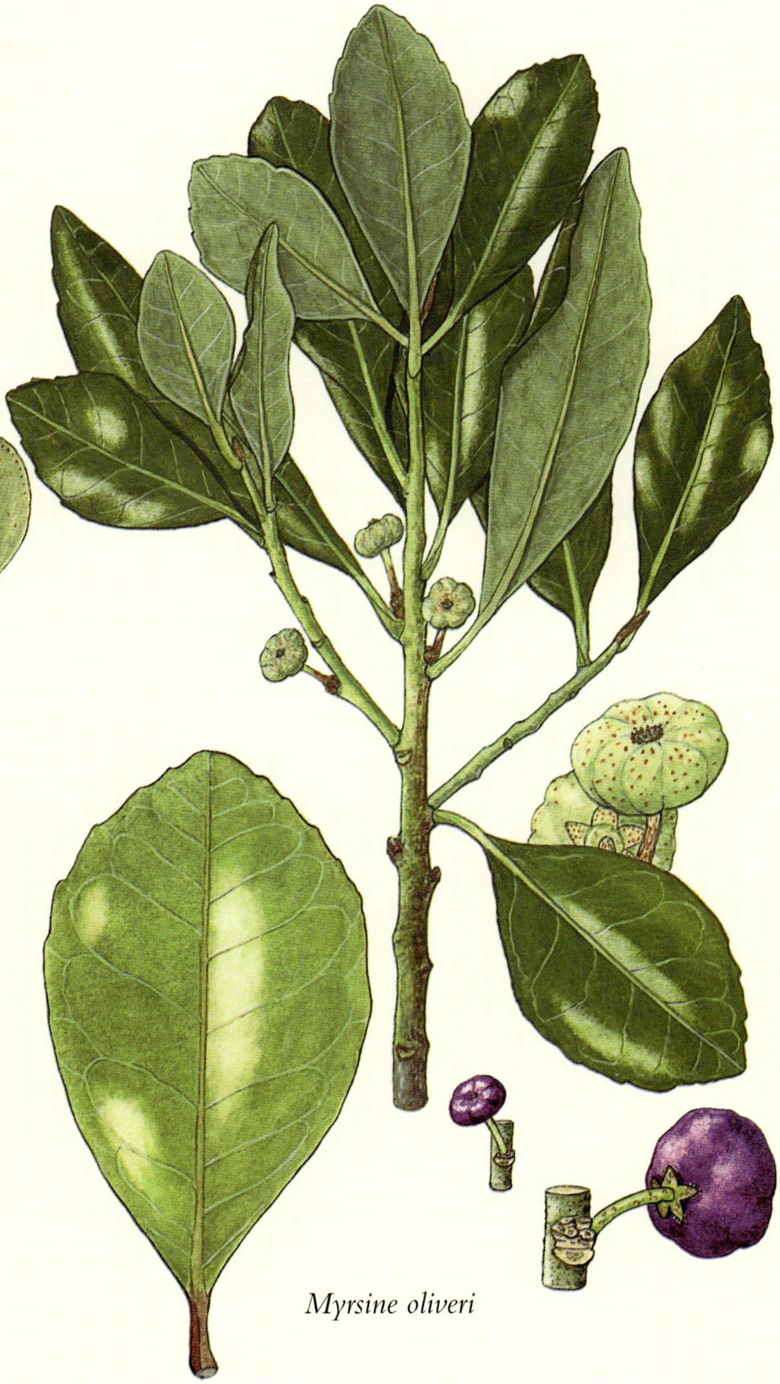

Myrsine oliveri

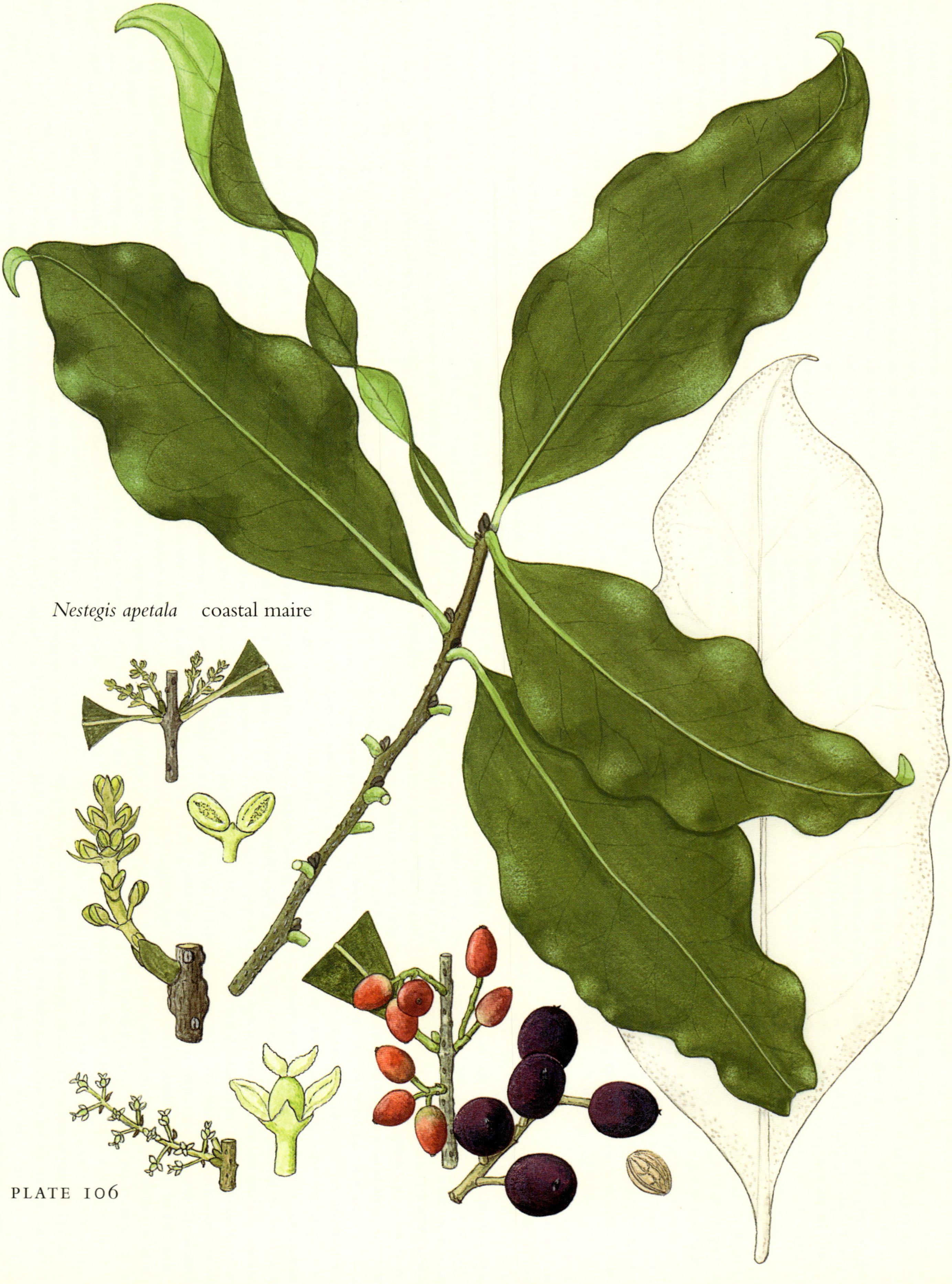

Nestegis apetala coastal maire

PLATE 106

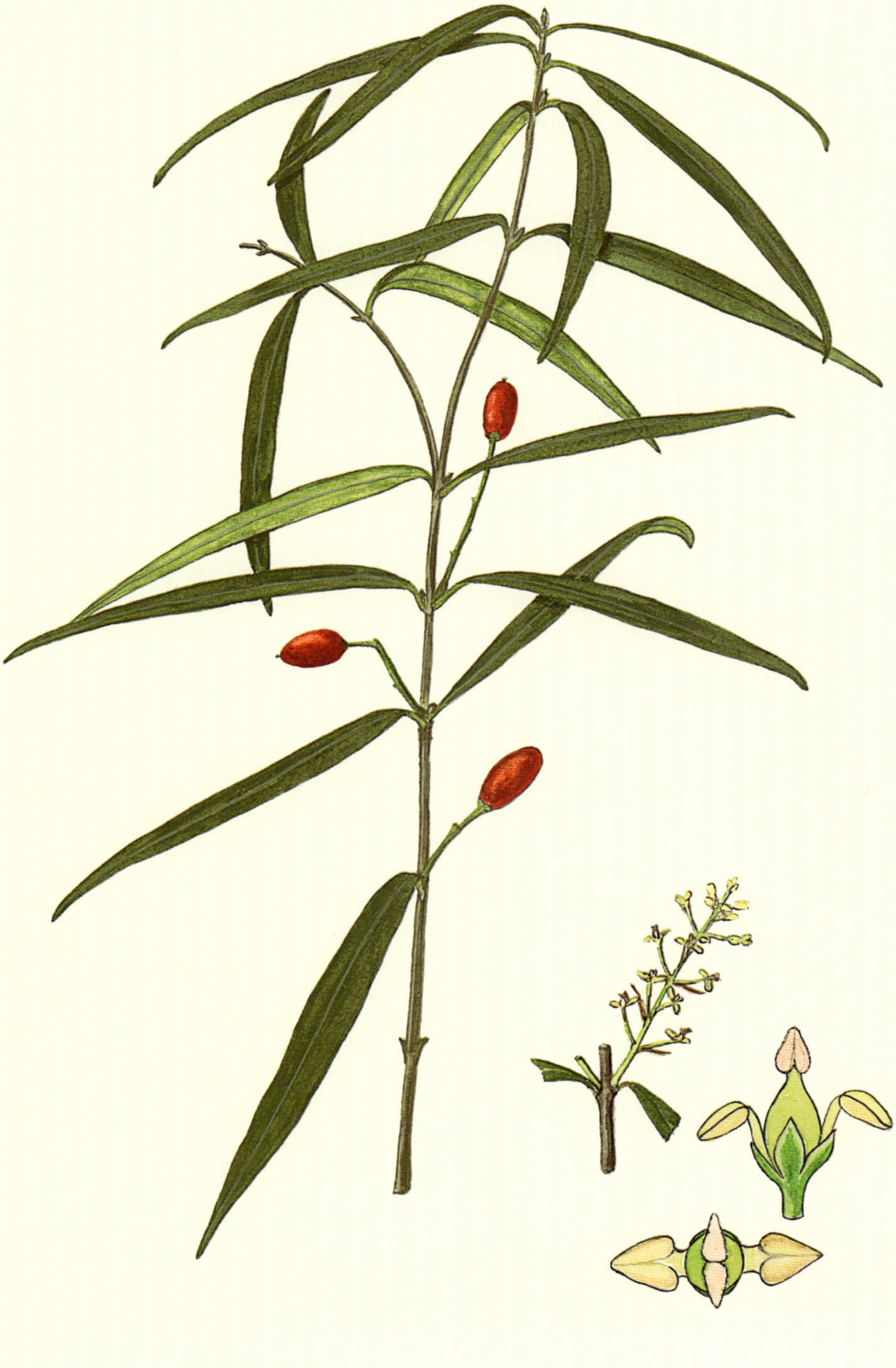

Nestegis montana rōroro, maire roro,
narrow-leaved maire

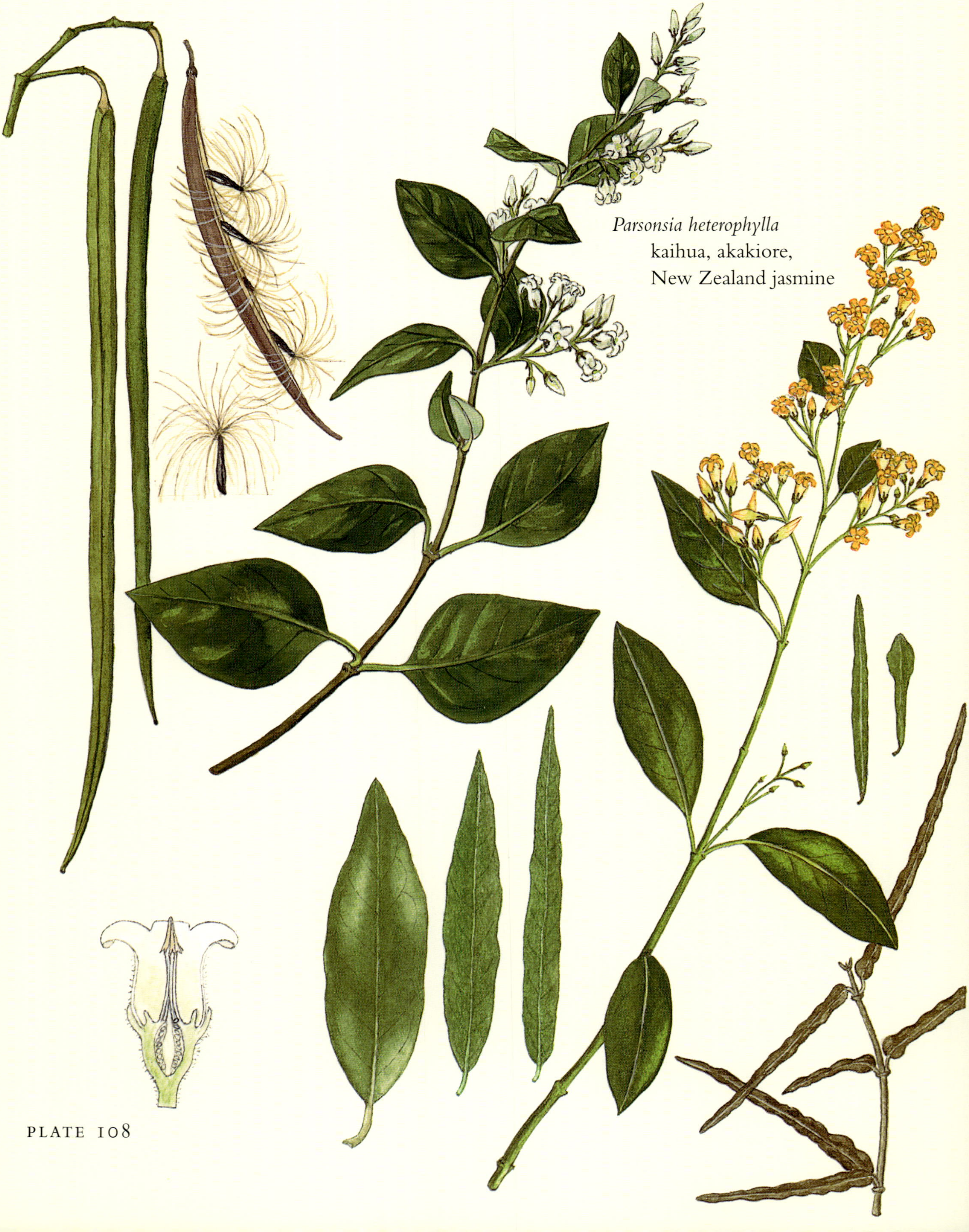

Parsonsia heterophylla
kaihua, akakiore,
New Zealand jasmine

PLATE 108

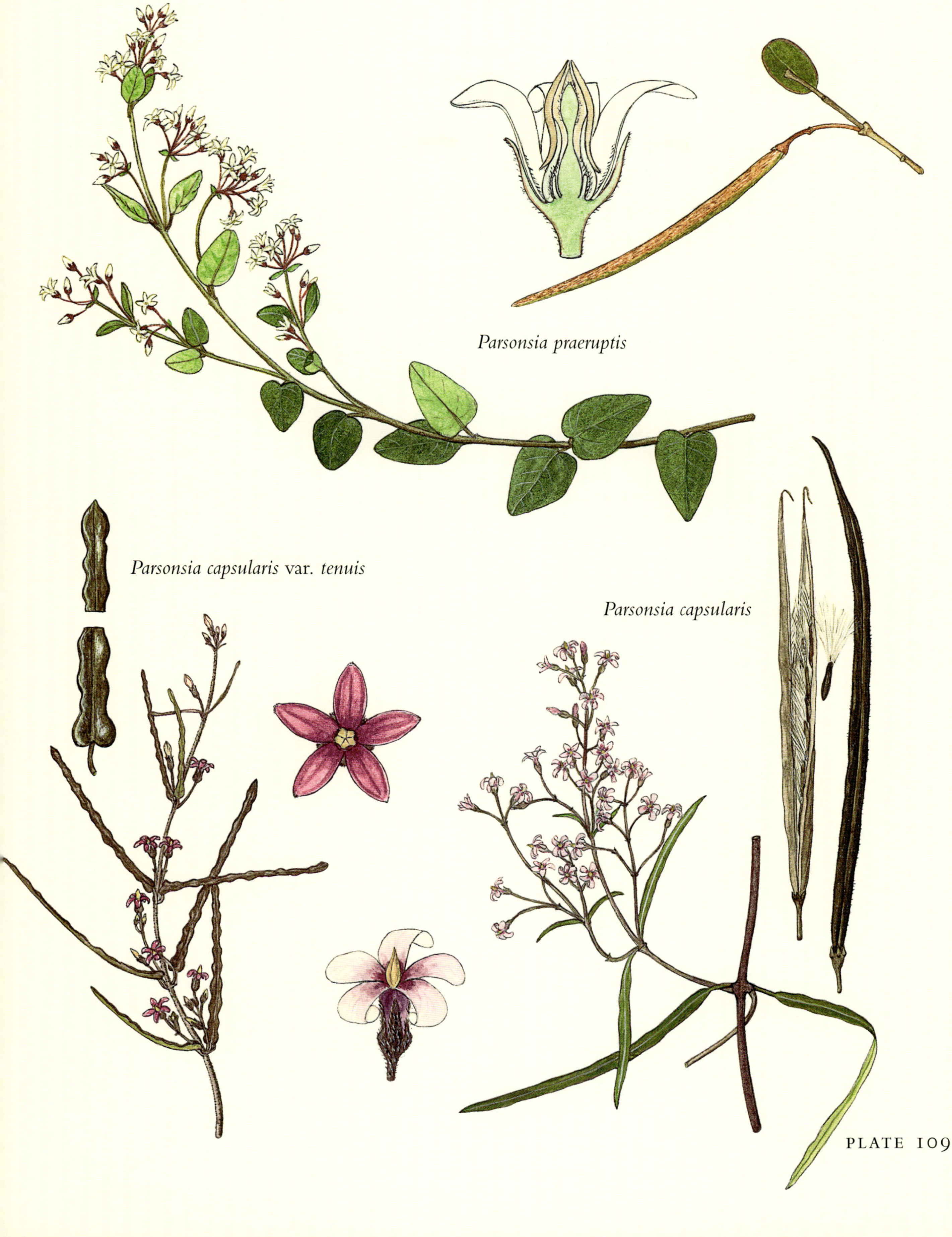
Parsonsia praeruptis
Parsonsia capsularis var. tenuis
Parsonsia capsularis

PLATE 109

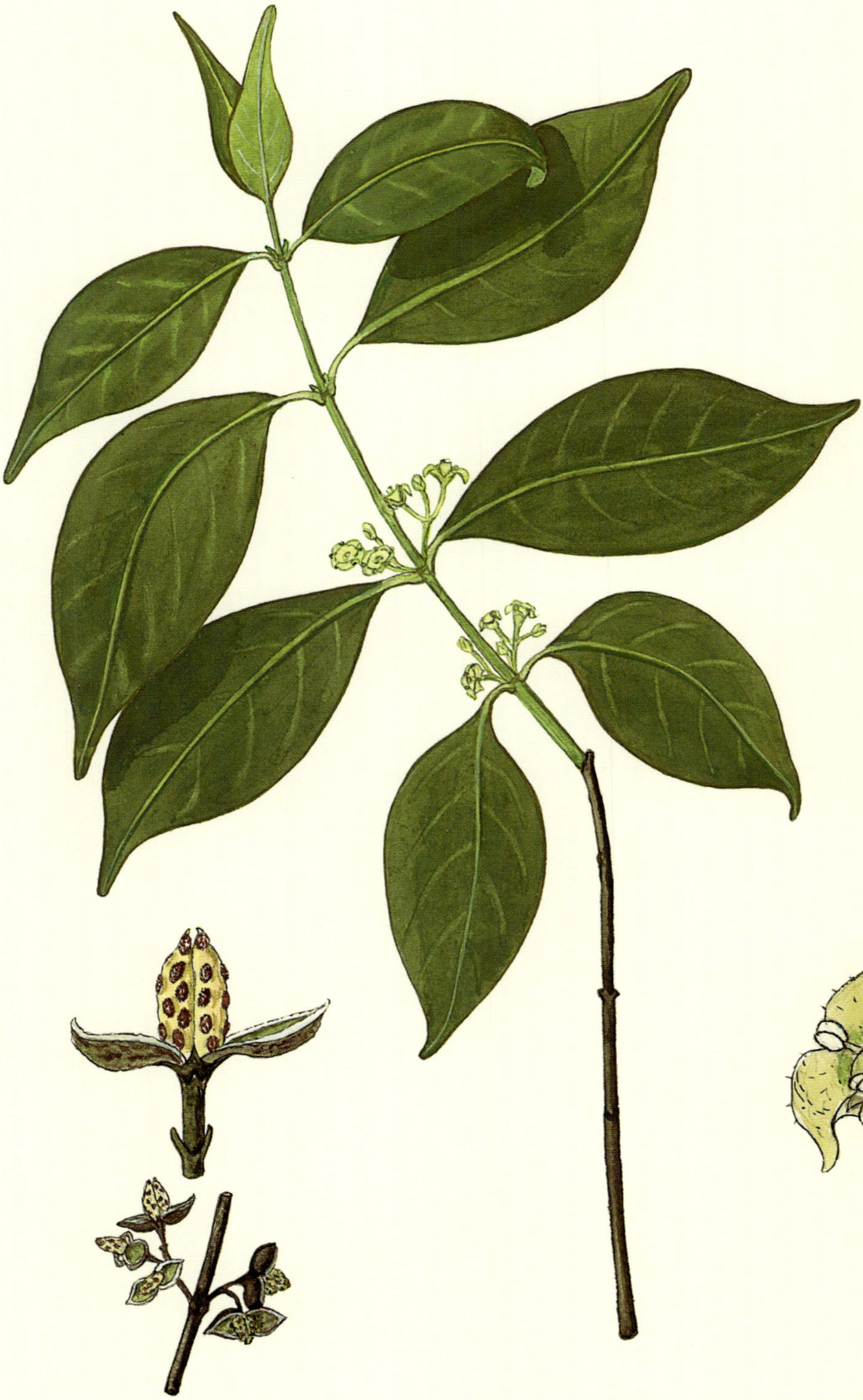

Geniostoma rupestre var. *ligustrifolium* hangehange,
New Zealand privet

PLATE 110

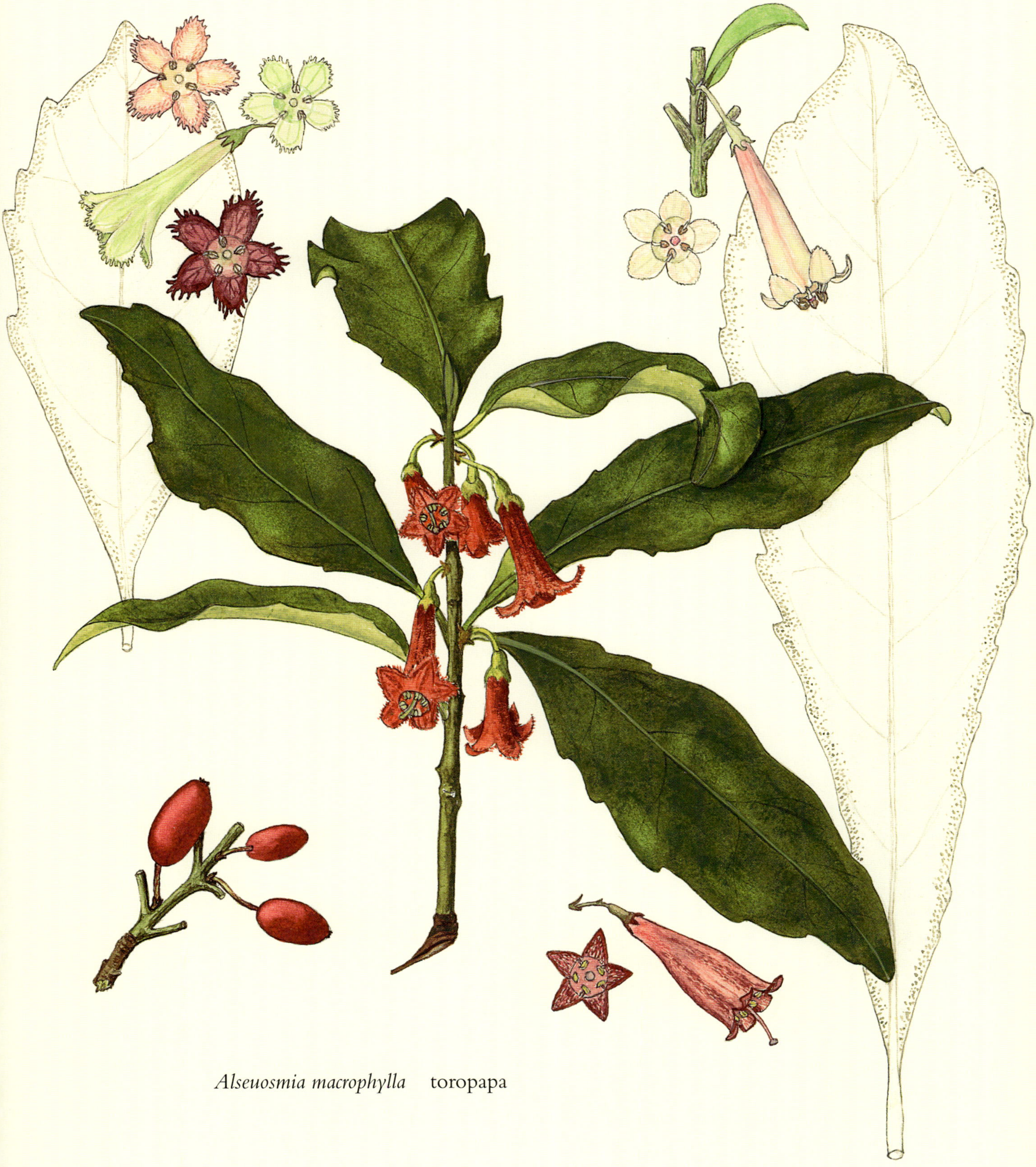

Alseuosmia macrophylla toropapa

PLATE III

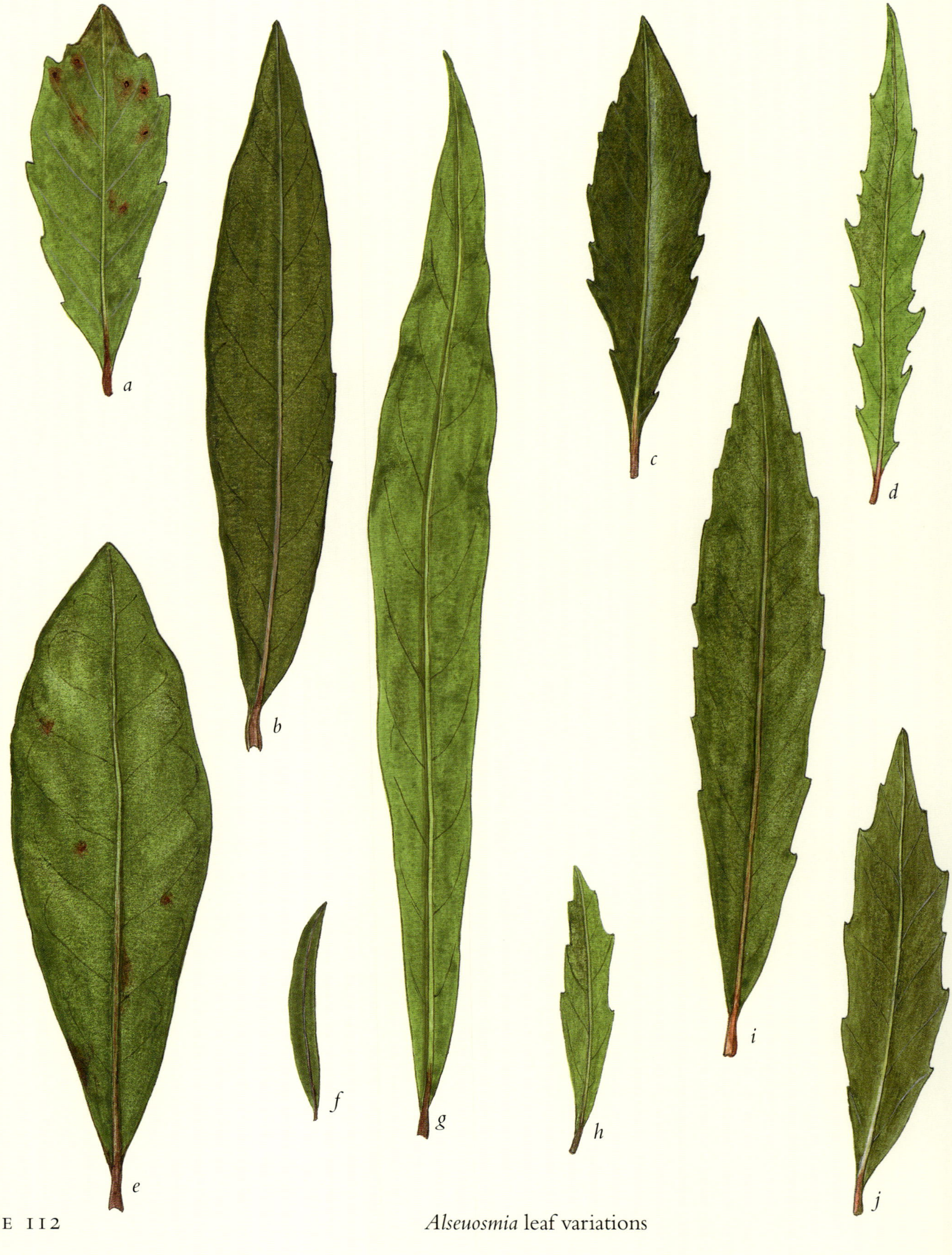

PLATE 112

Alseuosmia leaf variations

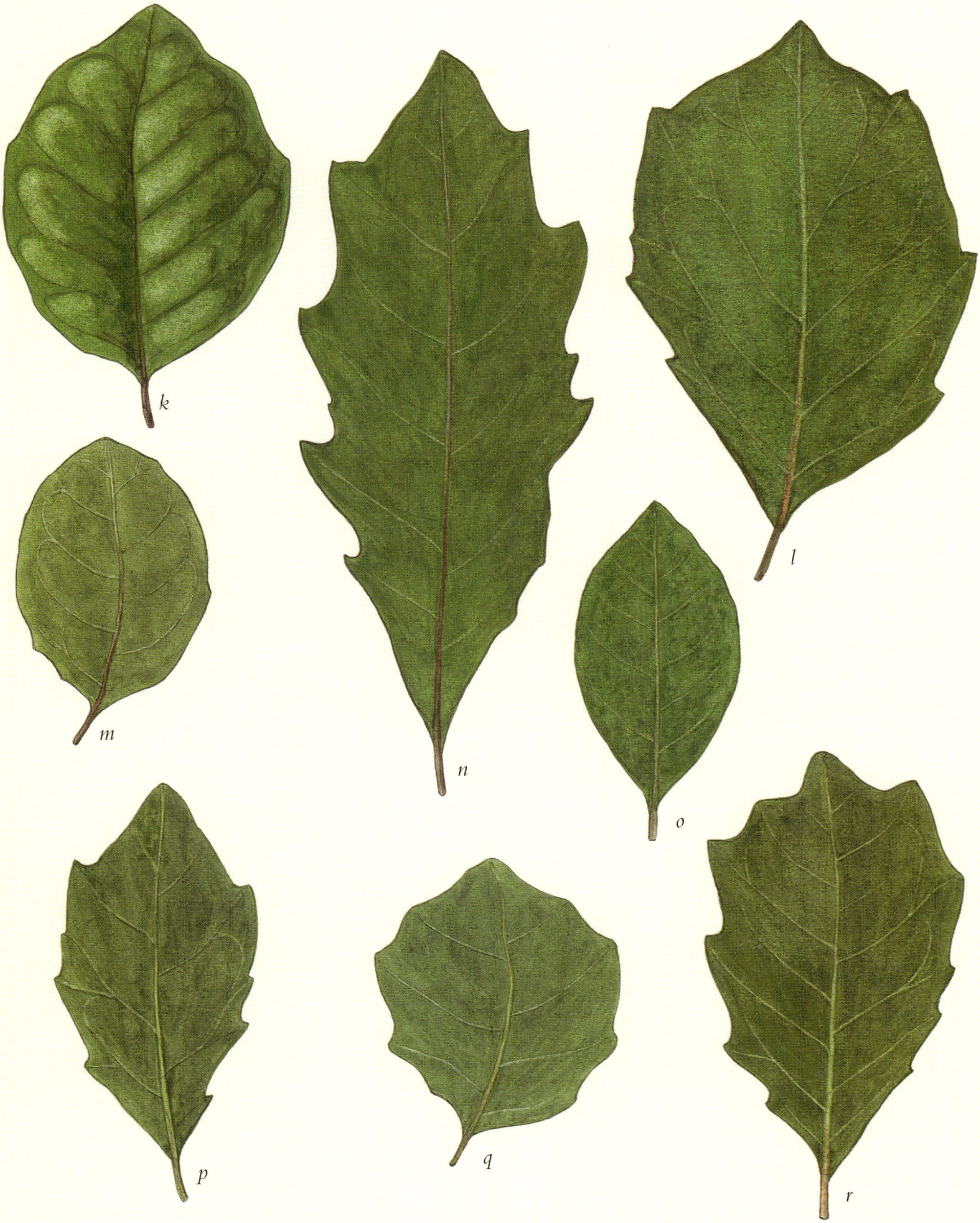

Alseuosmia leaf variations

PLATE 113

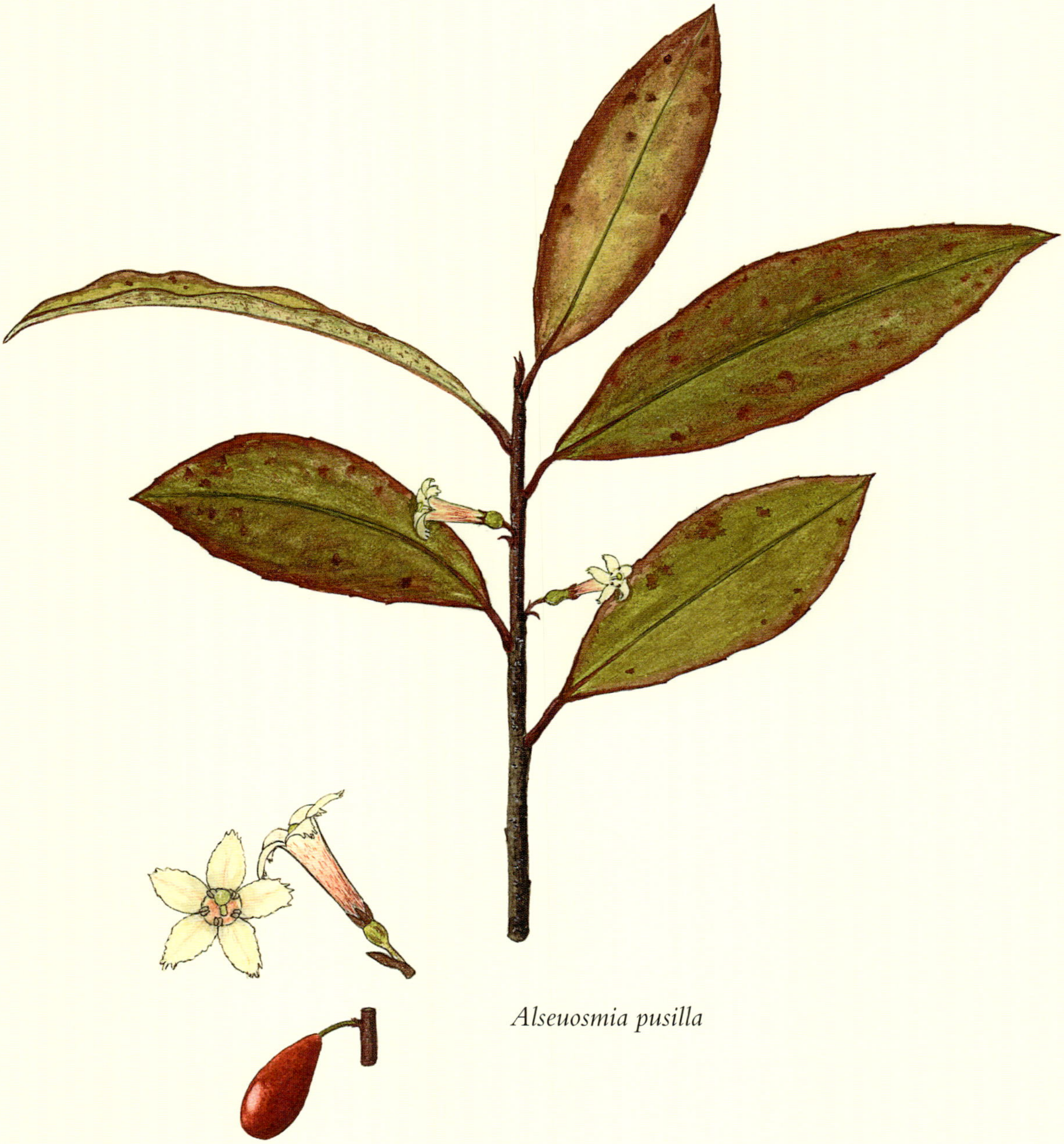

Alseuosmia pusilla

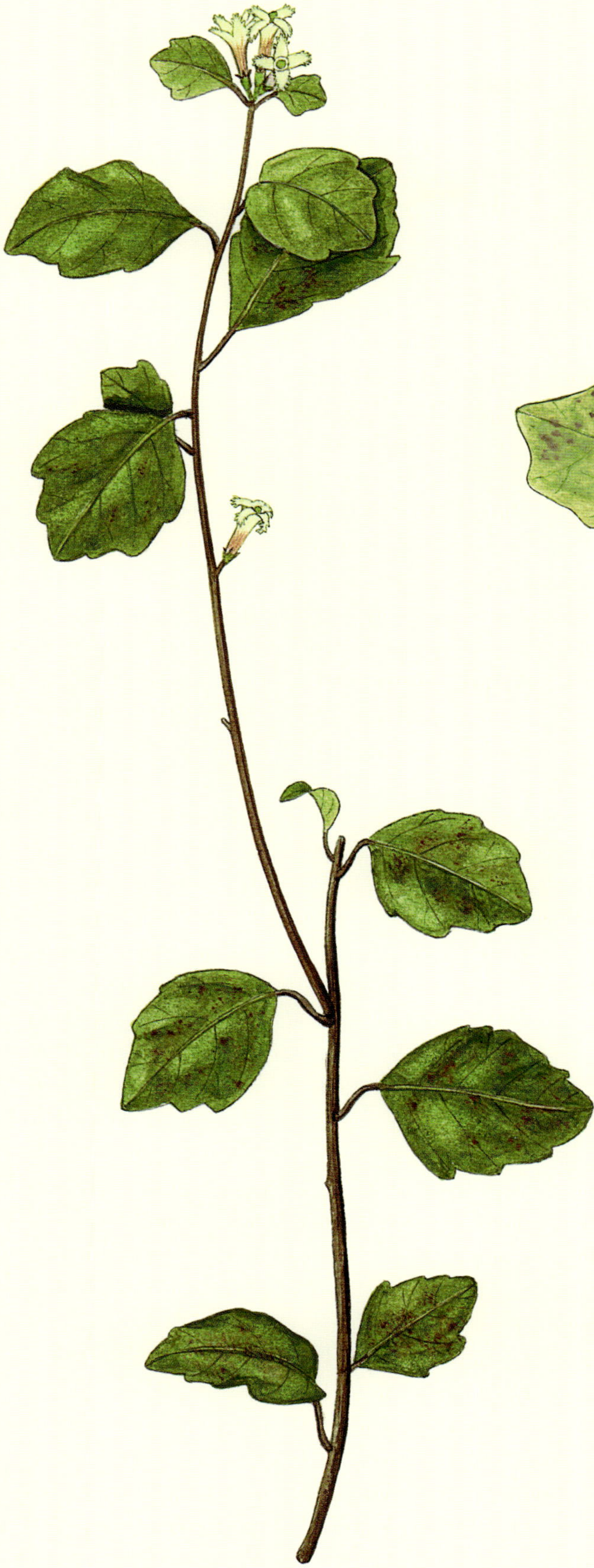

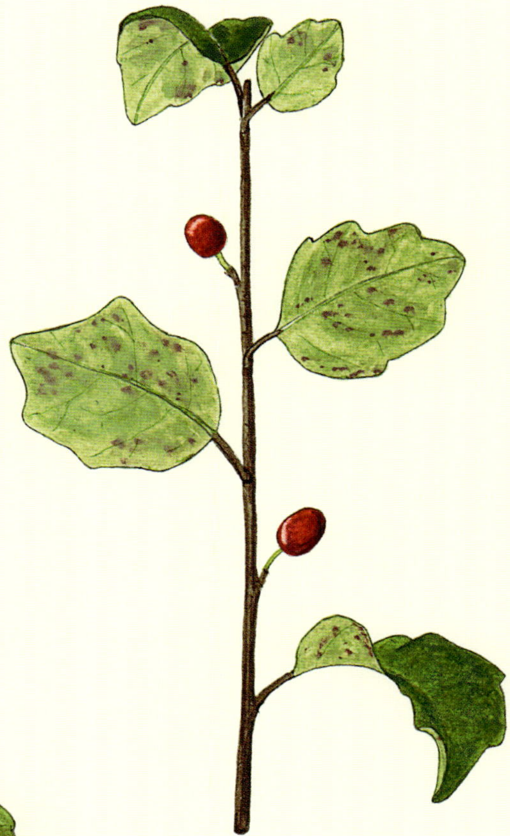

Alseuosmia banksii

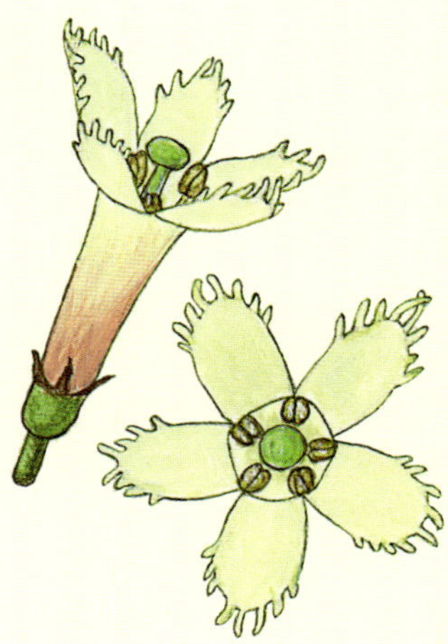

PLATE 115

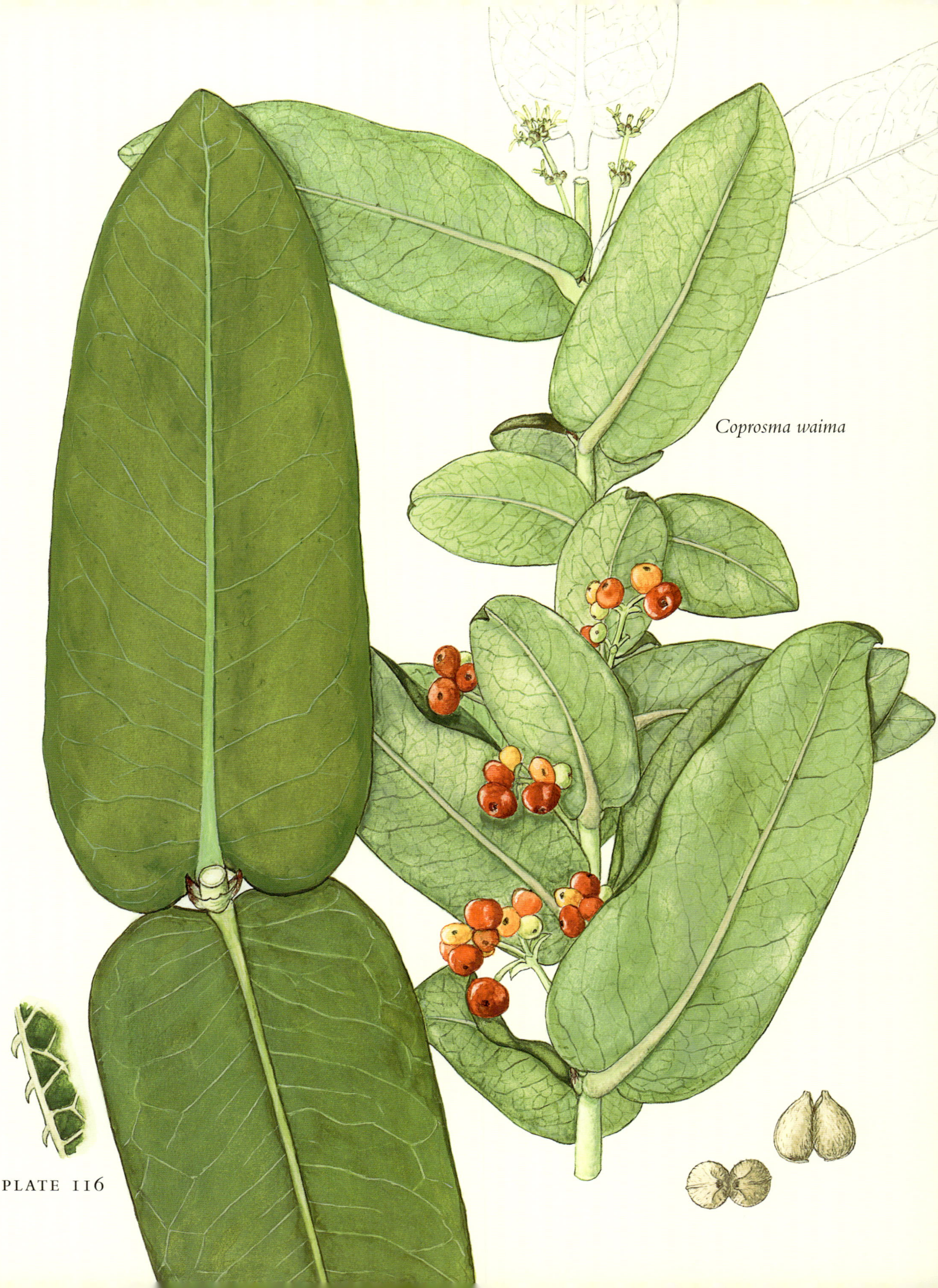

Coprosma waima

PLATE 116

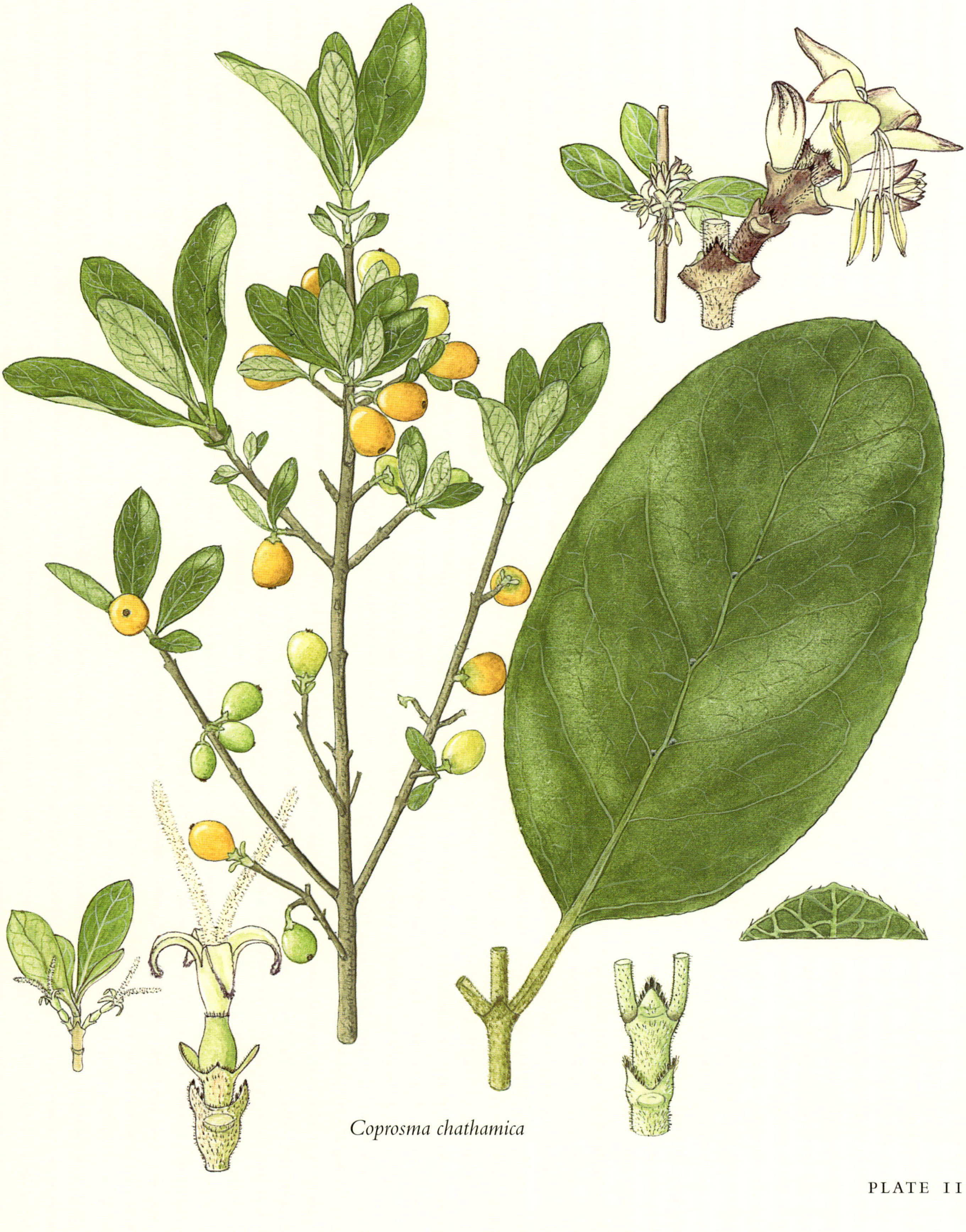

Coprosma chathamica

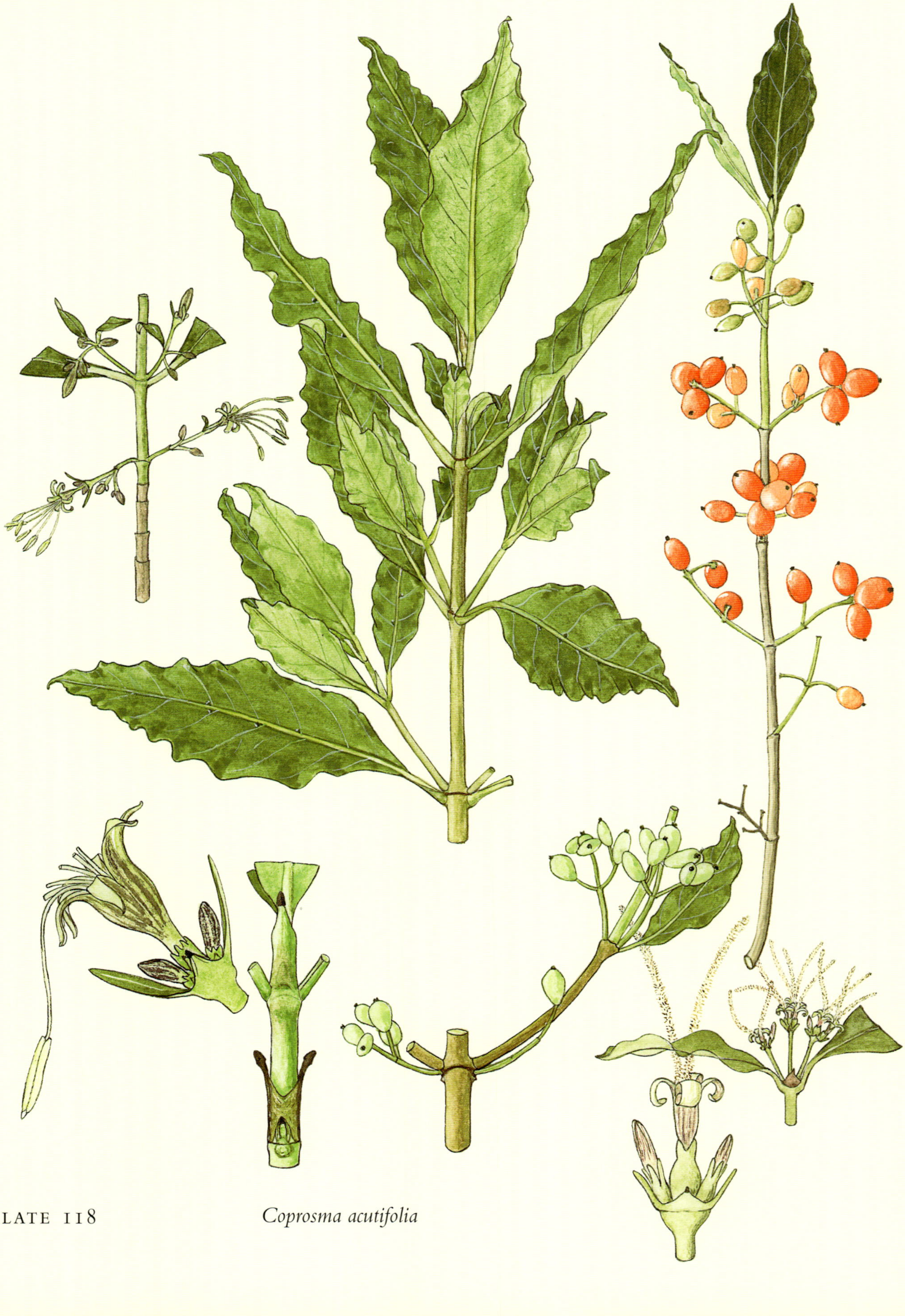

PLATE 118 *Coprosma acutifolia*

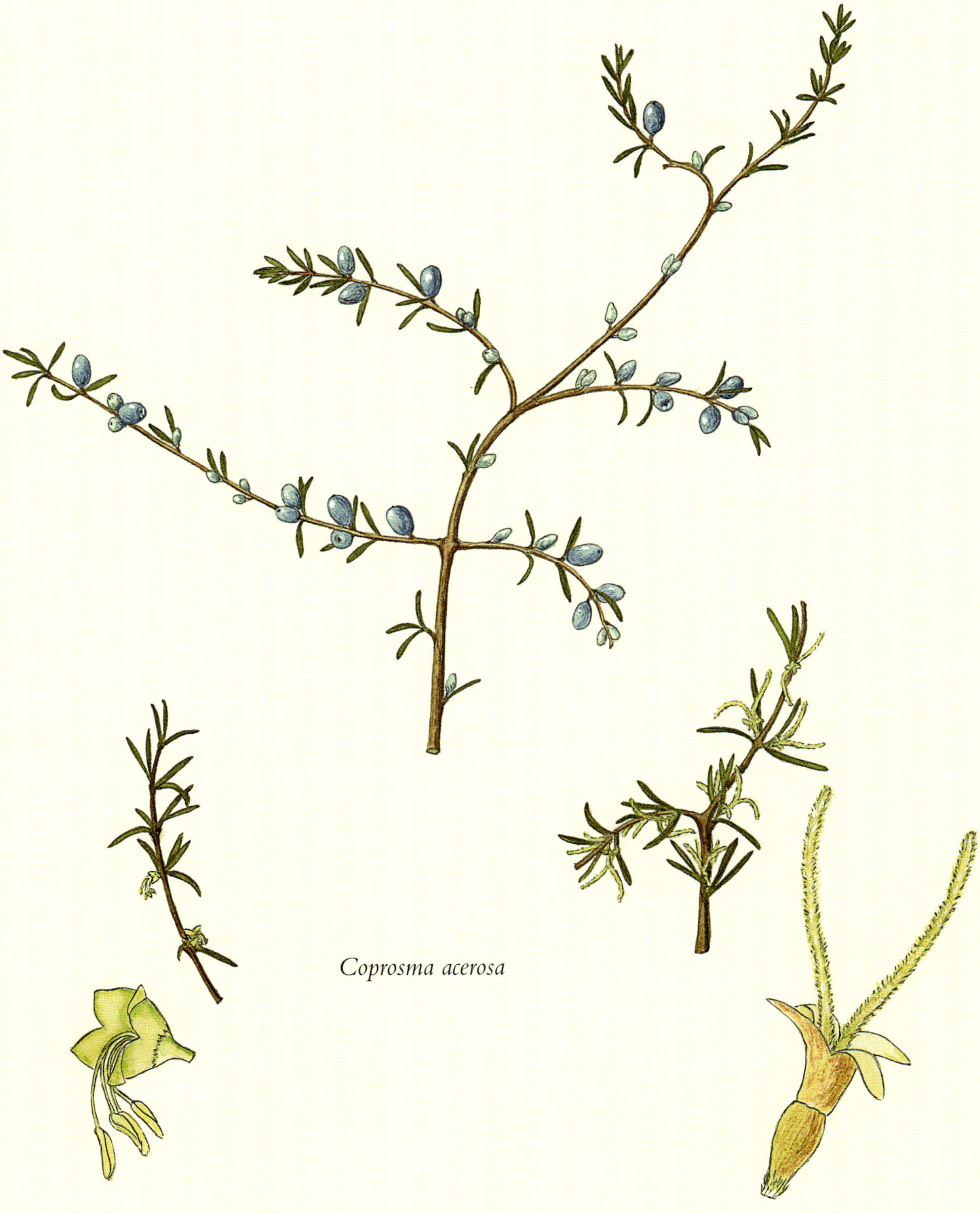

Coprosma acerosa

Coprosma spathulata subspecies *hikuruana*

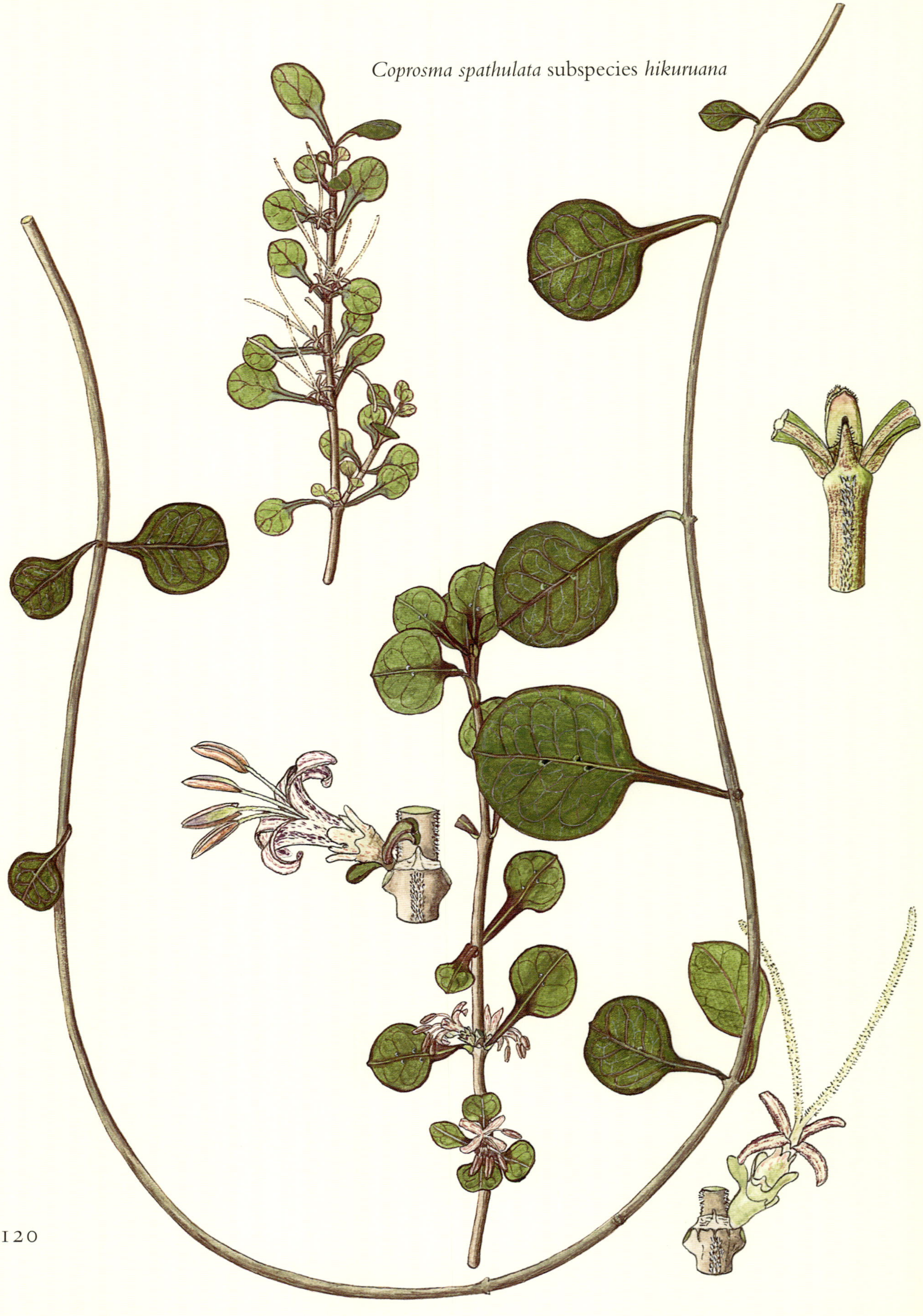

PLATE 120

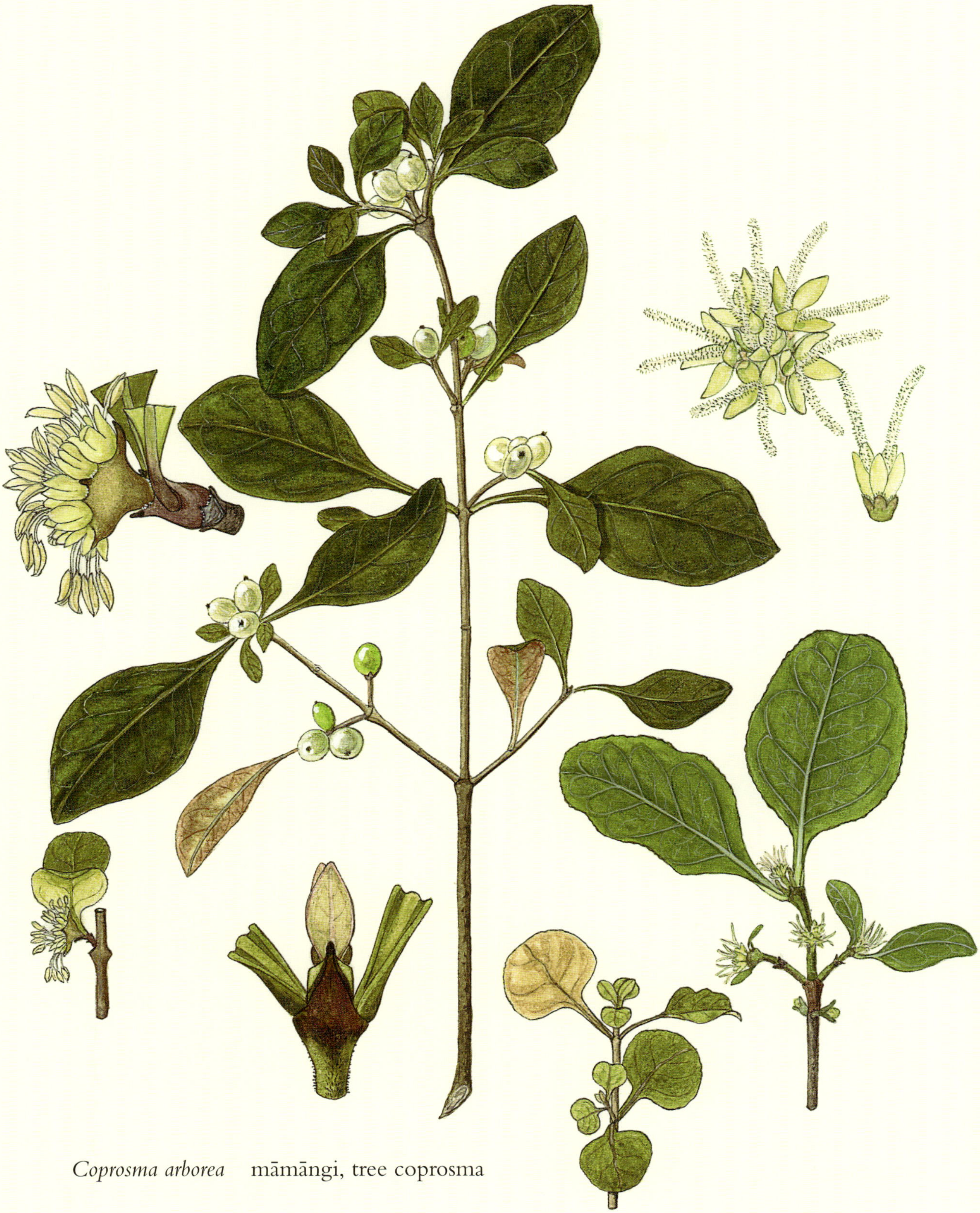

Coprosma arborea māmāngi, tree coprosma

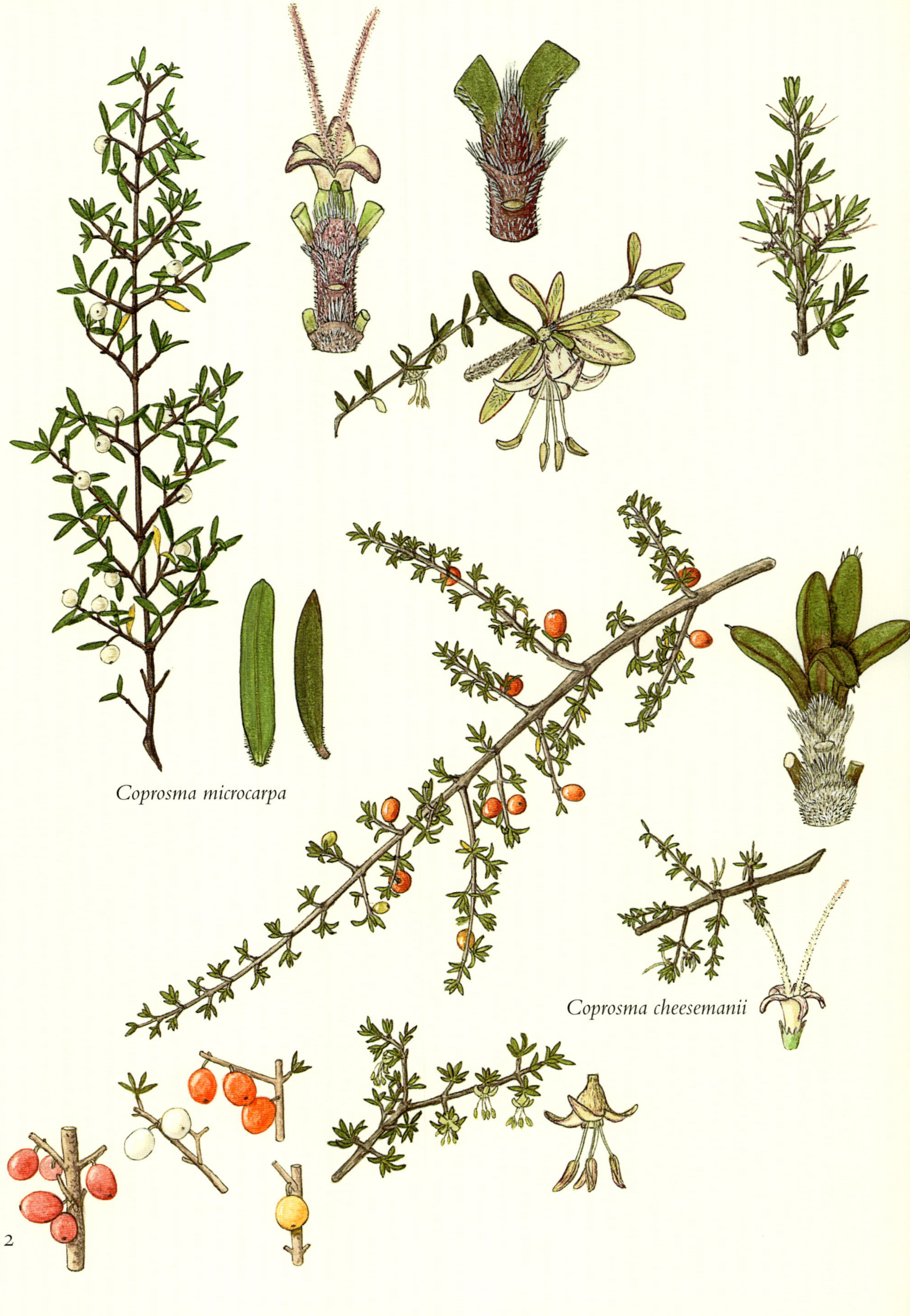

Coprosma microcarpa

Coprosma cheesemanii

PLATE 122

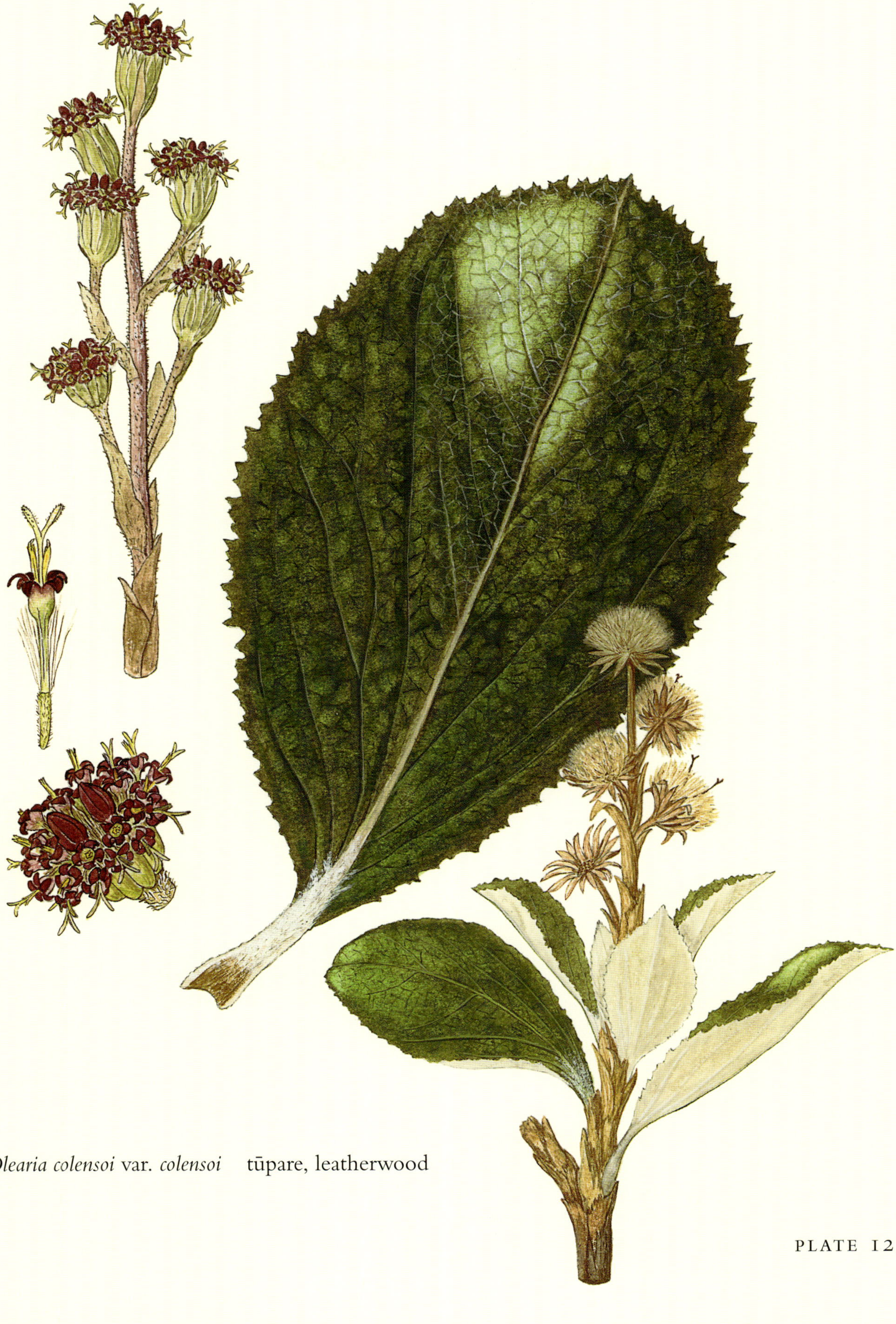

Olearia colensoi var. *colensoi* tūpare, leatherwood

PLATE 123

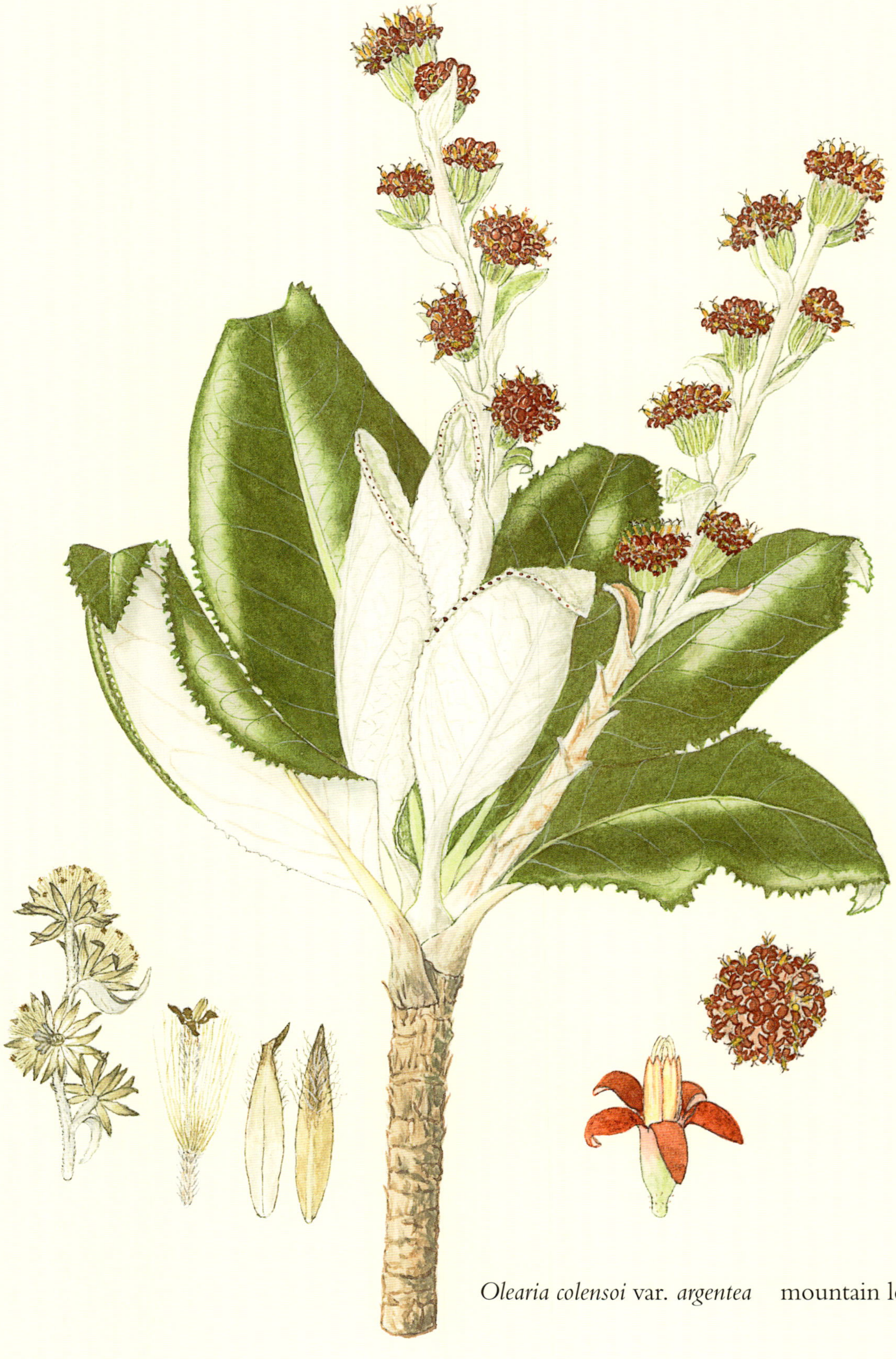

Olearia colensoi var. *argentea* mountain leatherwood

PLATE 124

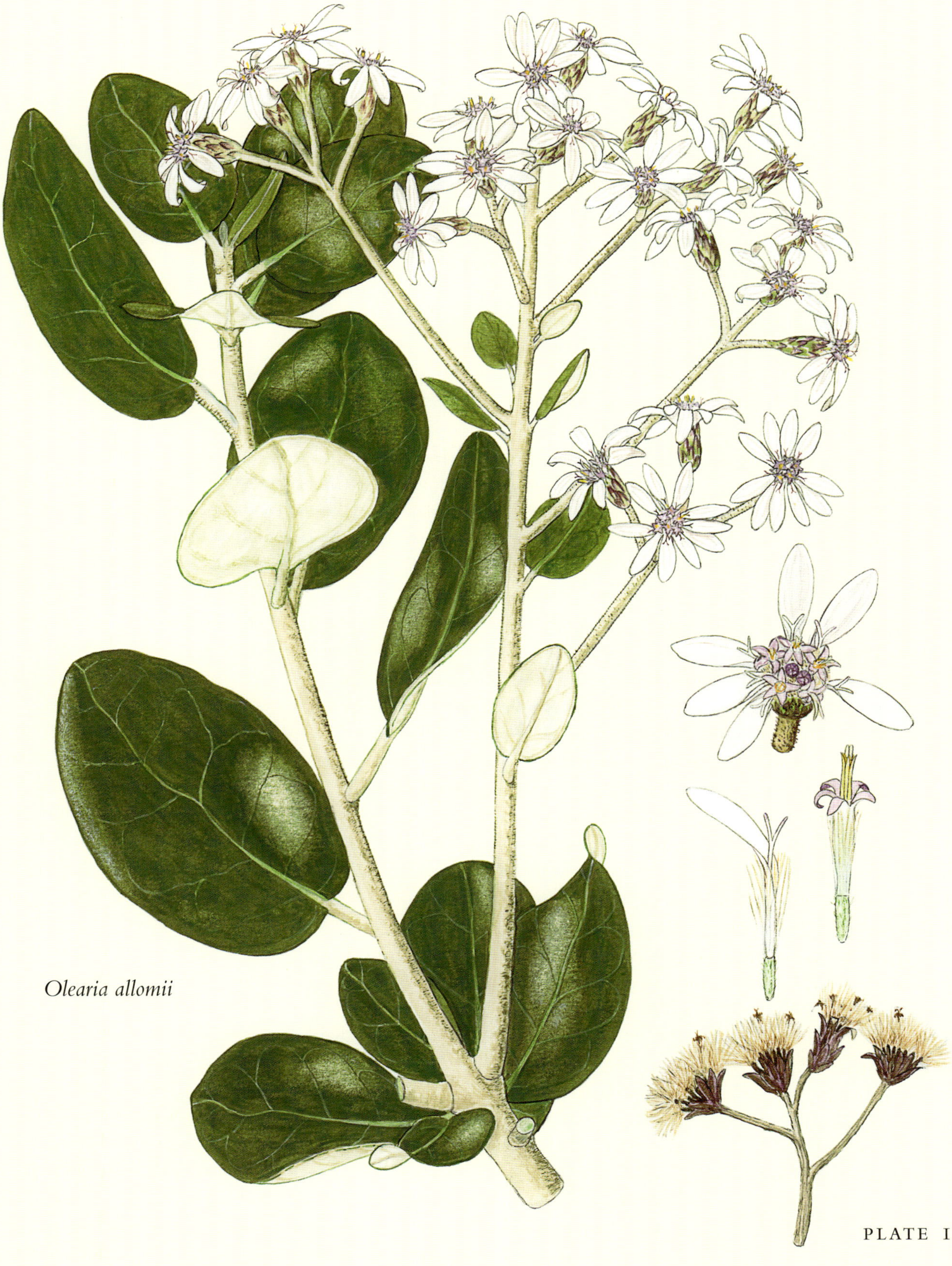

Olearia allomii

PLATE 125

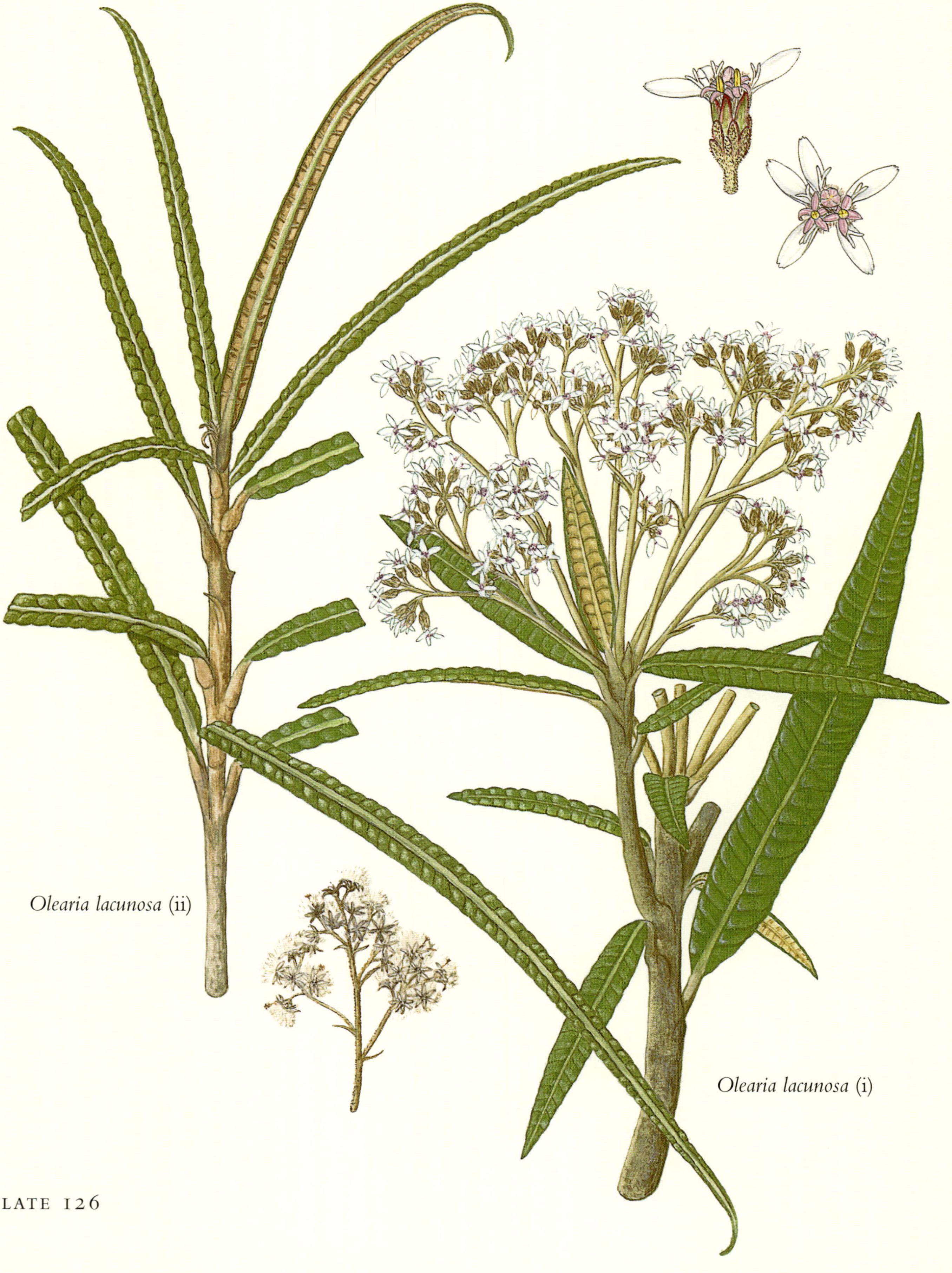
Olearia lacunosa (ii)
Olearia lacunosa (i)

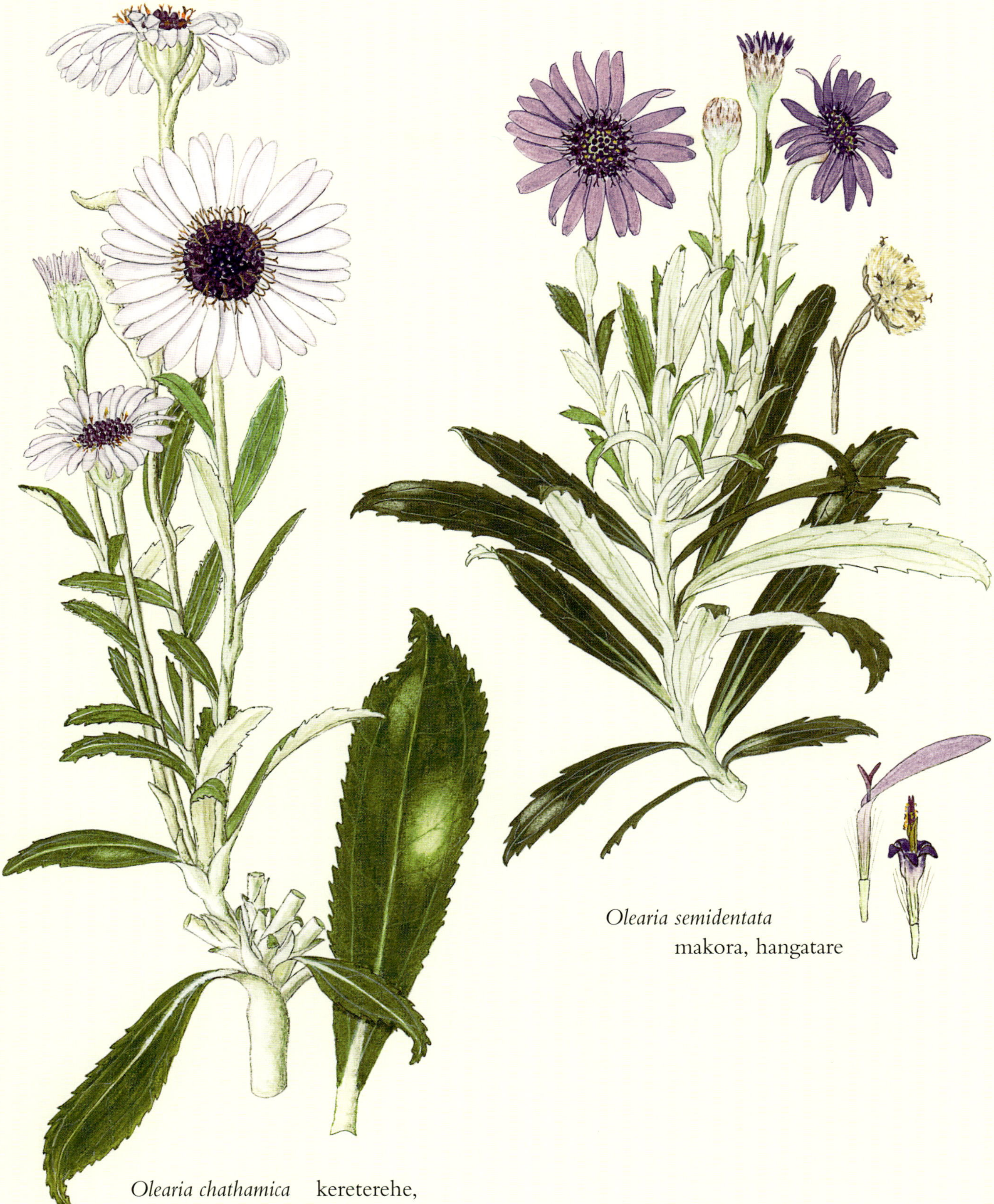

Olearia semidentata
makora, hangatare

Olearia chathamica kereterehe,
Chatham Island tree daisy

PLATE 127

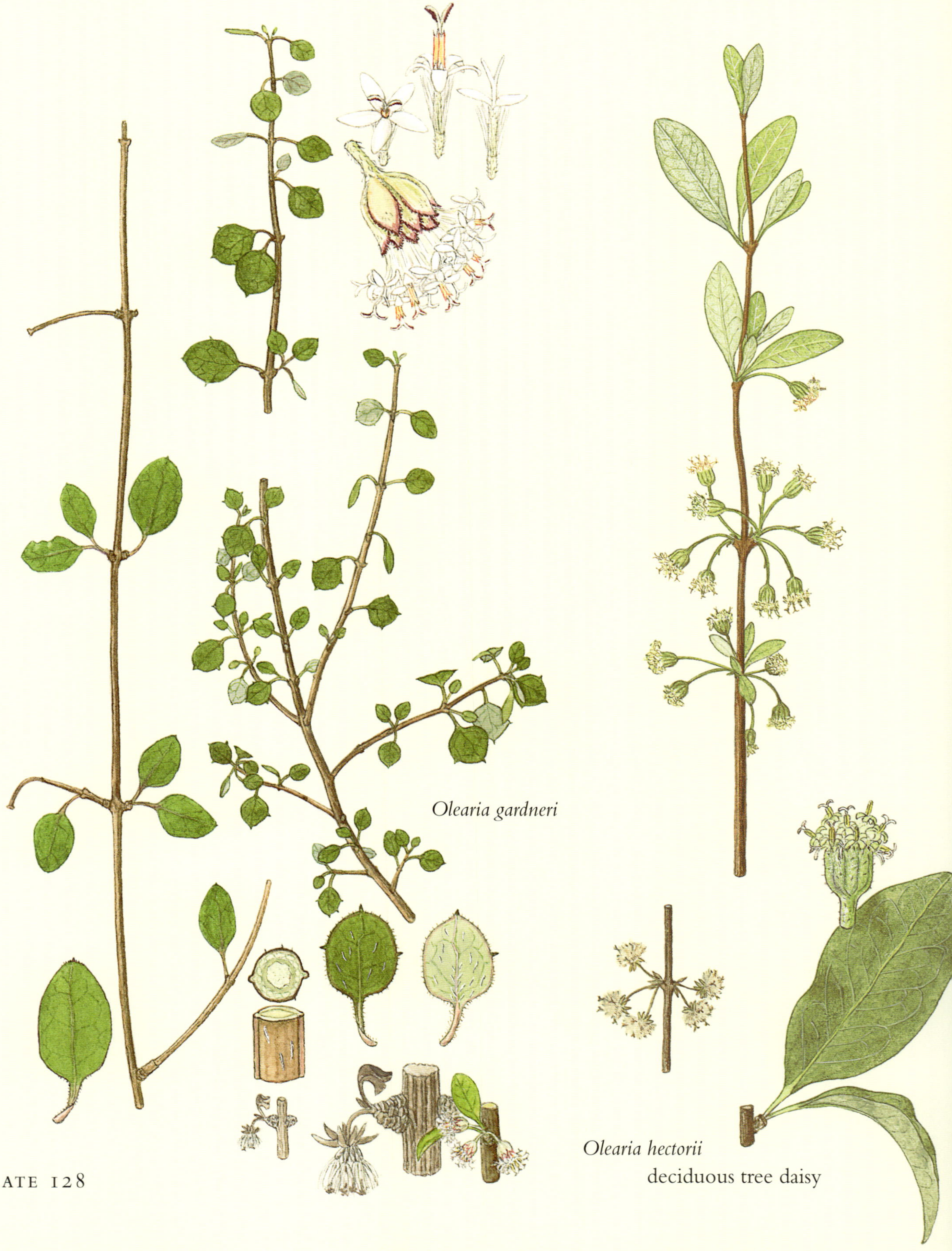

PLATE 128

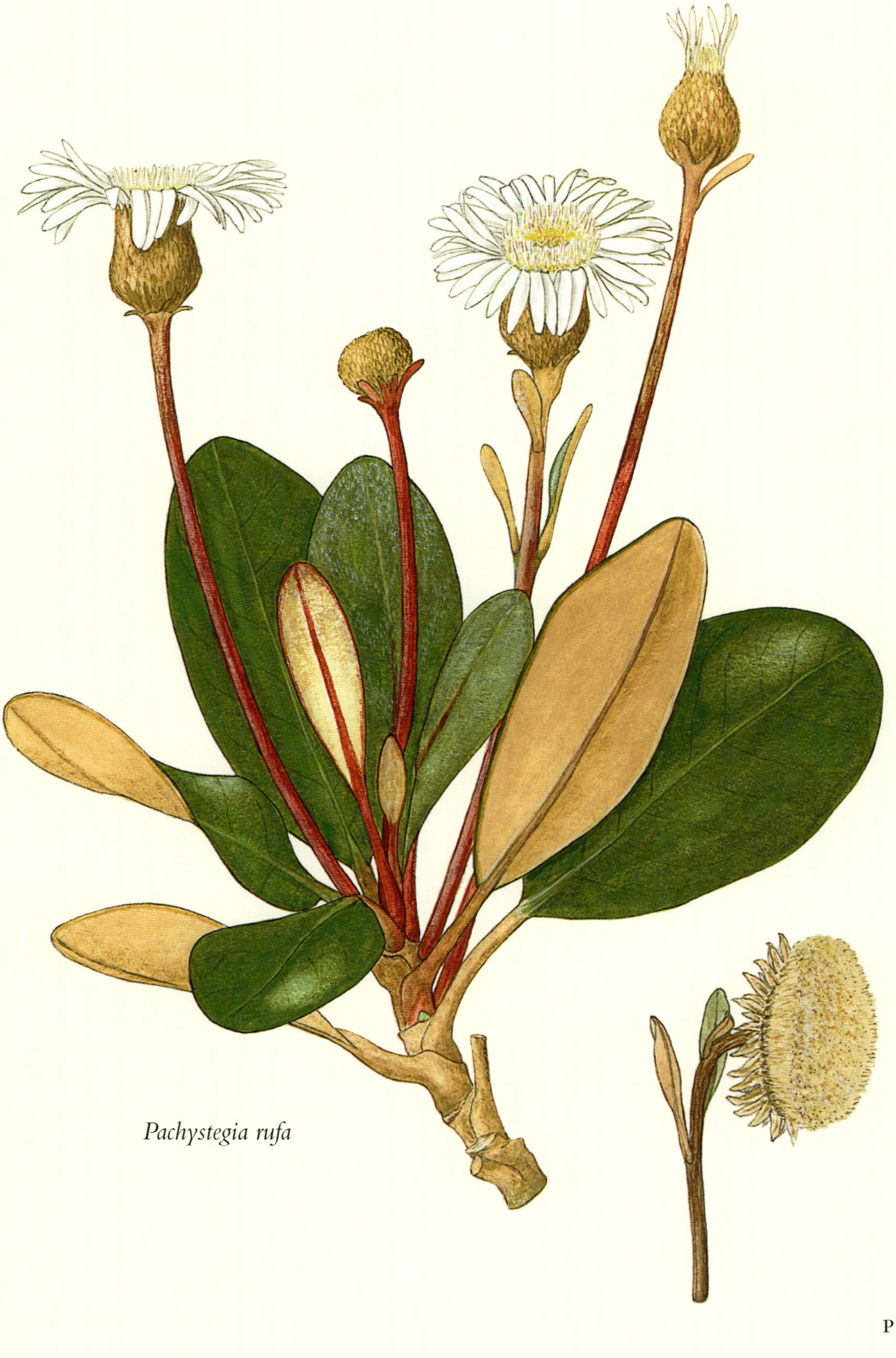

Pachystegia rufa

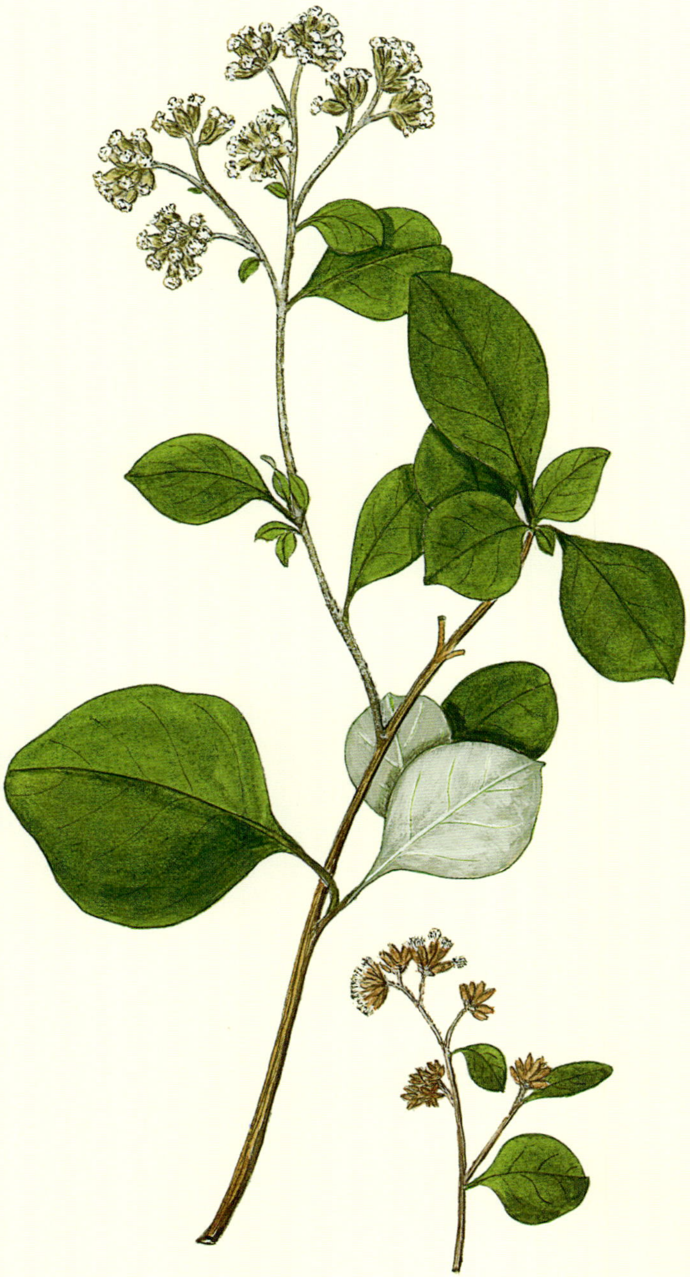

Helichrysum lanceolatum (i)

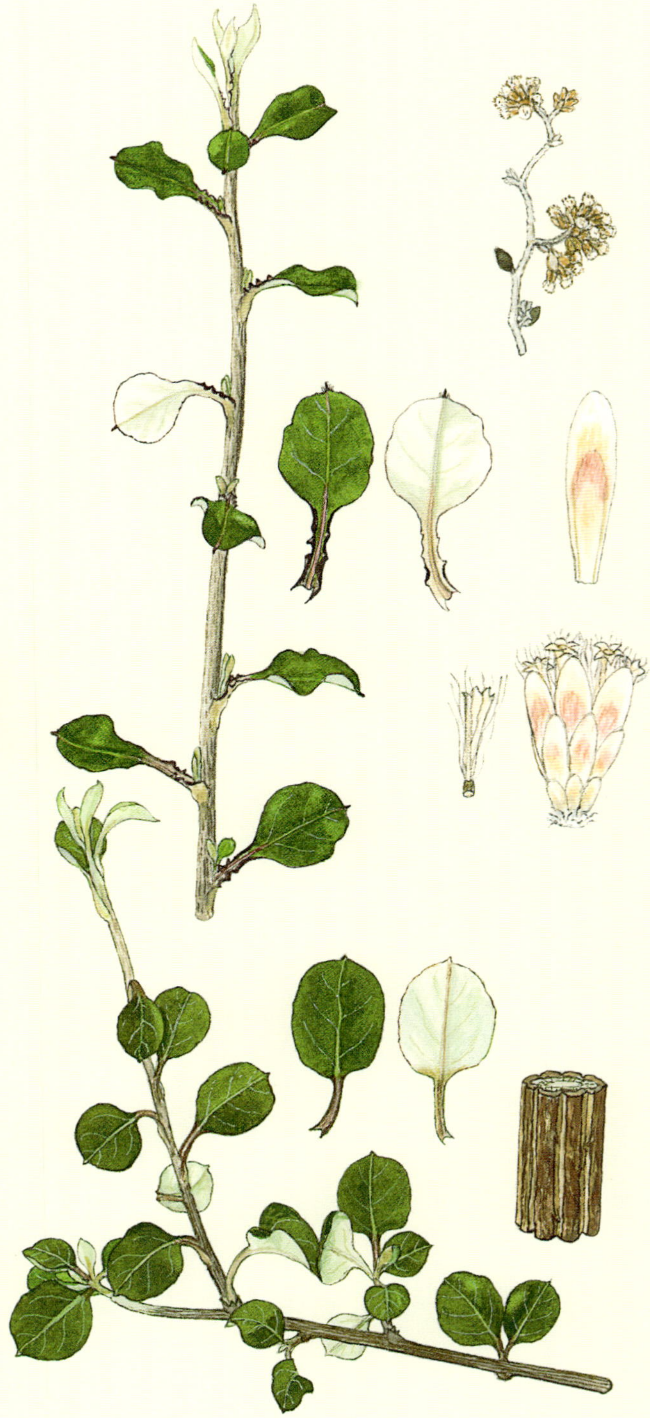

Helichrysum lanceolatum (ii)

PLATE 130

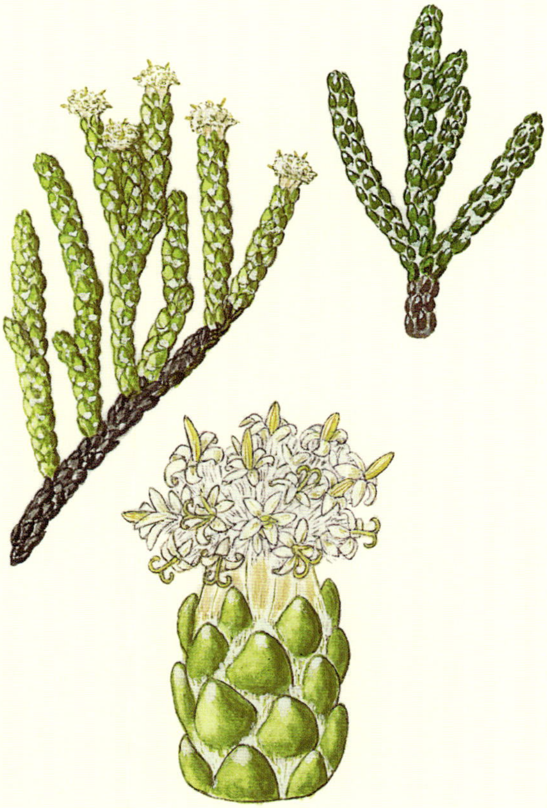

Helichrysum intermedium (iii)

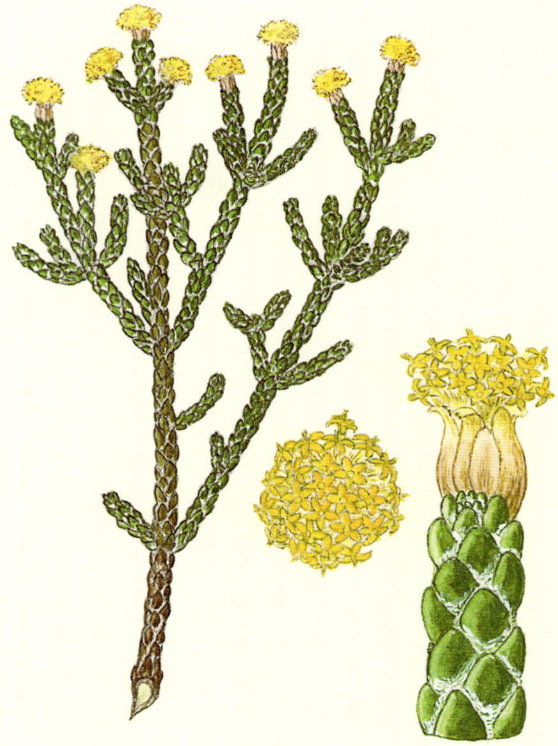

Helichrysum intermedium (iv)

Helichrysum intermedium (vi)

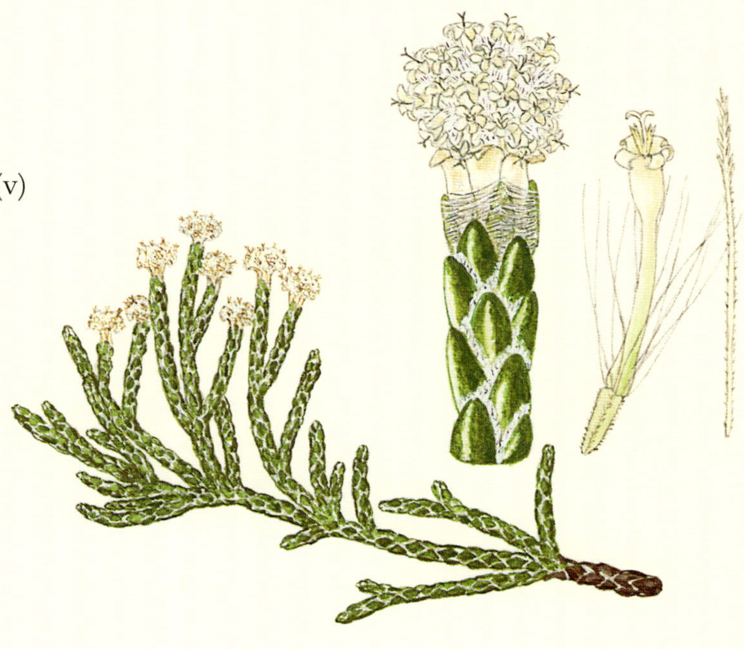

Helichrysum intermedium (v)

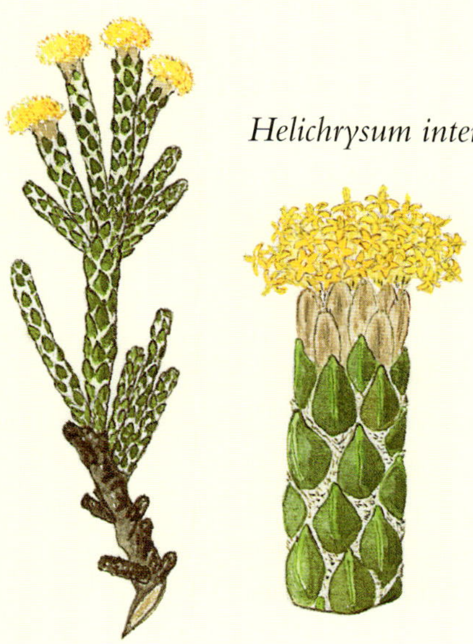

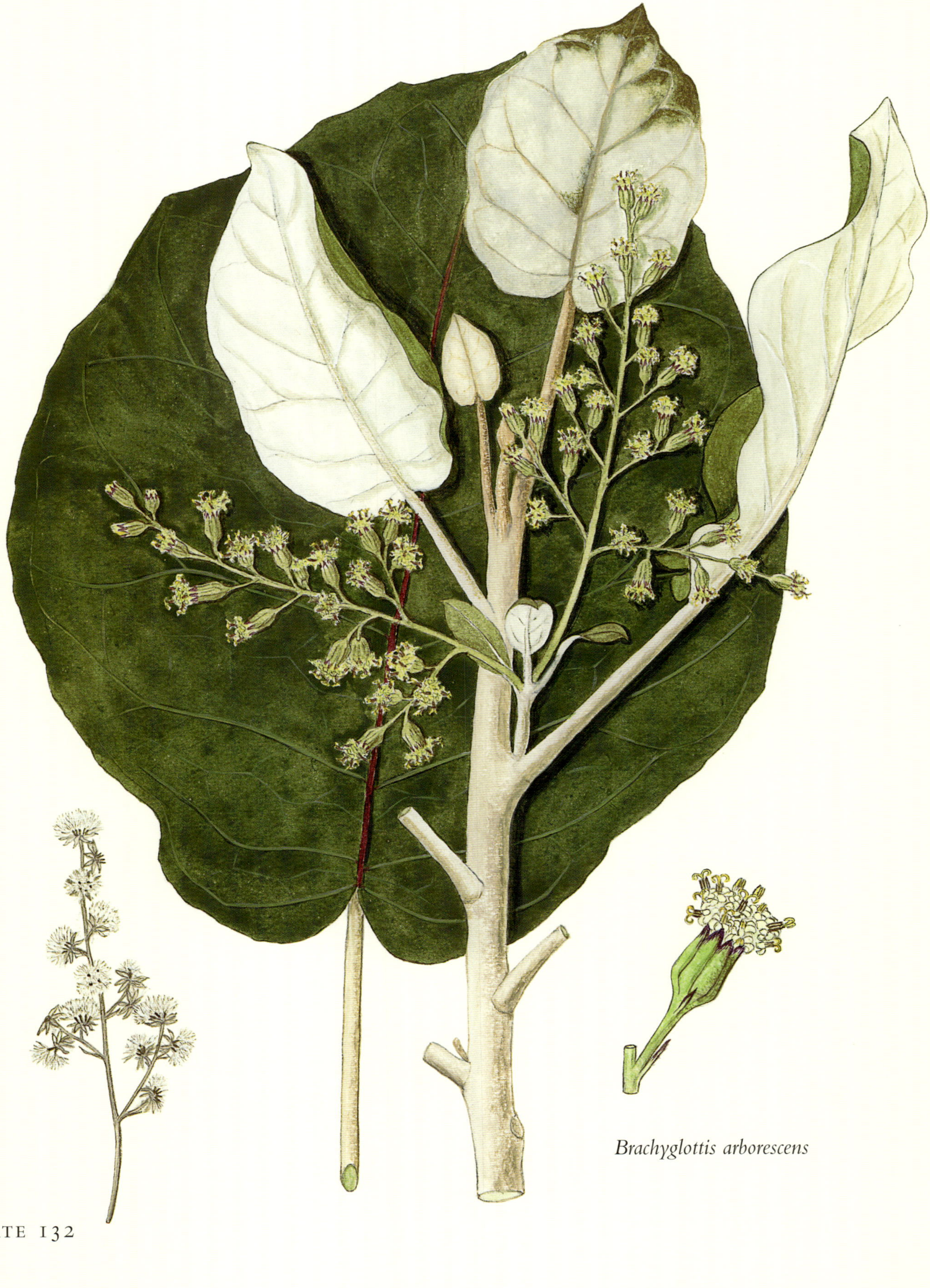

Brachyglottis arborescens

PLATE 132

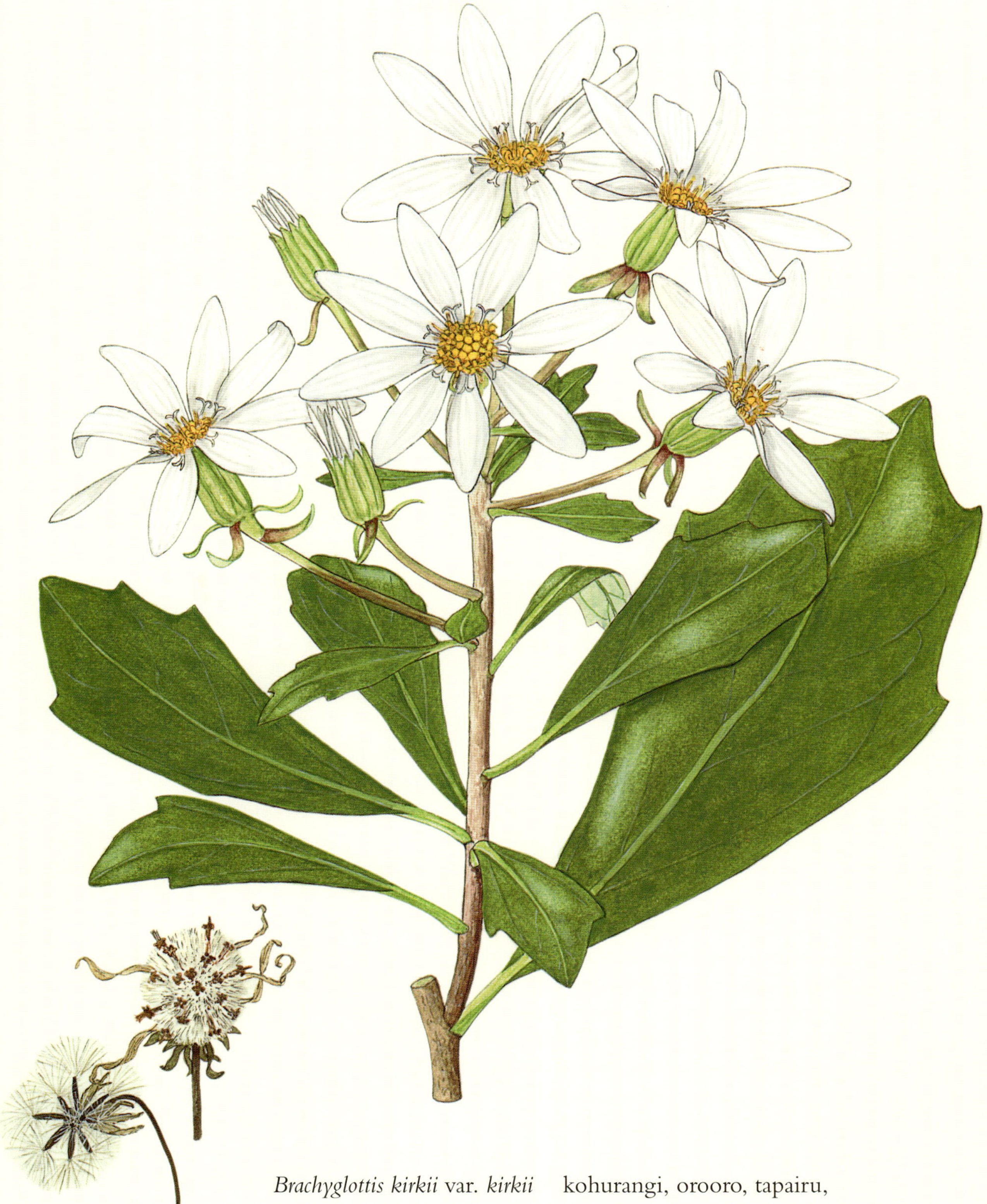

Brachyglottis kirkii var. *kirkii* kohurangi, orooro, tapairu, Kirk's tree daisy

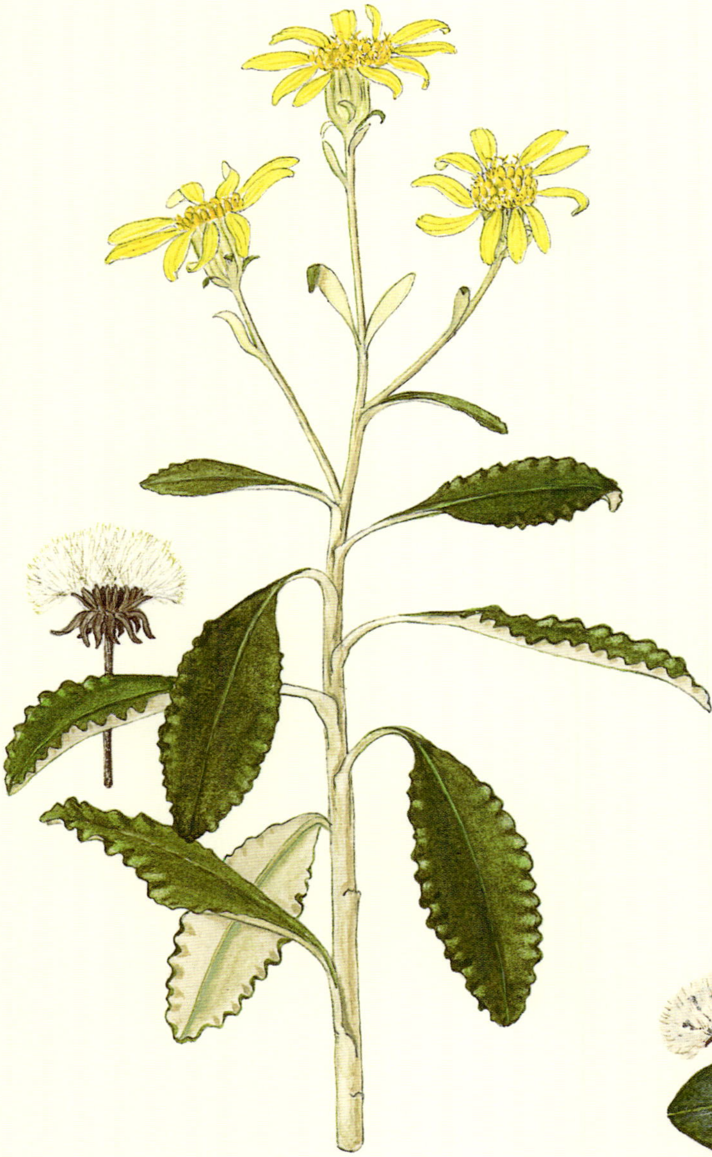

Brachyglottis monroi

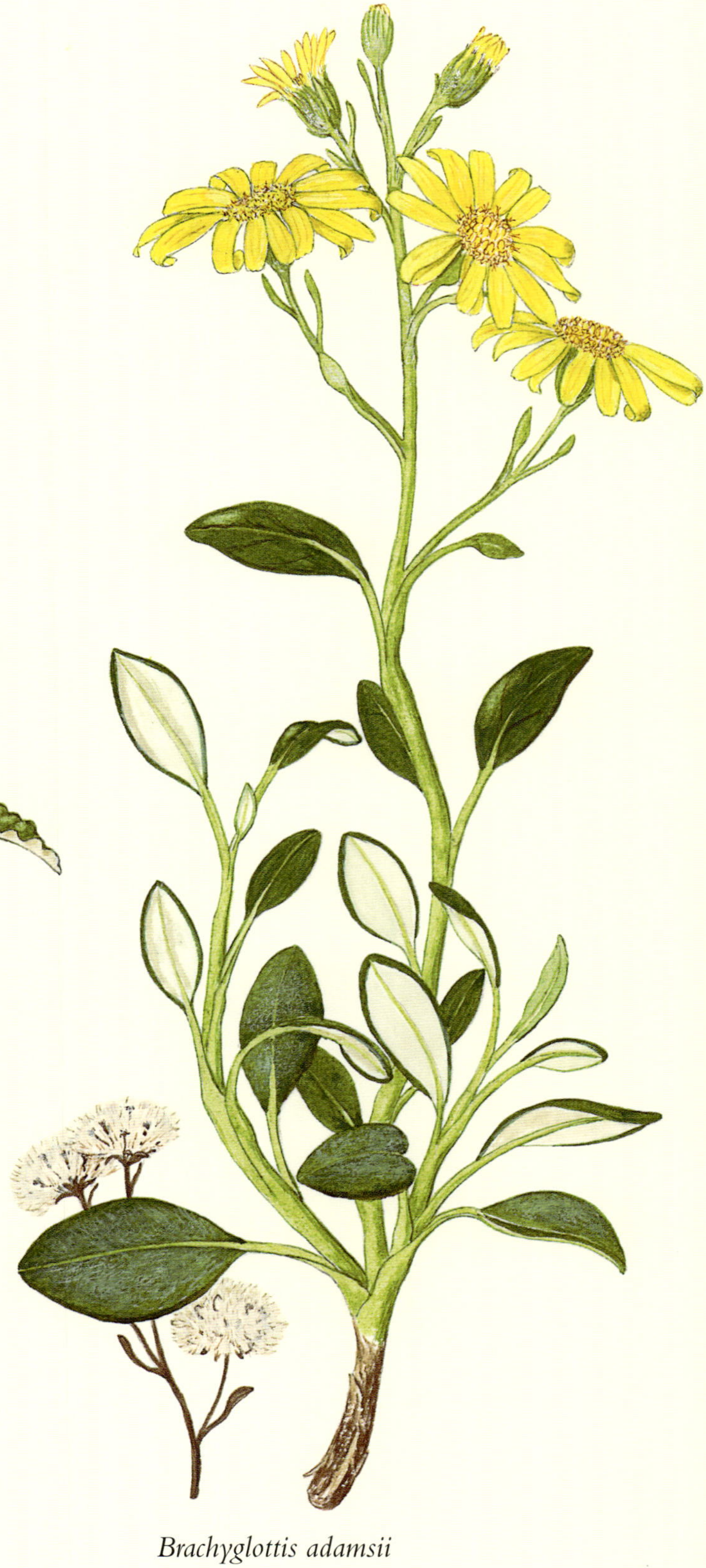

Brachyglottis adamsii

PLATE 134

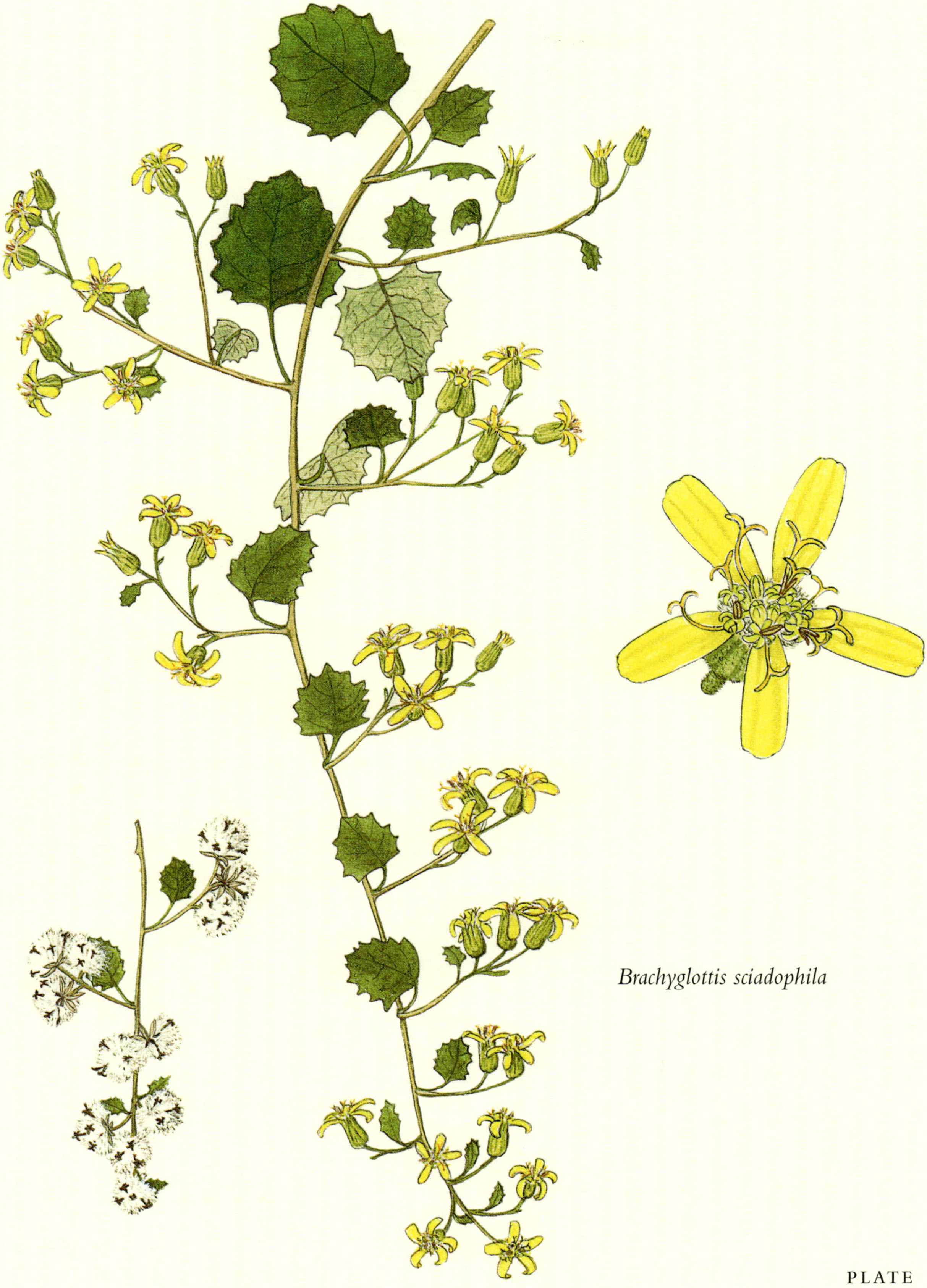

Brachyglottis sciadophila

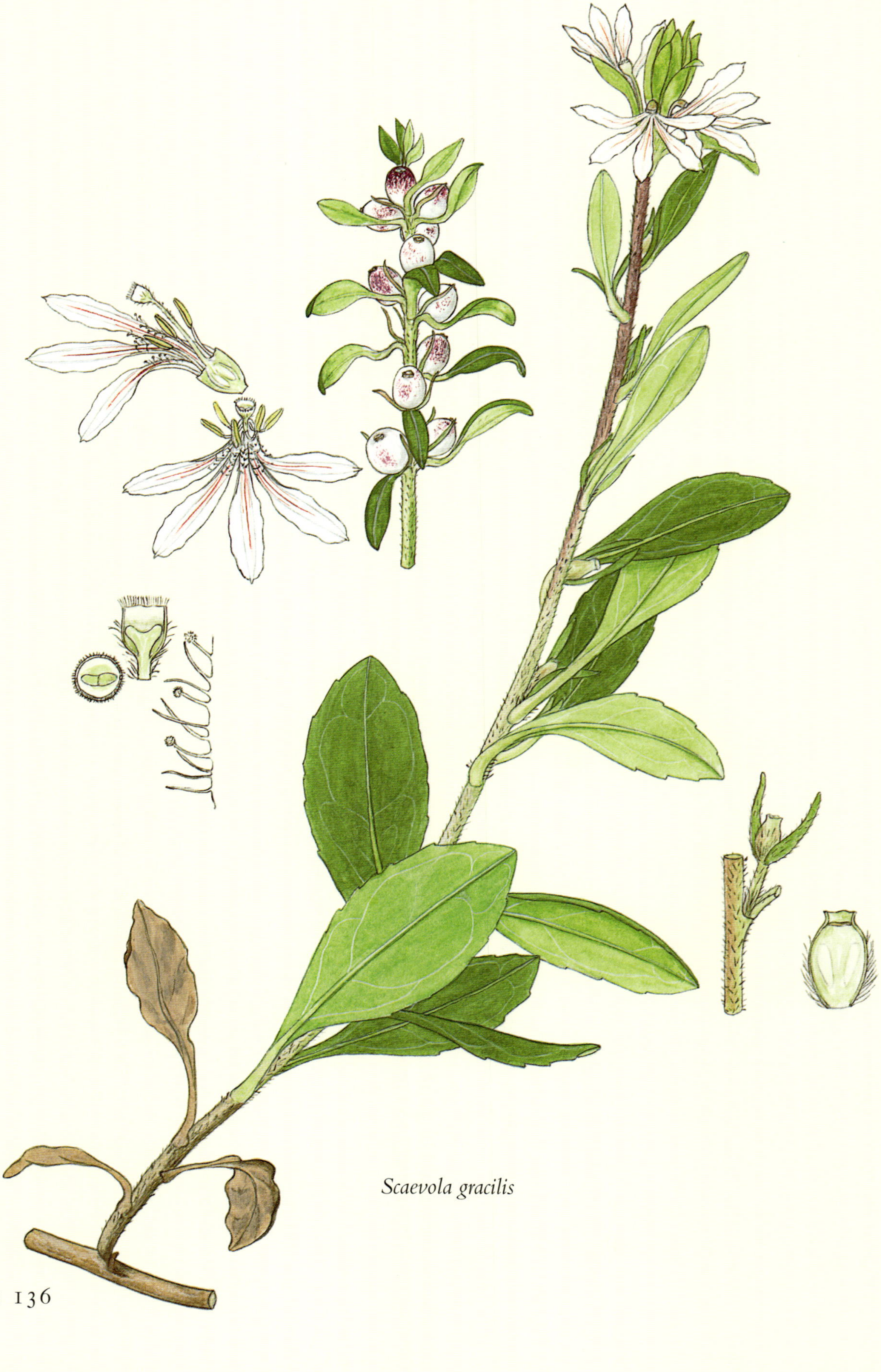

Scaevola gracilis

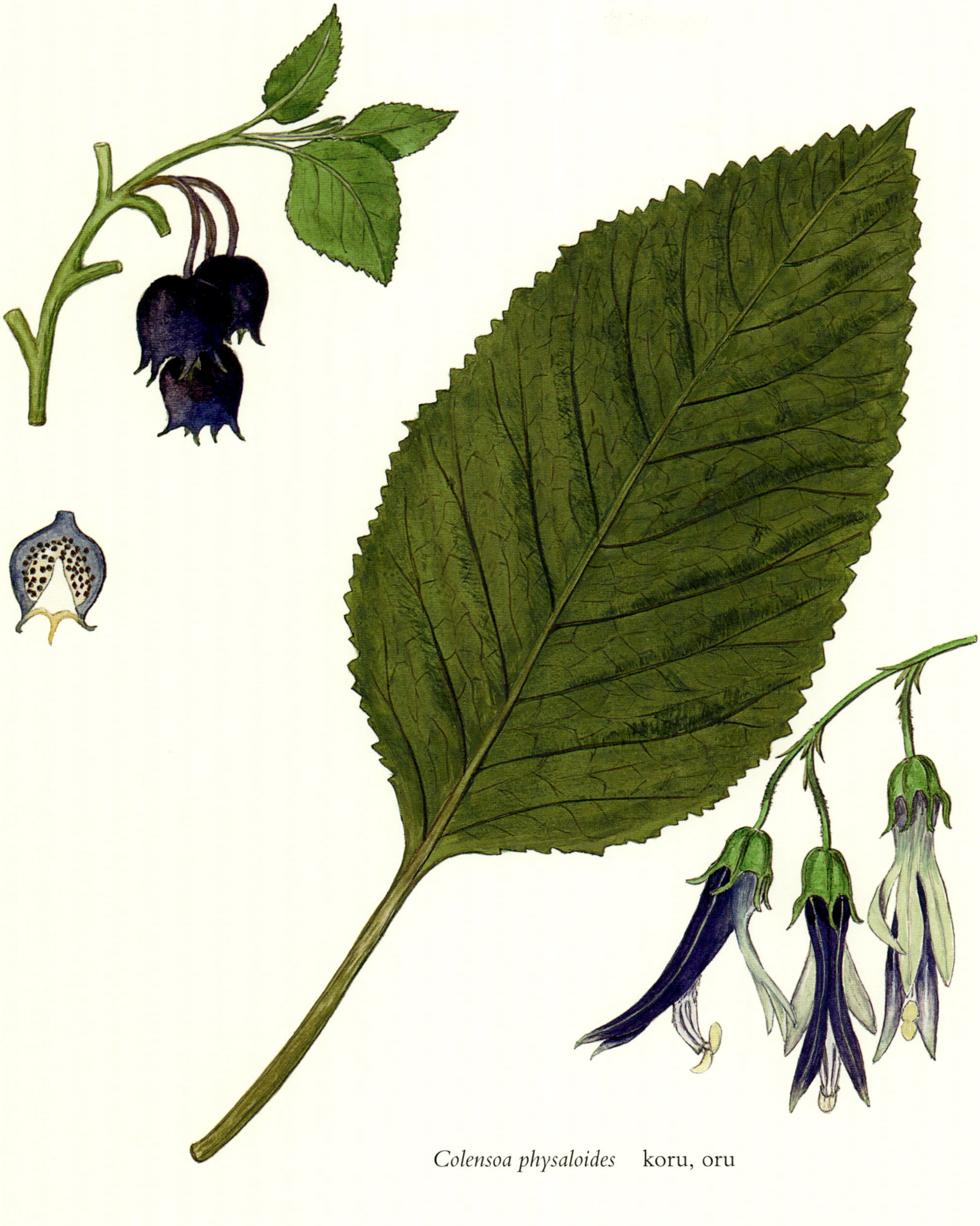

Colensoa physaloides koru, oru

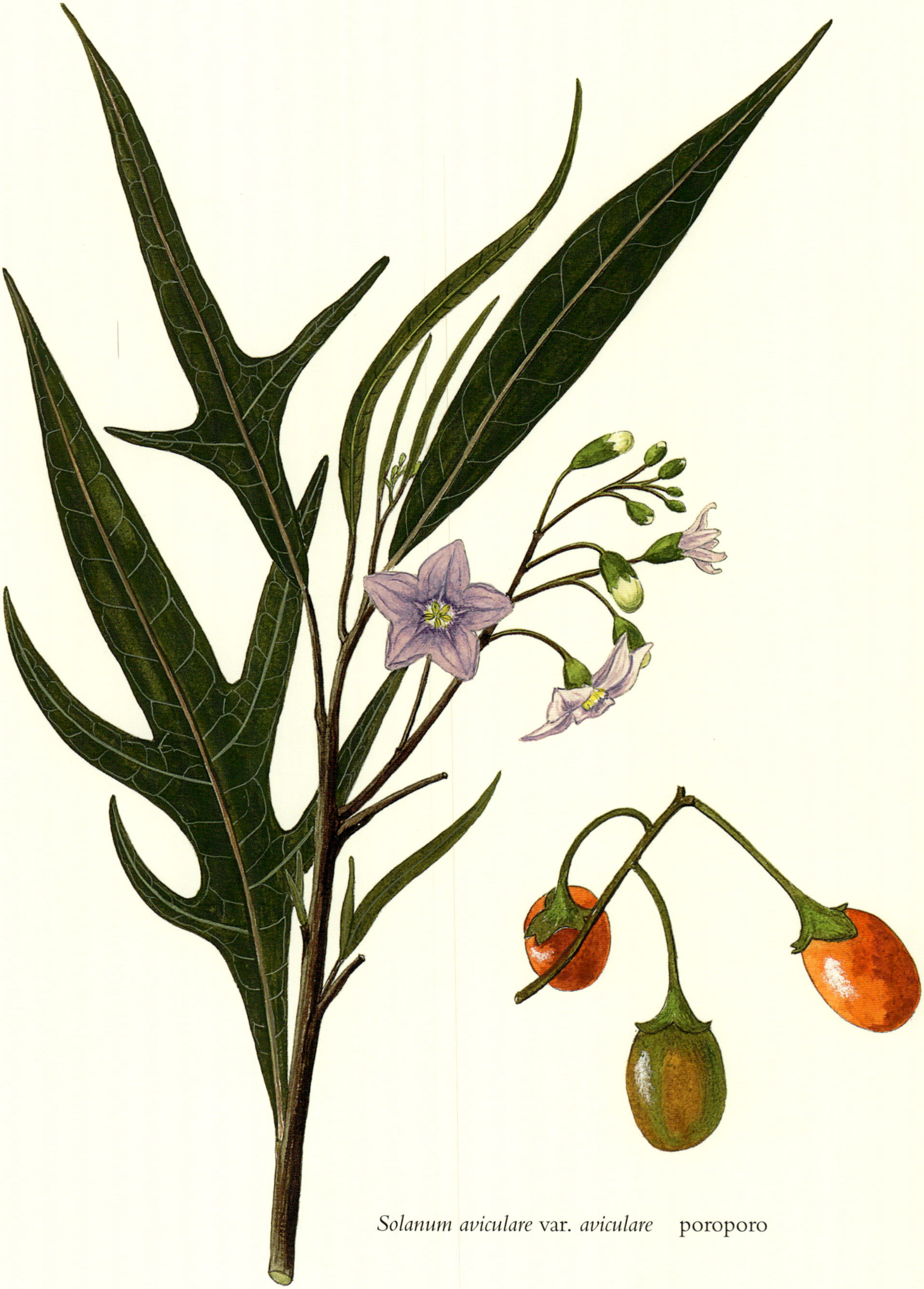

Solanum aviculare var. *aviculare* poroporo

PLATE 138

Solanum aviculare var. *latifolium*

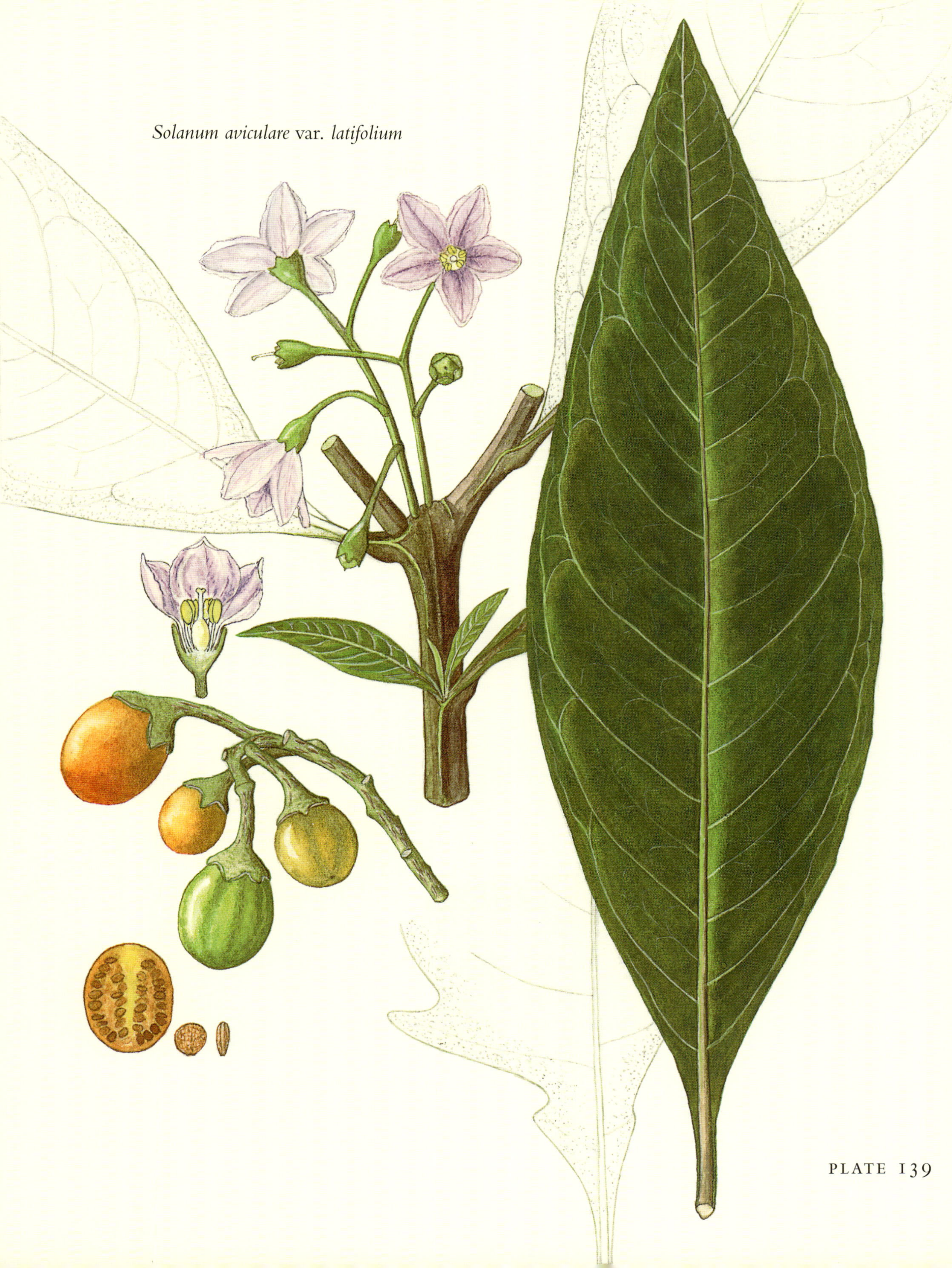

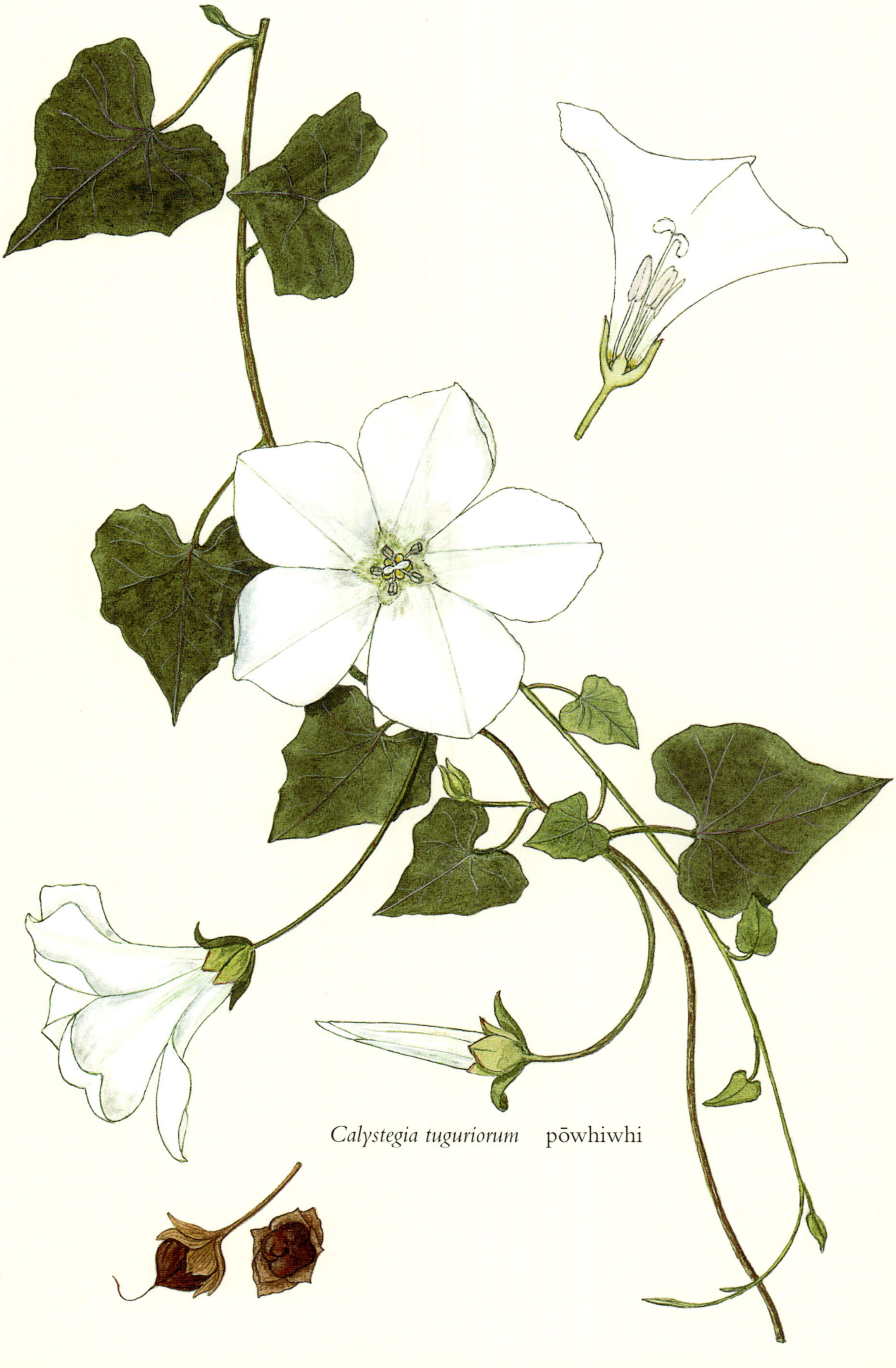

Calystegia tuguriorum pōwhiwhi

PLATE 140

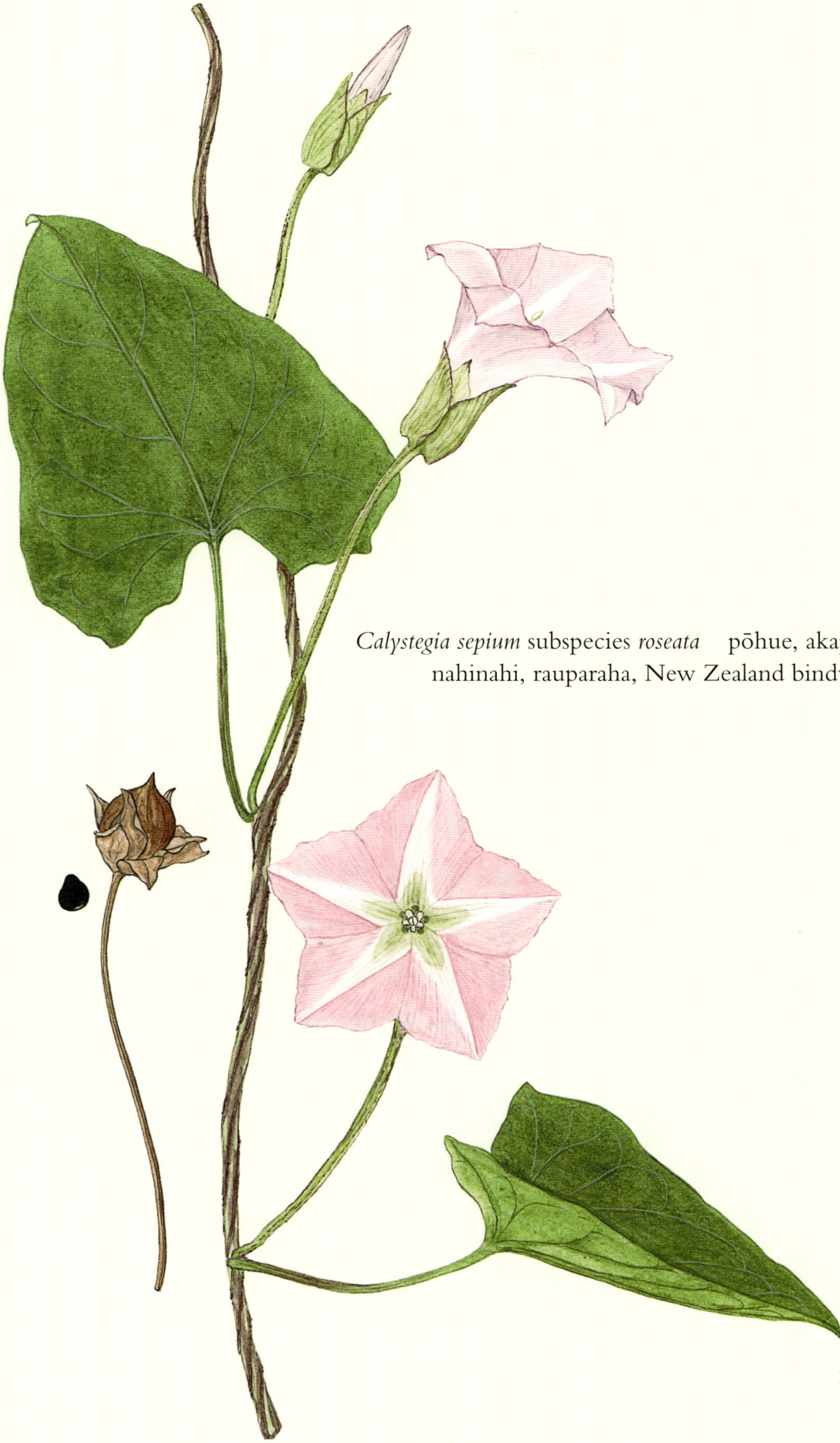

Calystegia sepium subspecies *roseata* pōhue, akapōhue, nahinahi, rauparaha, New Zealand bindweed

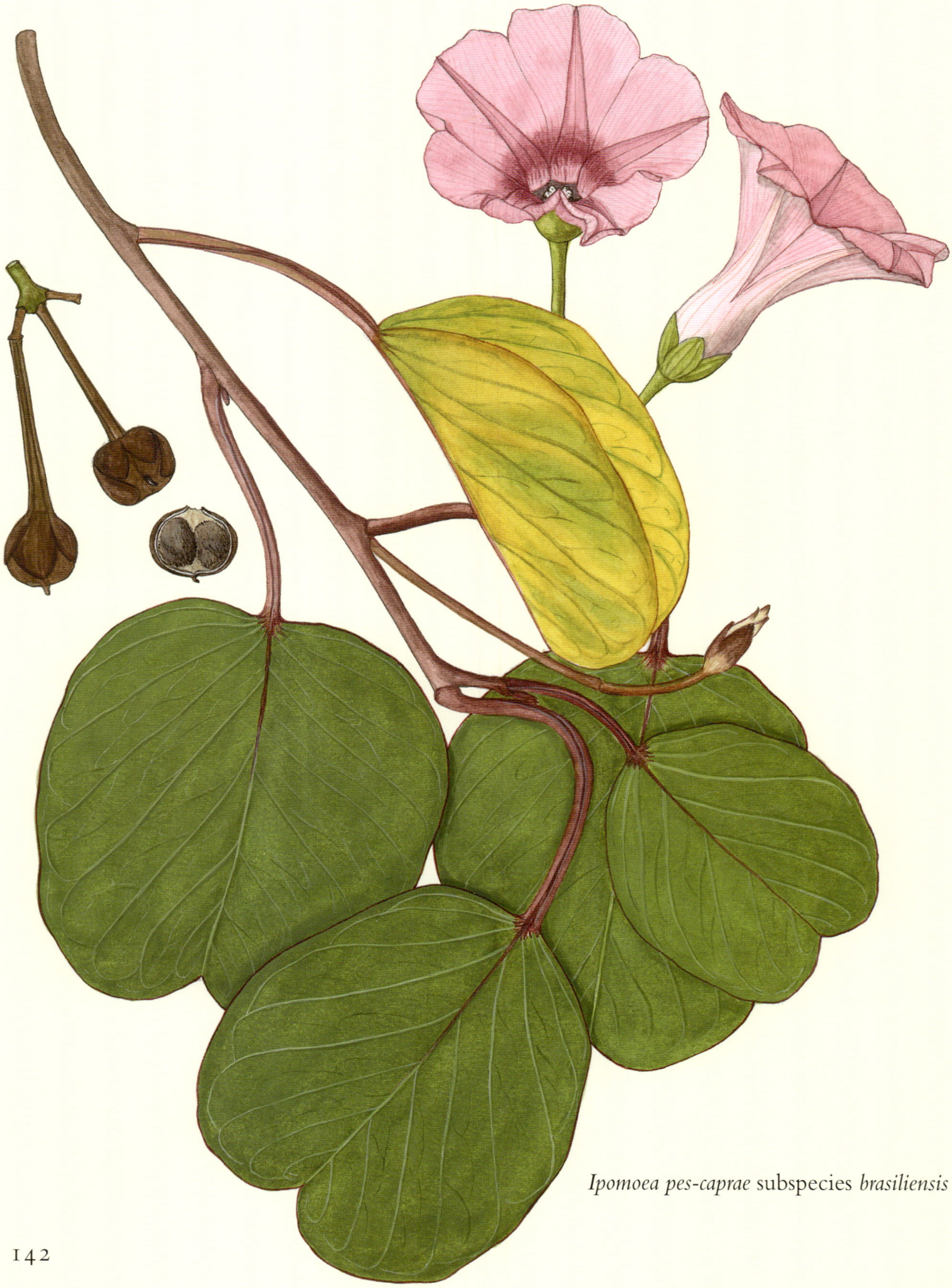

Ipomoea pes-caprae subspecies *brasiliensis*

PLATE 142

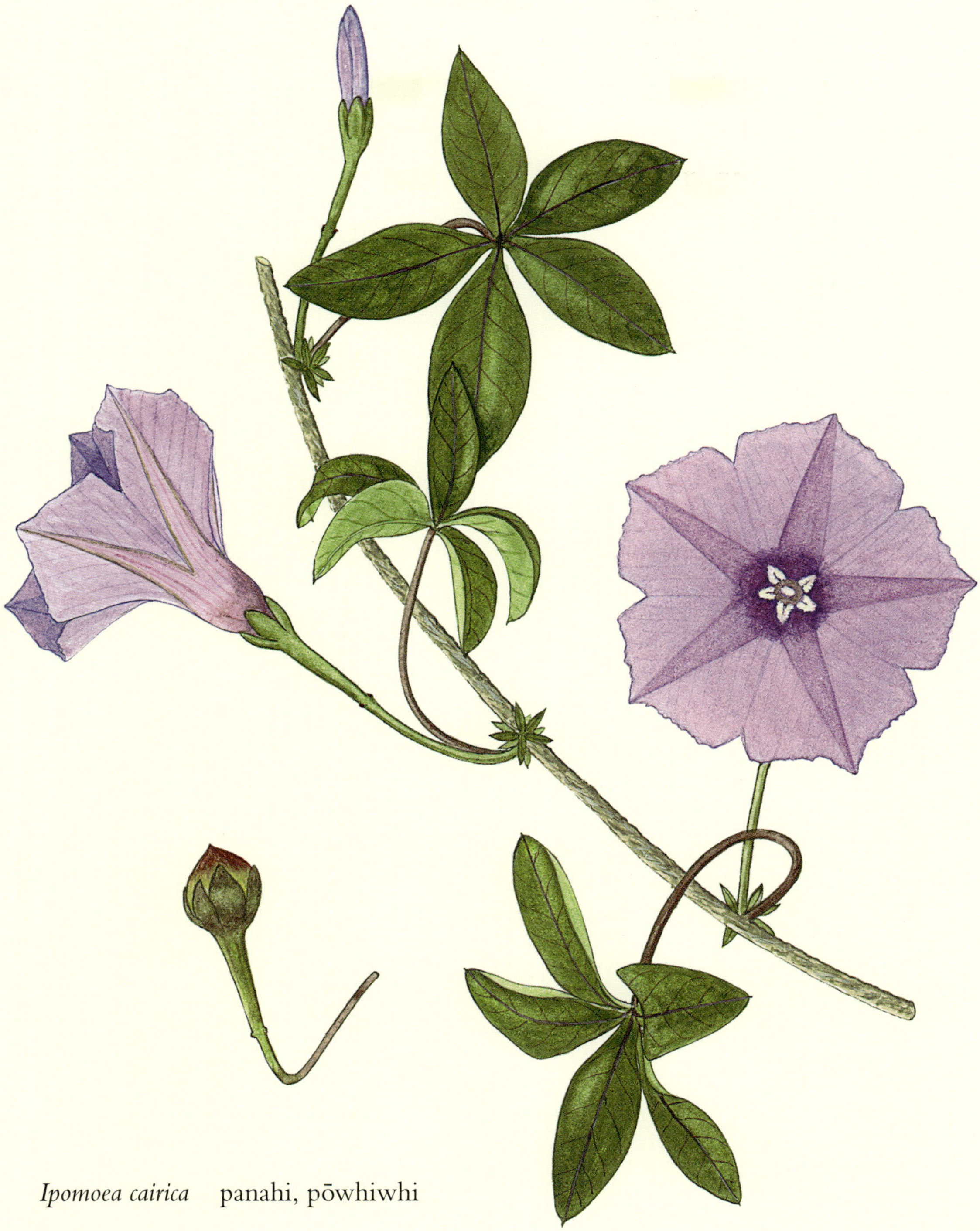

Ipomoea cairica panahi, pōwhiwhi

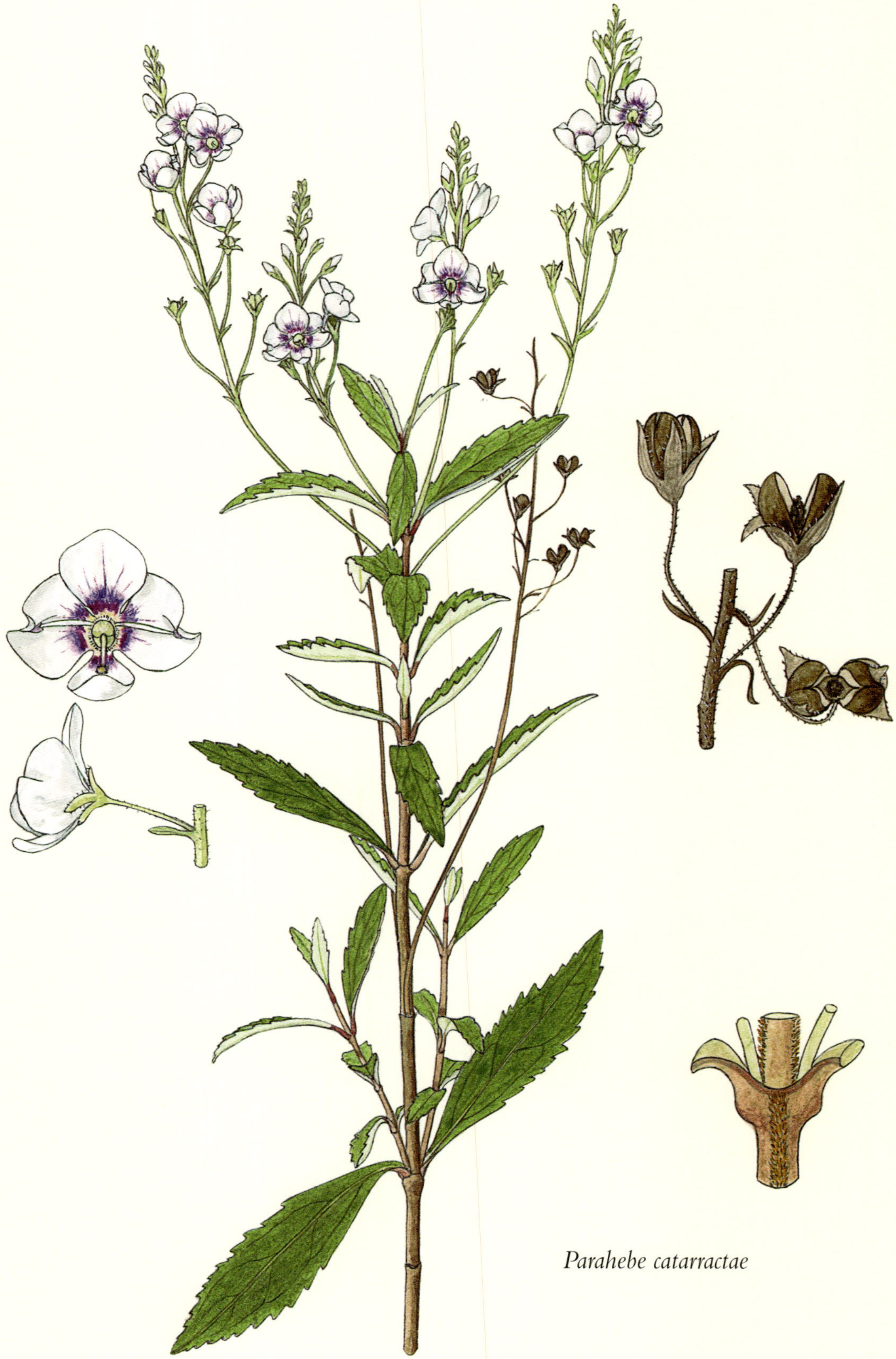

Parahebe catarractae

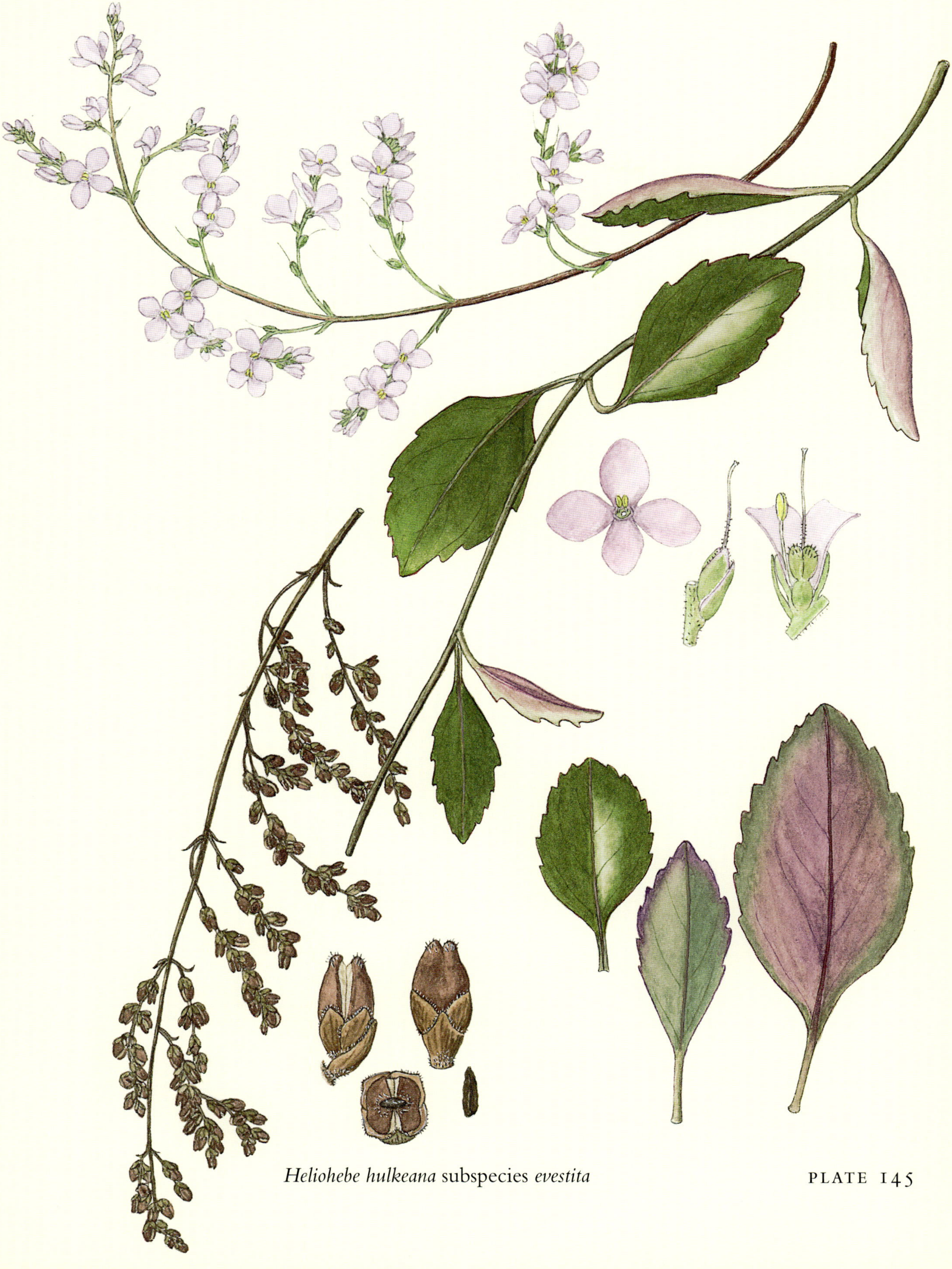

Heliohebe hulkeana subspecies *evestita*

PLATE 145

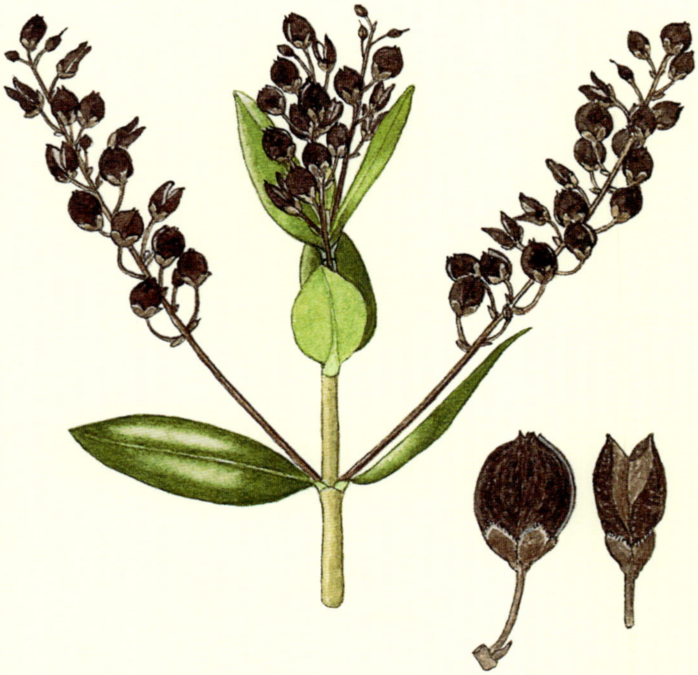

Hebe aff. *brevifolia*

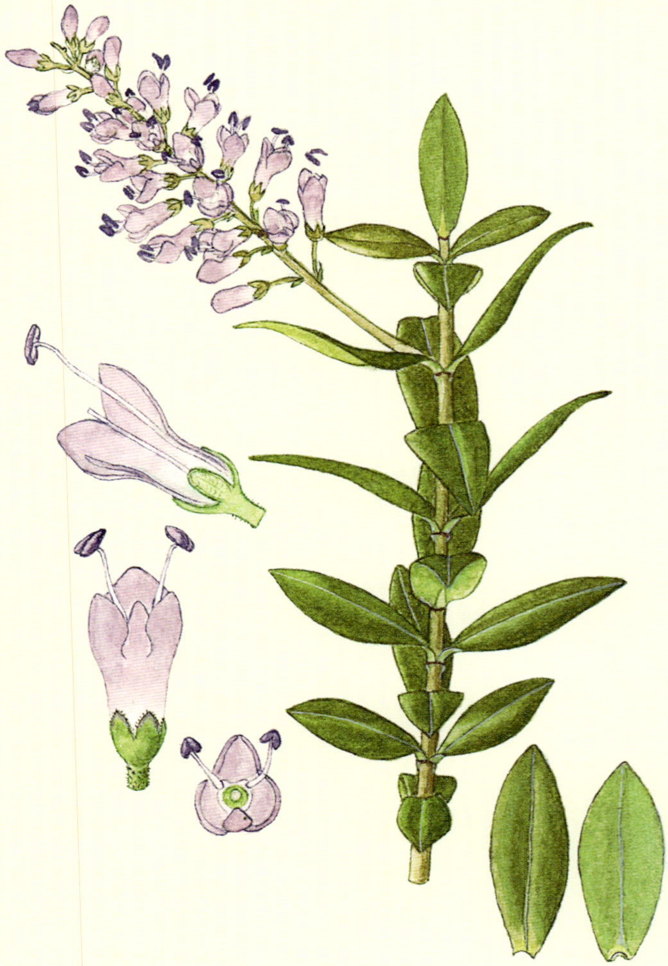

Heliohebe acuta

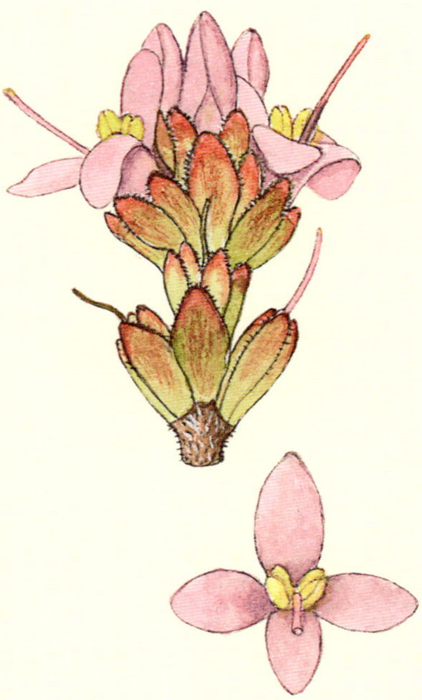

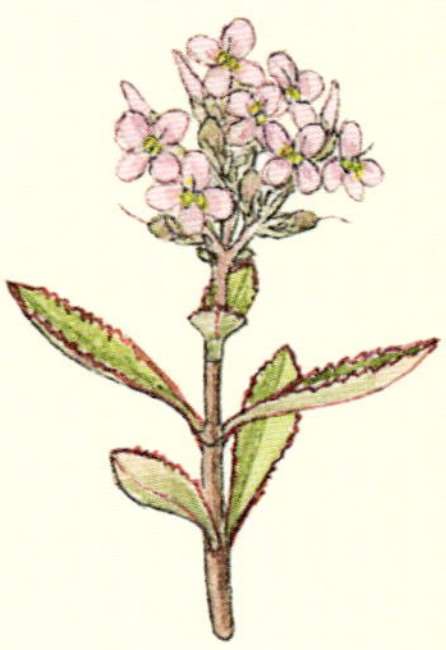

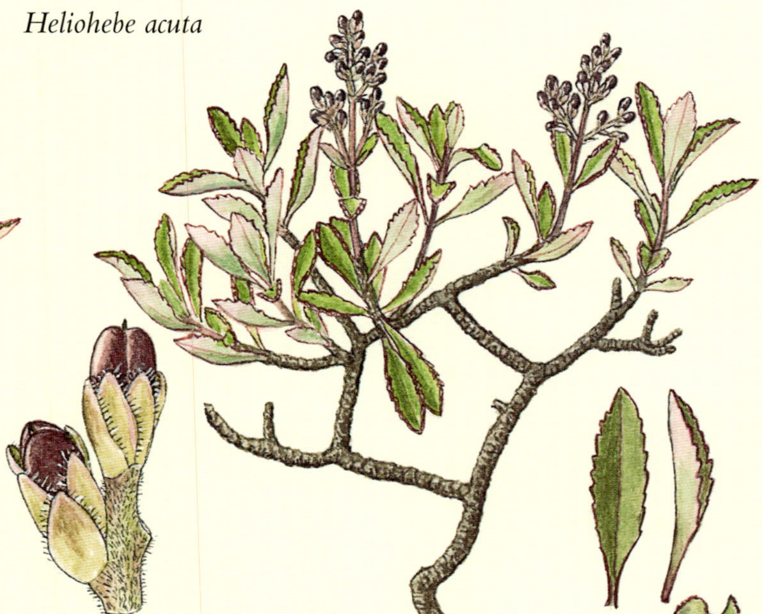

Hebe speciosa tītīrangi, napuka

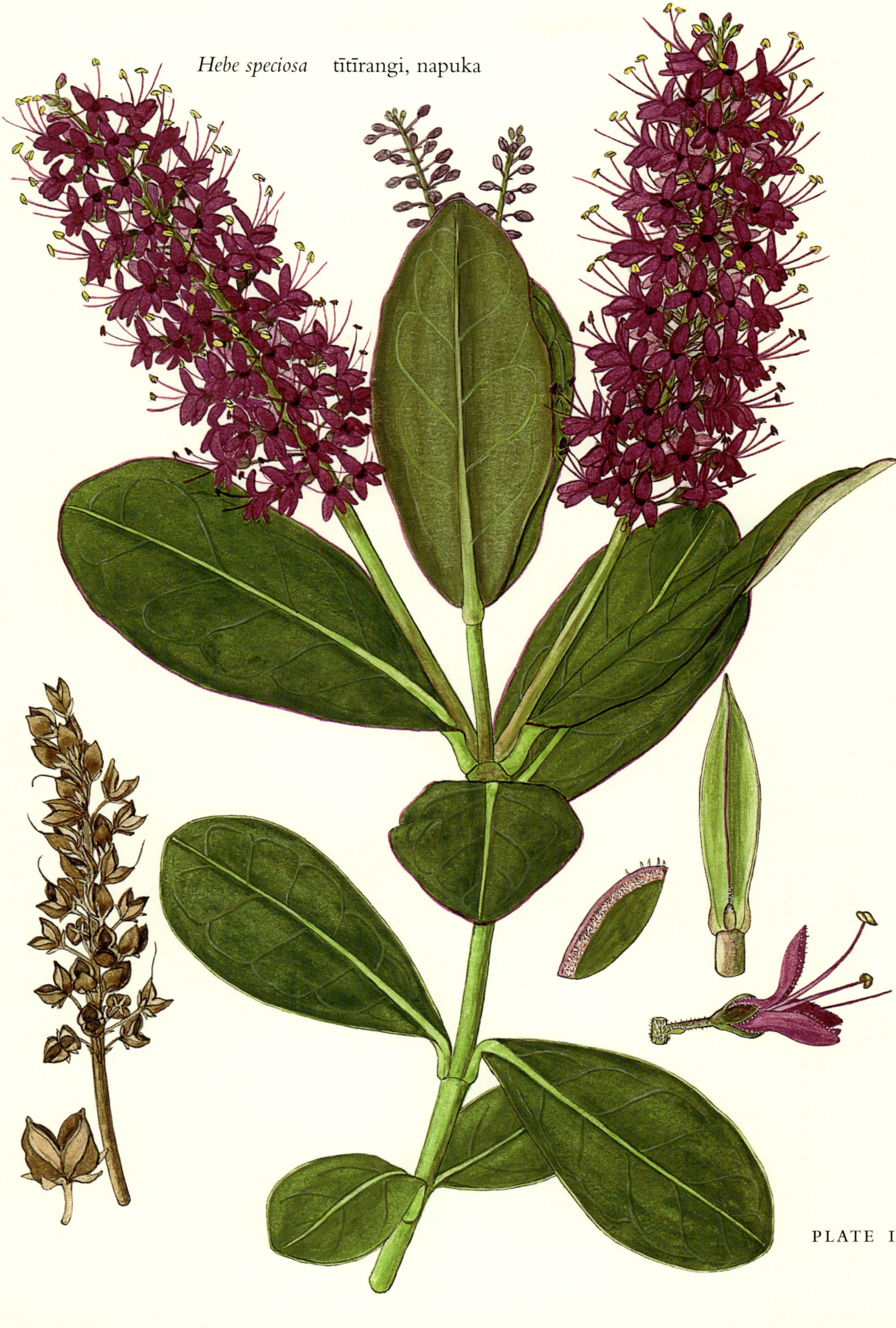

PLATE 147

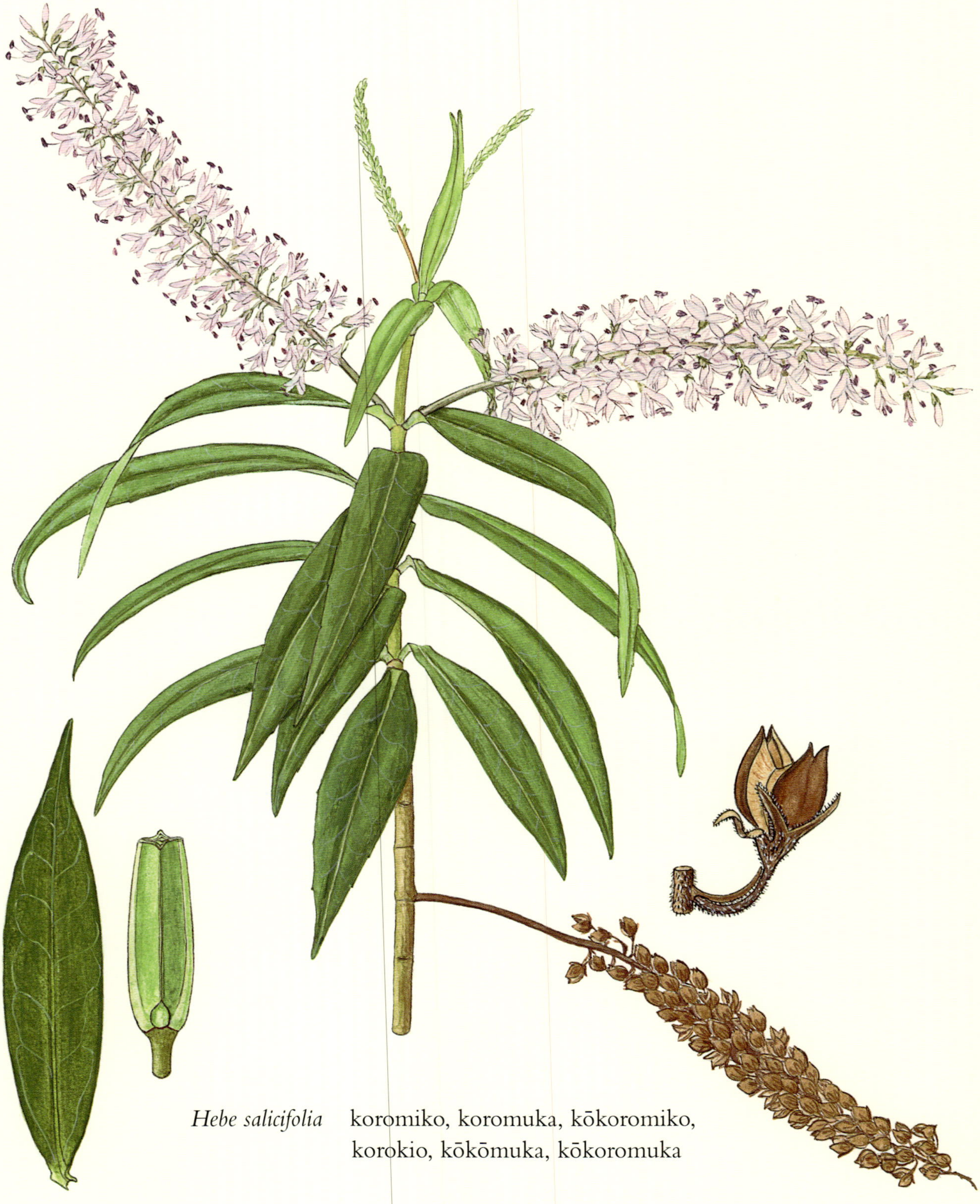

Hebe salicifolia koromiko, koromuka, kōkoromiko, korokio, kōkōmuka, kōkoromuka

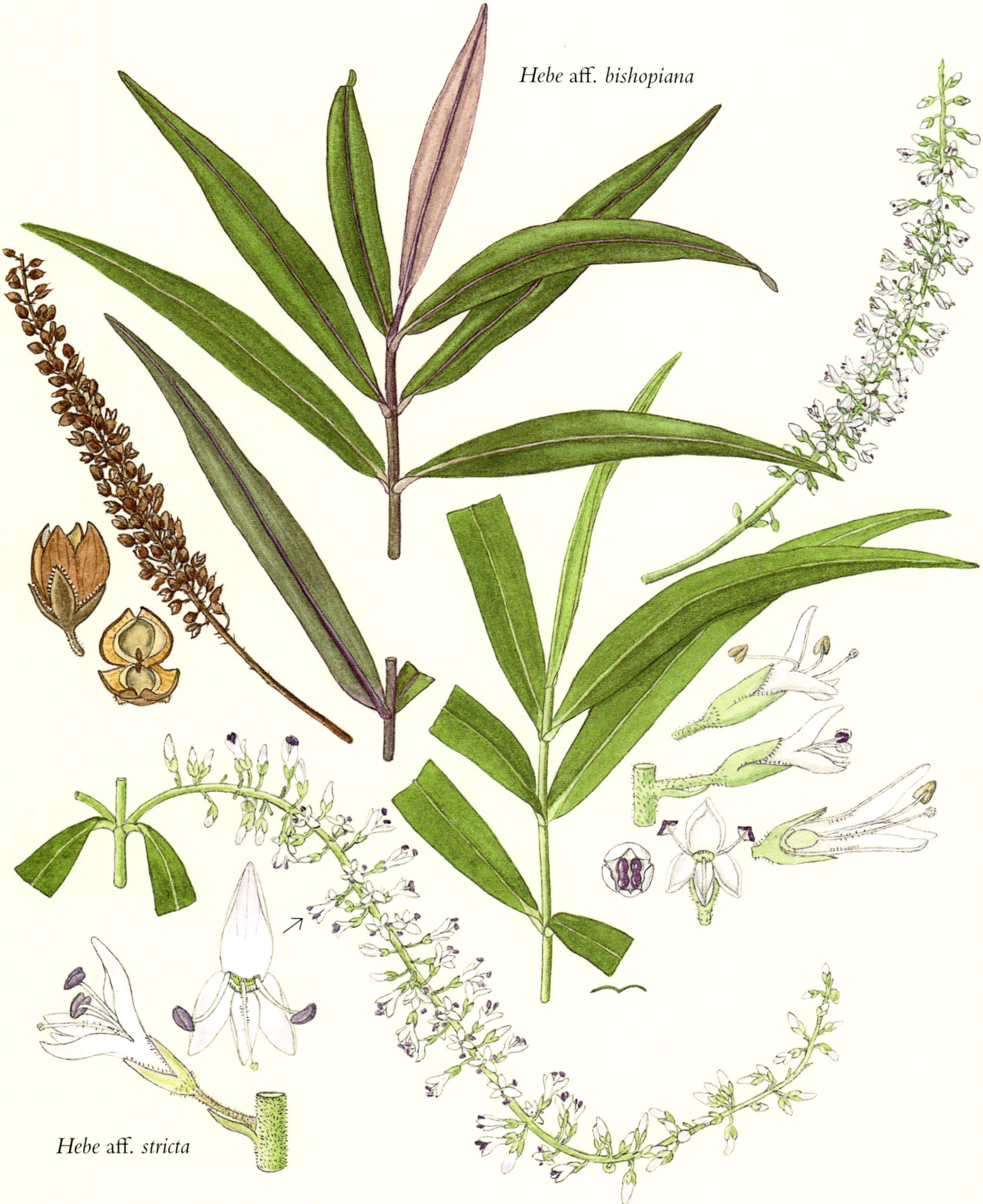

PLATE 149

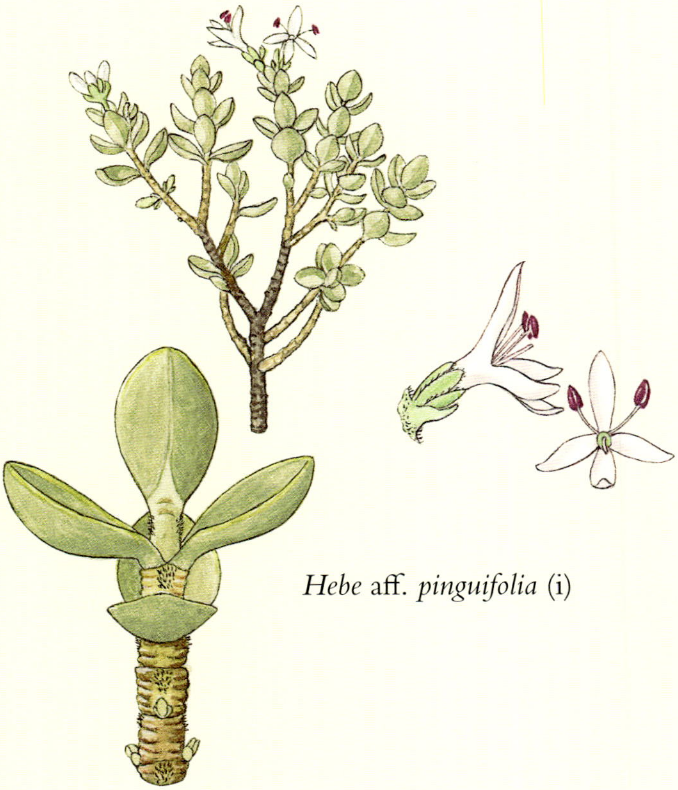

Hebe aff. *pinguifolia* (i)

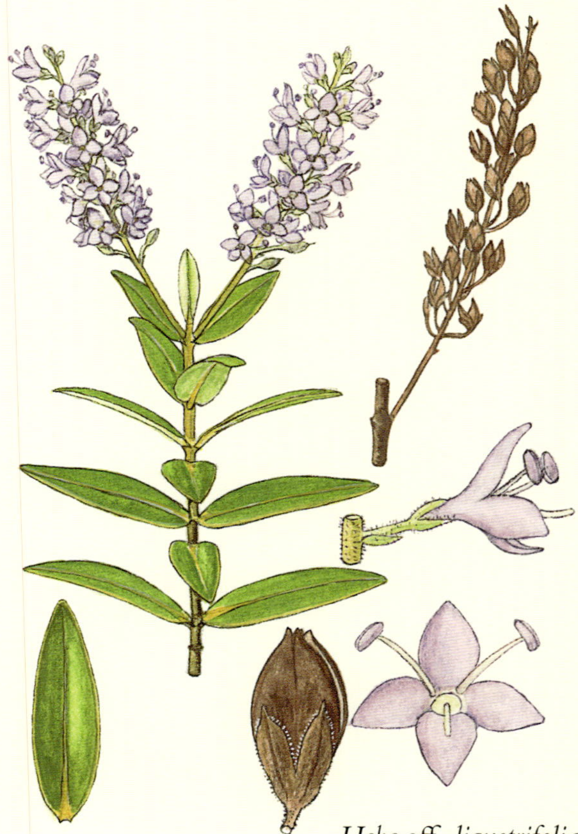

Hebe aff. *ligustrifolia*

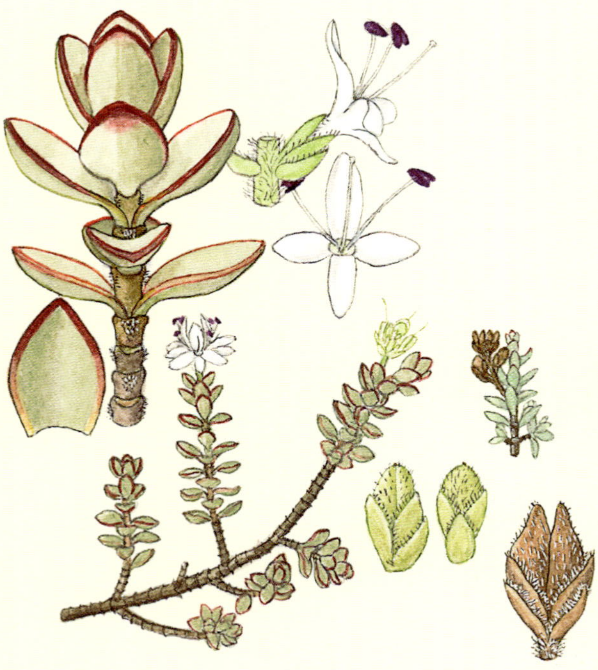

Hebe aff. *pinguifolia* (ii)

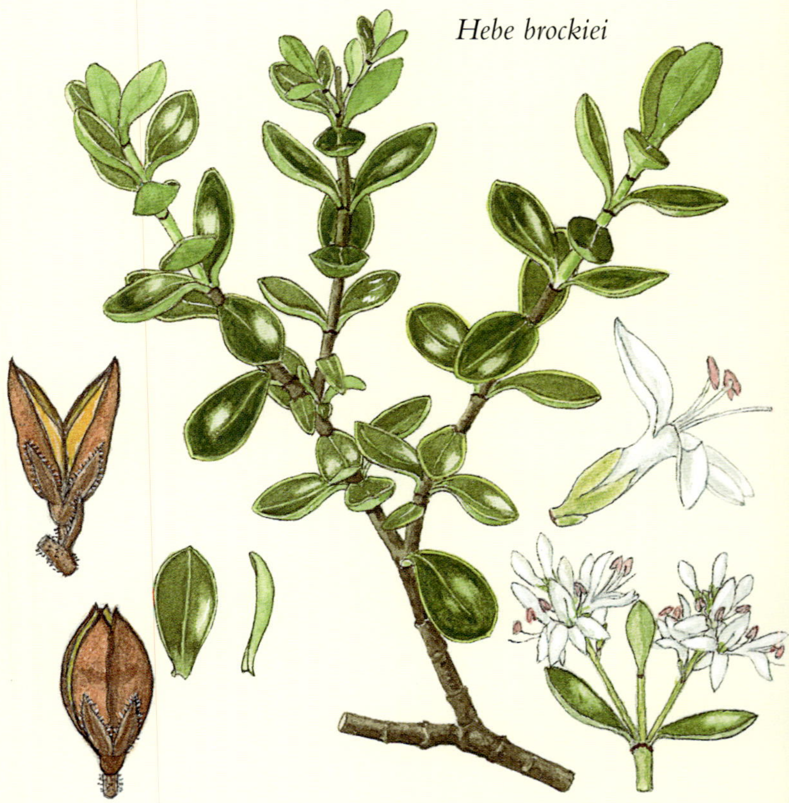

Hebe brockiei

PLATE 150

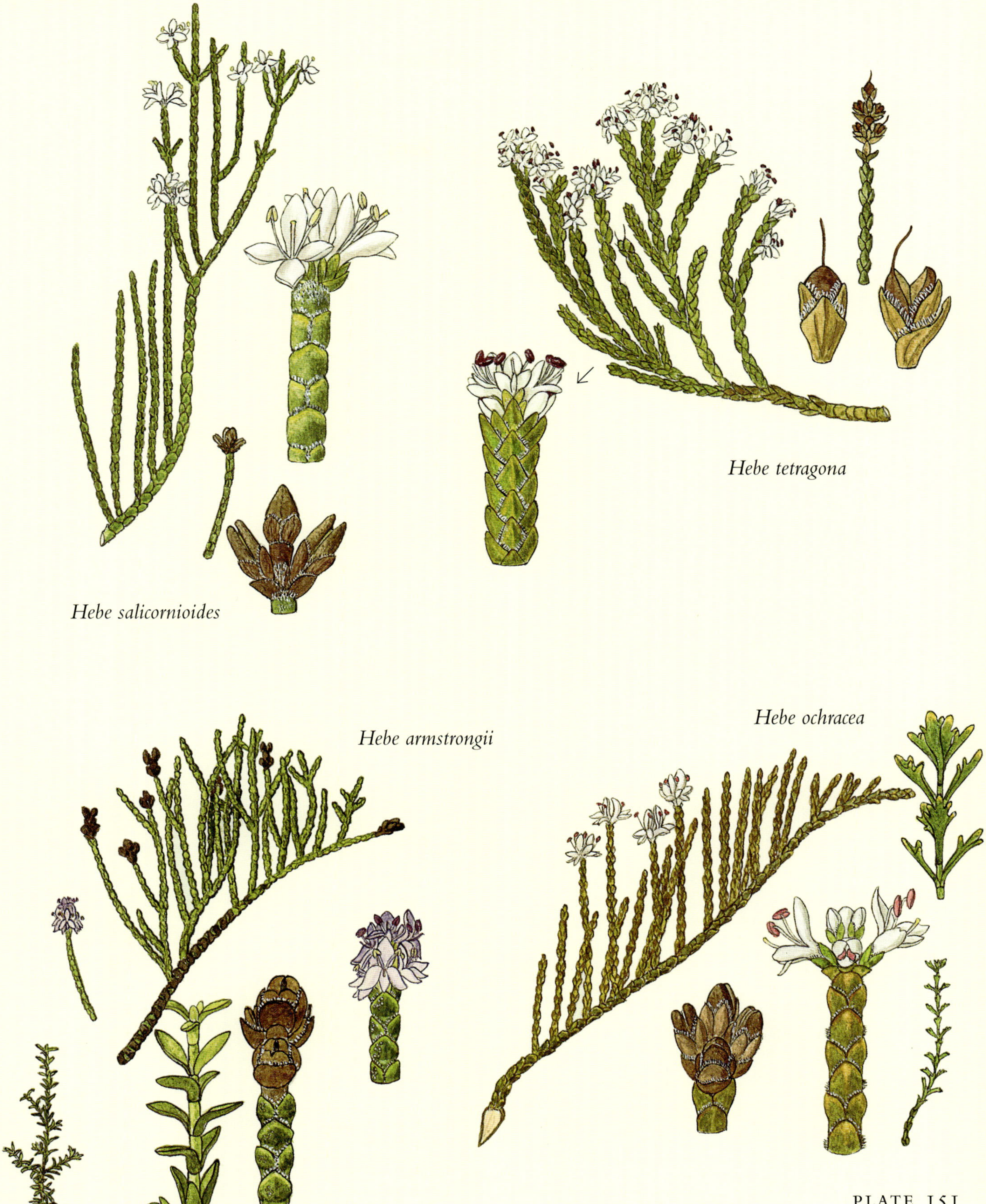
Hebe tetragona
Hebe salicornioides
Hebe armstrongii
Hebe ochracea

PLATE 151

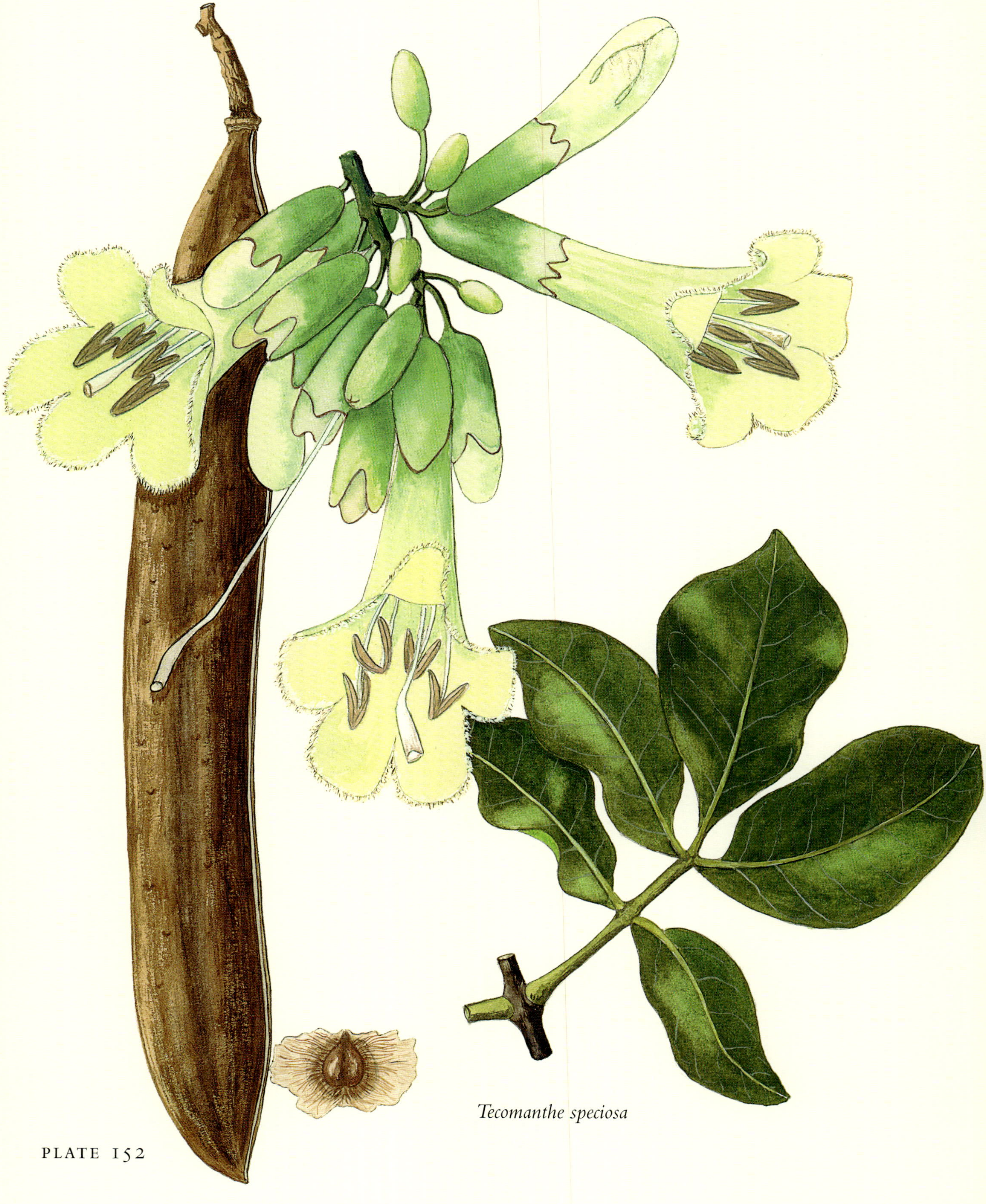

Tecomanthe speciosa

PLATE 152

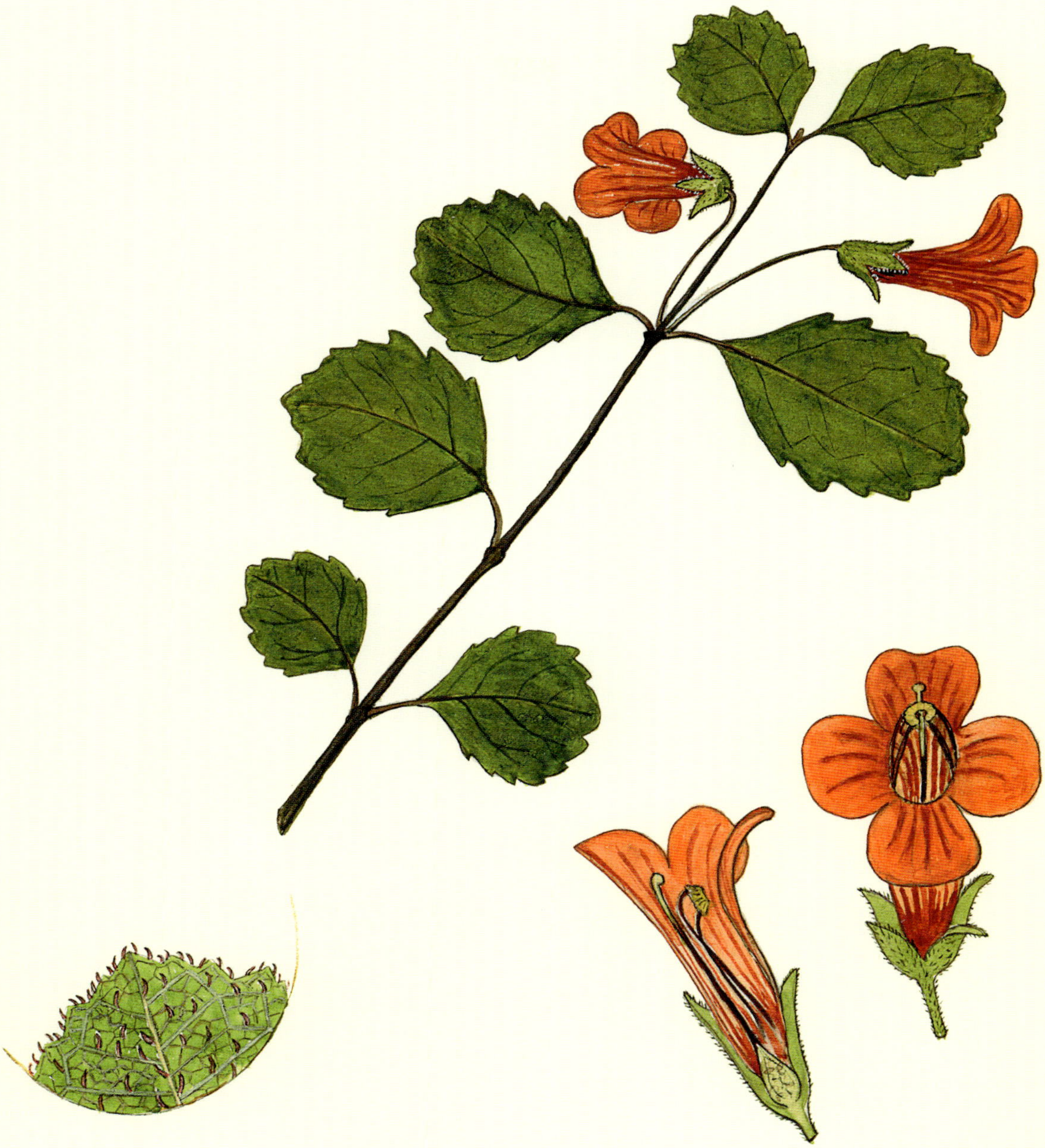

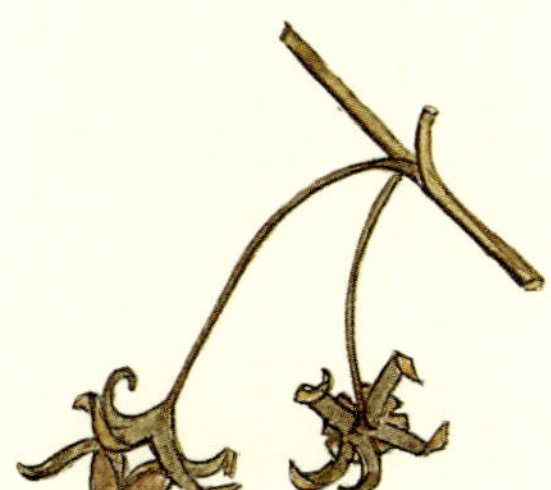

Rhabdothamnus solandri taurepo, waiūatua,
New Zealand gloxinia

PLATE 153

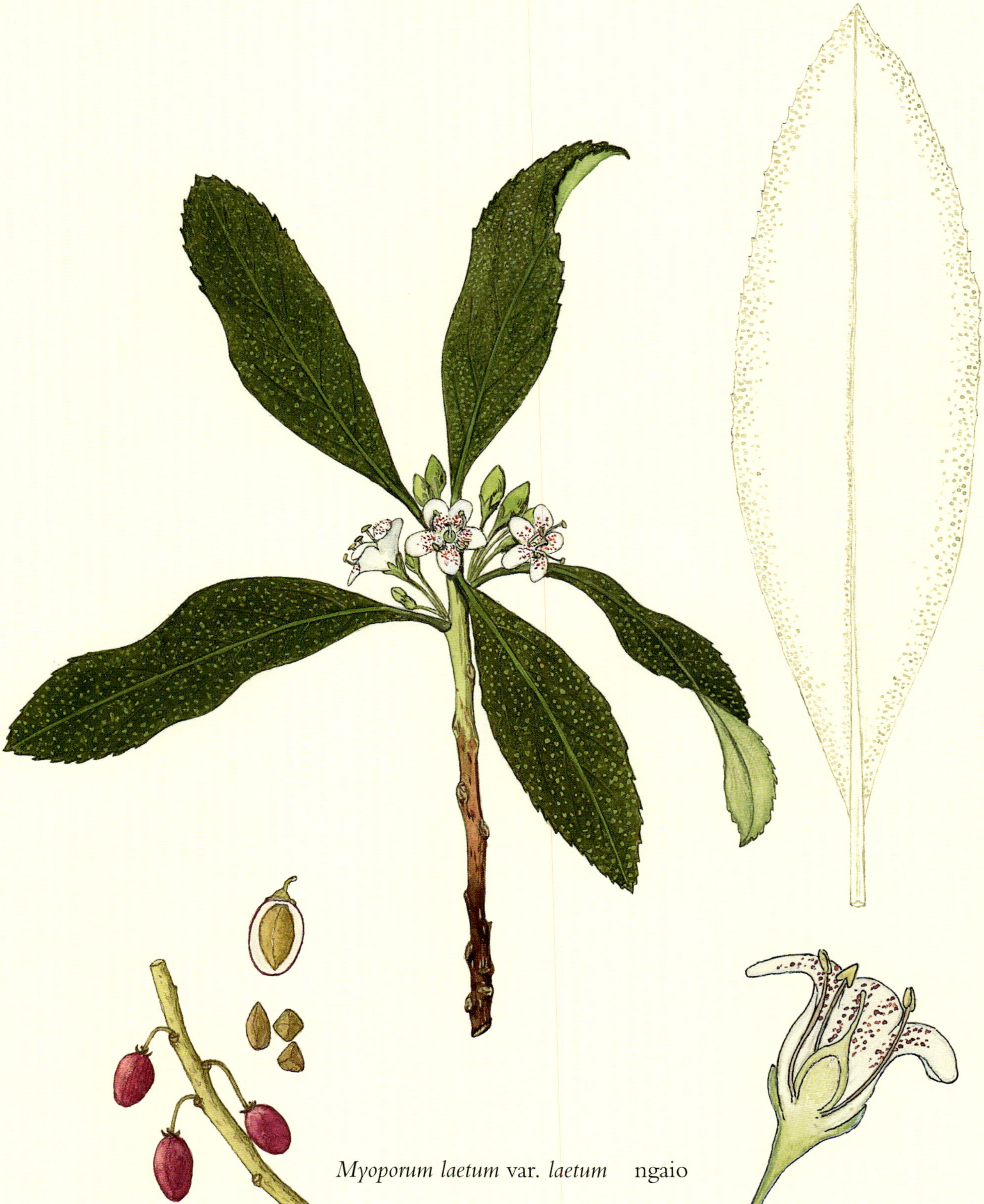

Myoporum laetum var. *laetum* ngaio

PLATE 154

Vitex lucens pūriri

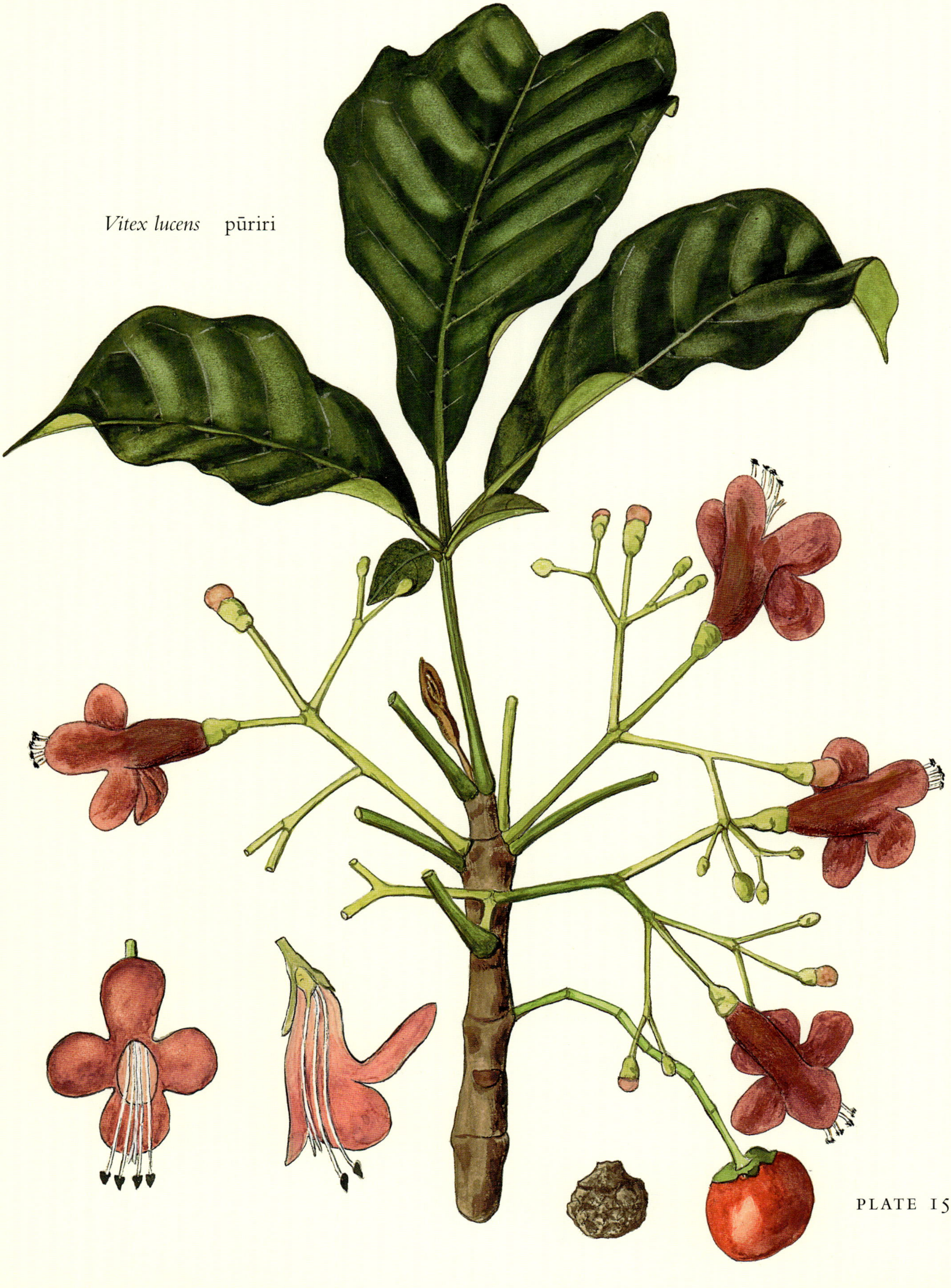

PLATE 155

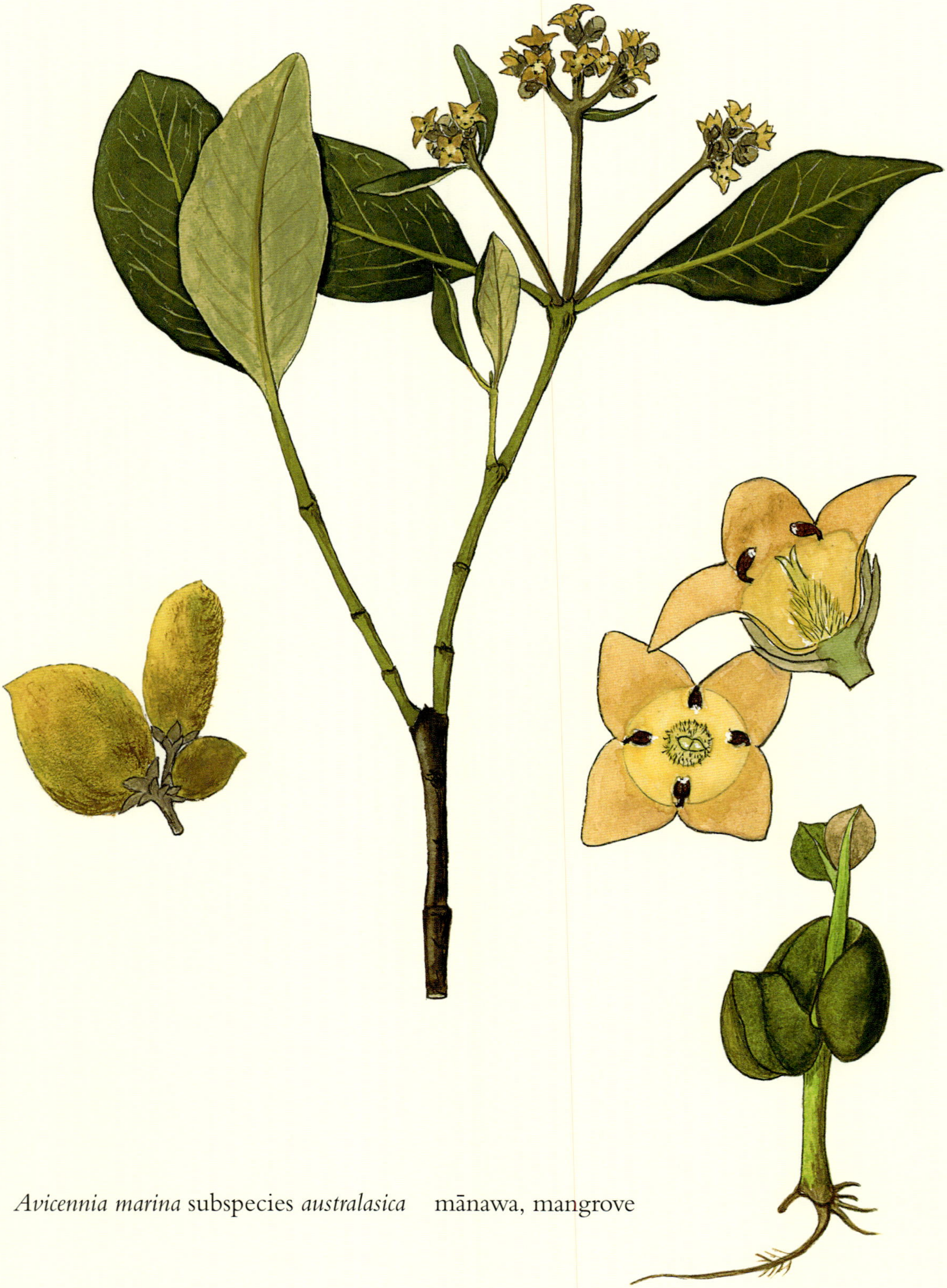

Avicennia marina subspecies *australasica* mānawa, mangrove

PLATE 156

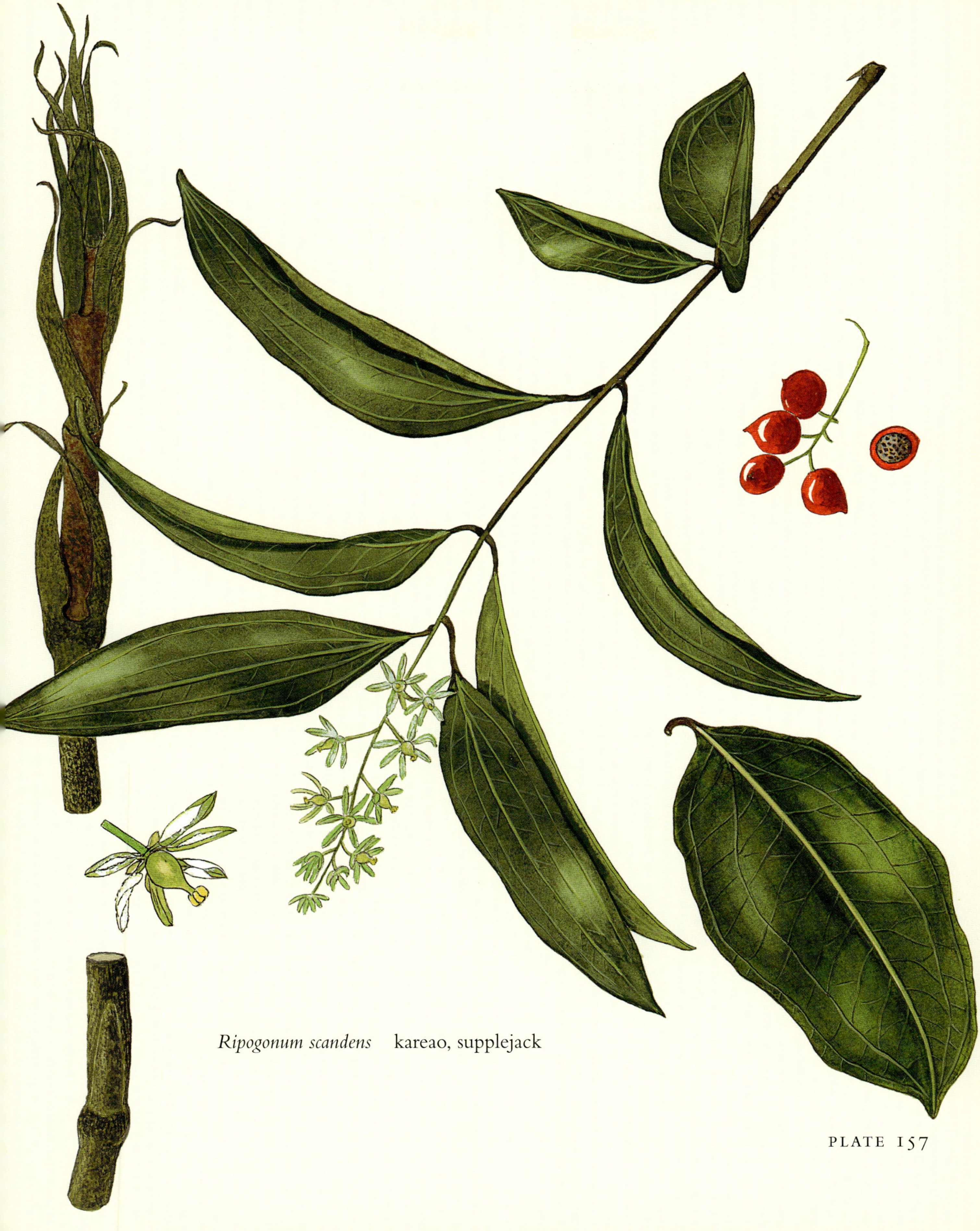

Ripogonum scandens kareao, supplejack

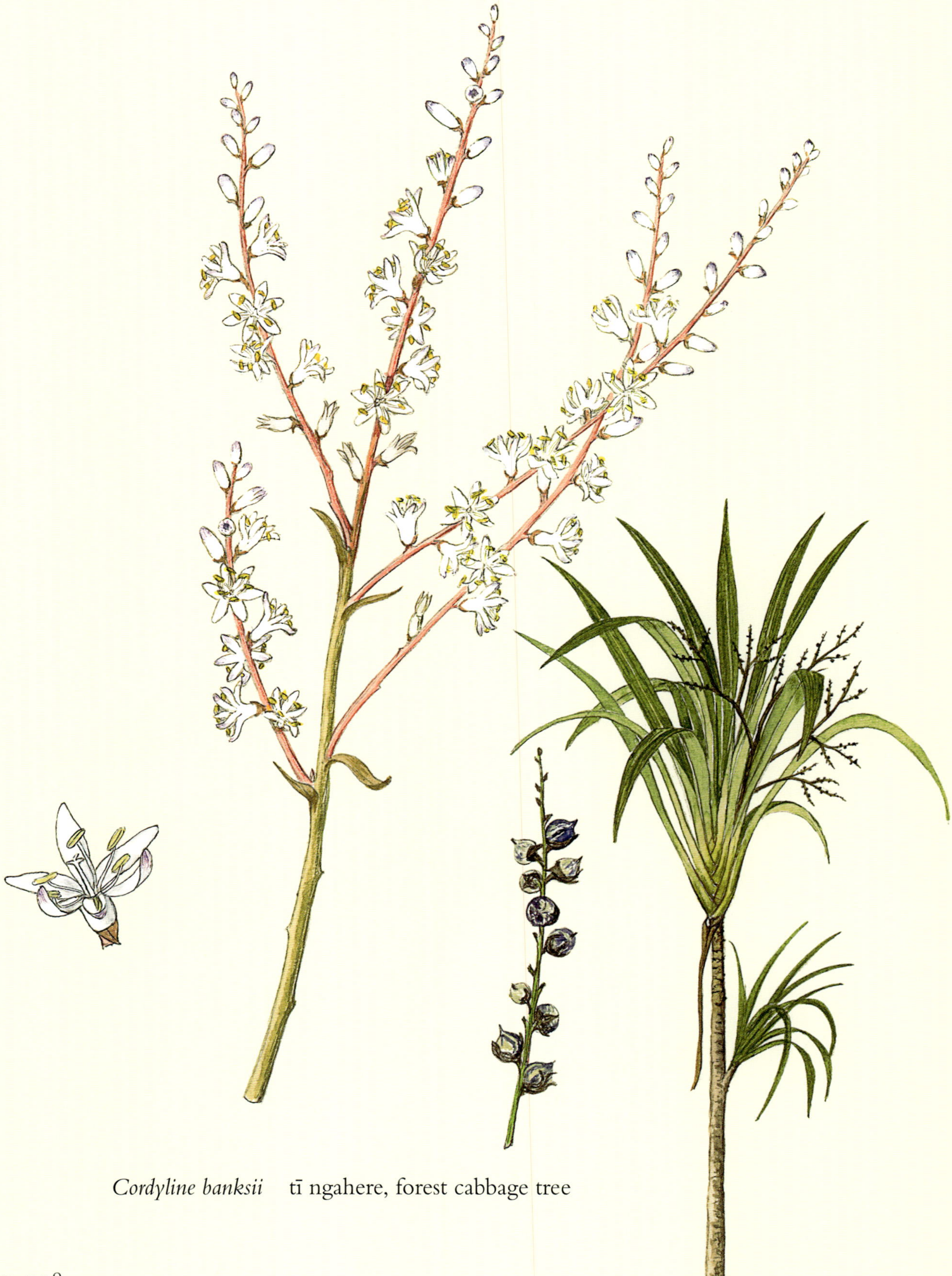

Cordyline banksii tī ngahere, forest cabbage tree

PLATE 158

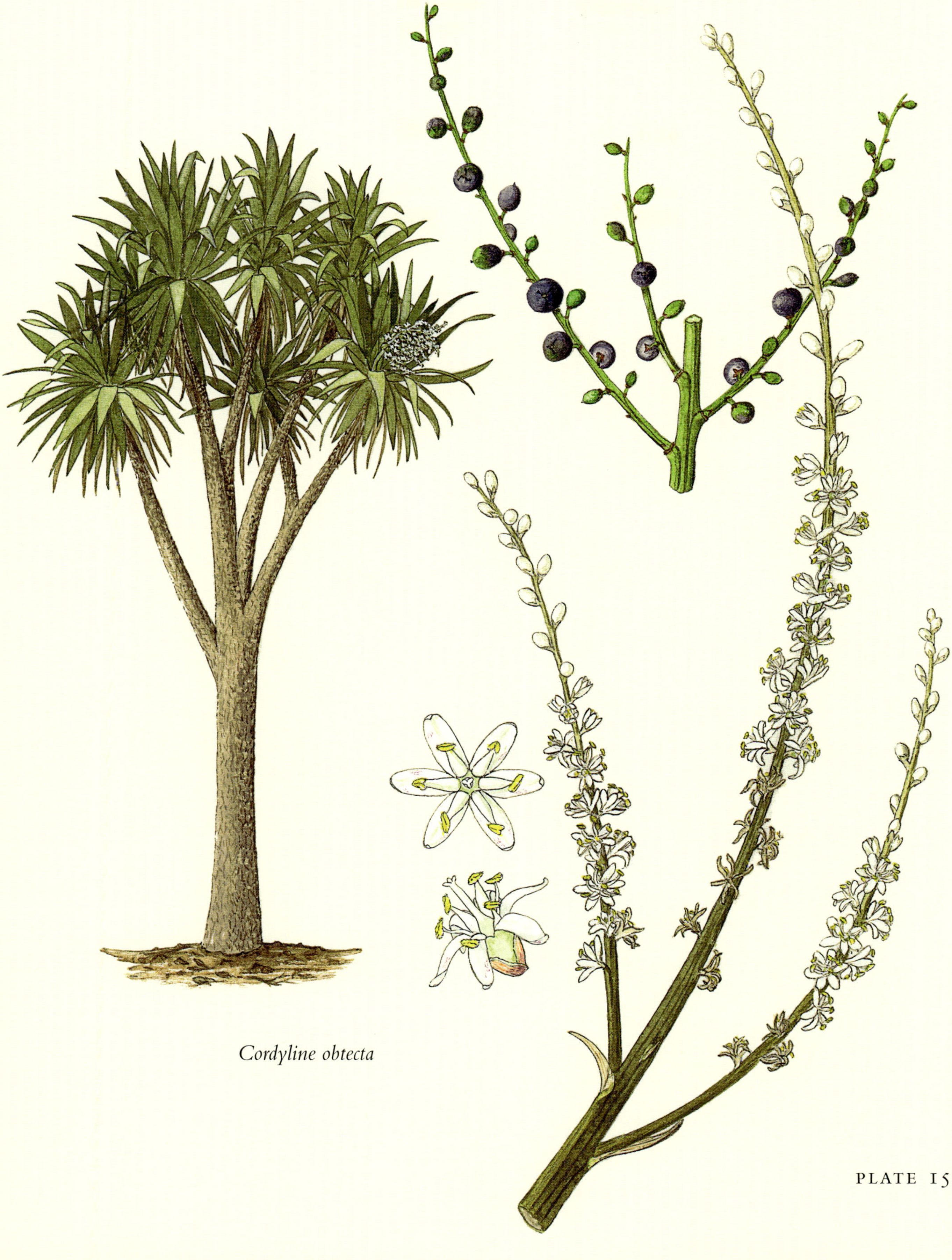

Cordyline obtecta

PLATE 159

Phormium tenax harakeke, kōrari (the flowering stem),
New Zealand flax

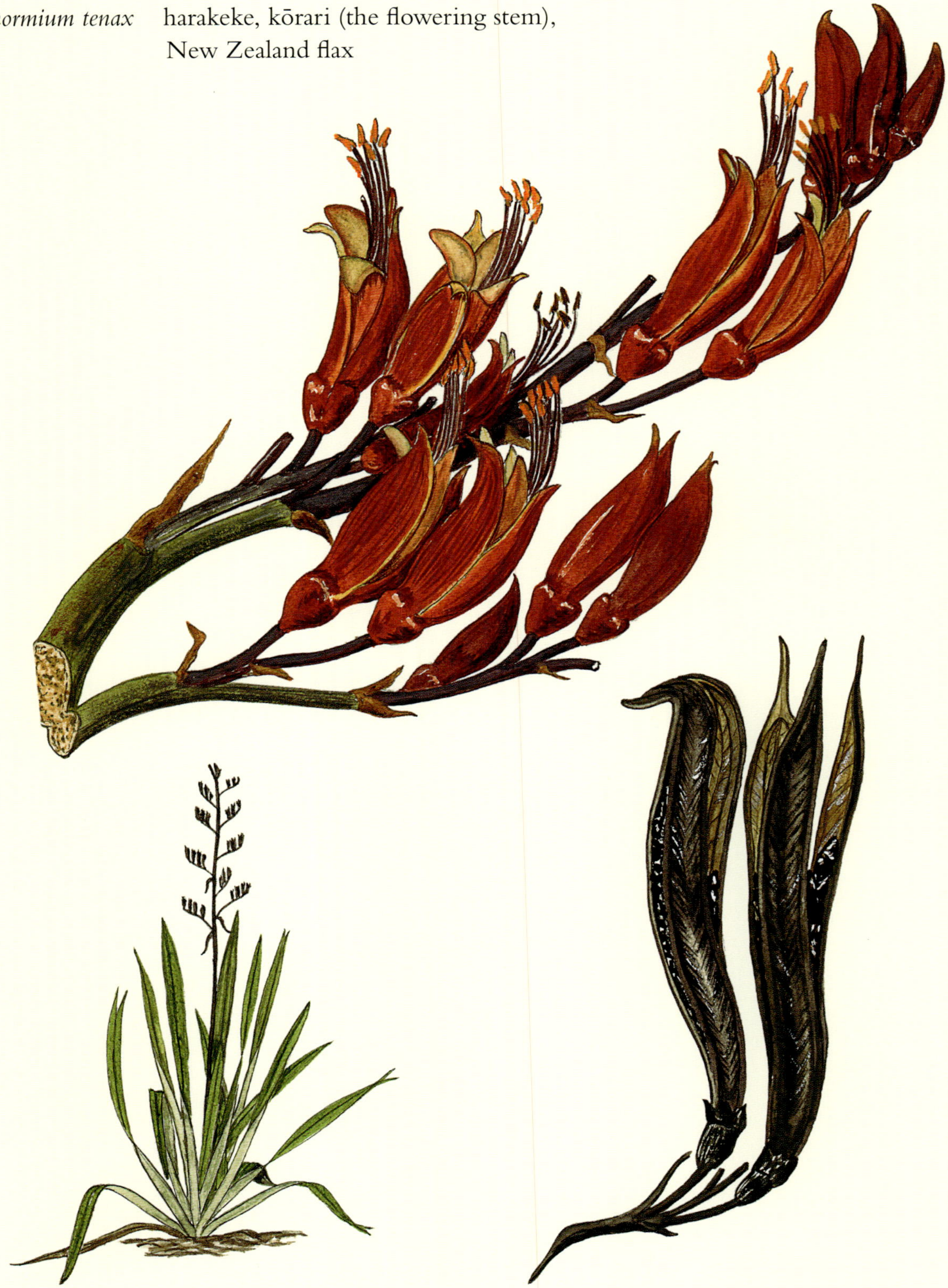

PLATE 160

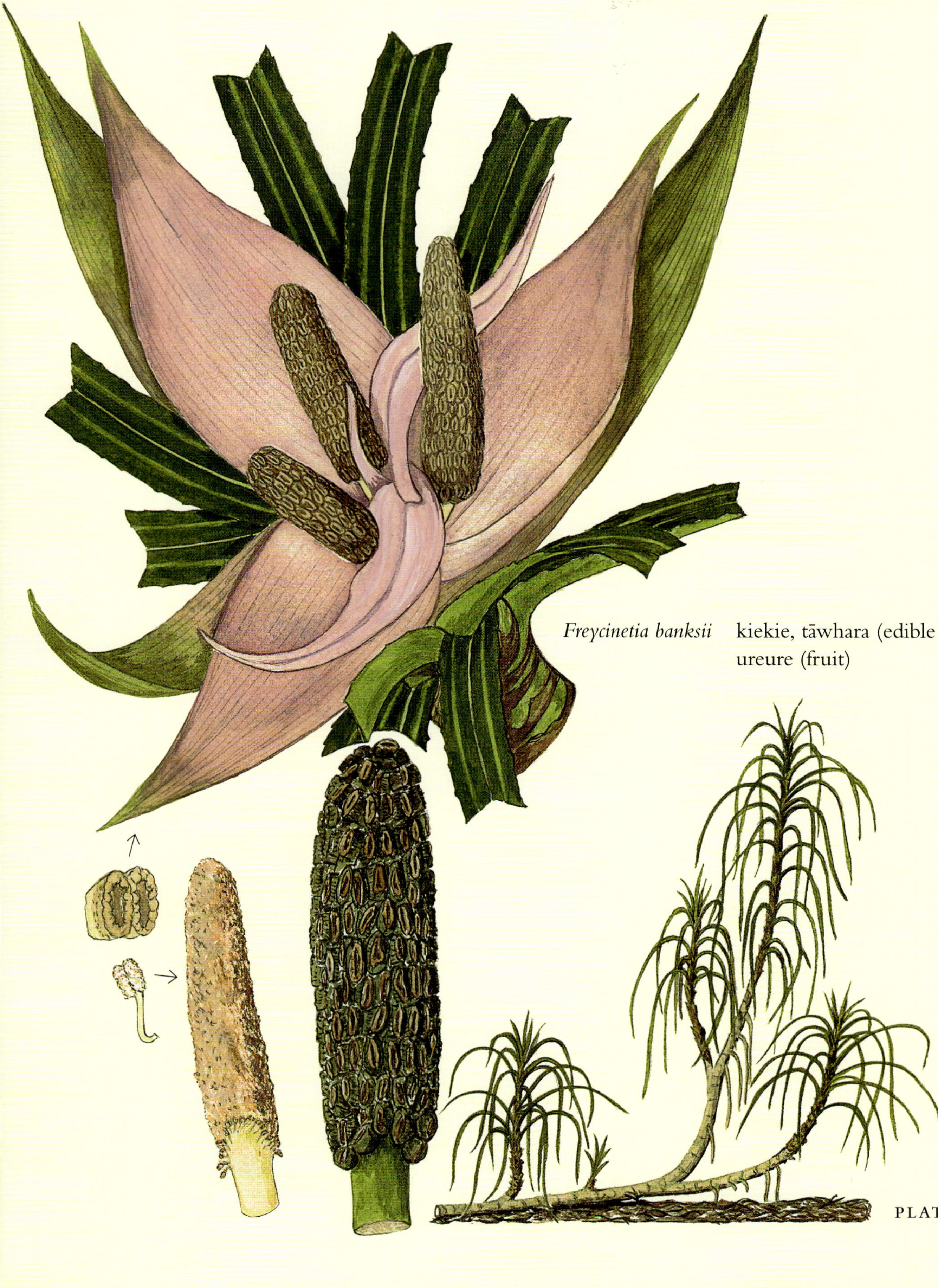

Freycinetia banksii kiekie, tāwhara (edible bracts), ureure (fruit)

List of Plates

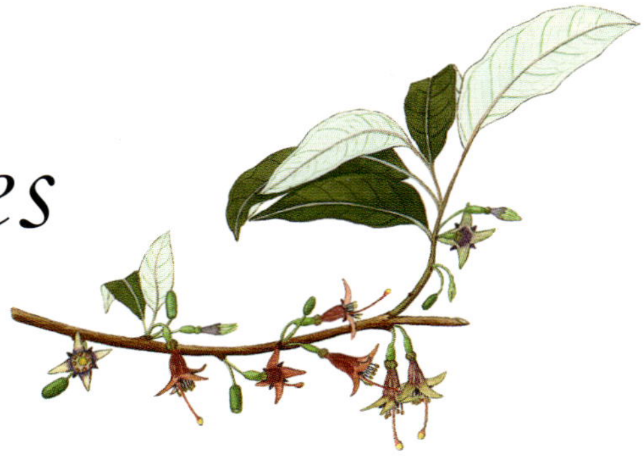

PODOCARPACEAE

PLATE 1 *Dacrycarpus dacrydioides* **kahikatea, white pine**

Dacrycarpus: weeping fruit
dacrydioides: like *Dacrydium*

One of New Zealand's tallest trees, occasionally to 43 m, normally 24–36 m. Trunk sometimes 2 m diameter, usually 0.5–1 m. Trees live for at least 600 years.

DISTRIBUTION: North, South and Stewart Islands. Lowland forest, common except Stewart Island, often dominant in swamp forest.

PLATE 2 *Prumnopitys ferruginea* **miro**

Prumnopitys: plum fruit, after the small, rounded, plum-like fruit produced by the trees
ferruginea: rusty red, apparently applied because of the colour of dried specimens

Tree to 25 m tall or more, trunk to 1 m diameter, occasionally larger. The fruit smells and tastes of turpentine. Trees may live at least 600 years.

DISTRIBUTION: North, South and Stewart Islands. Common in lowland and montane forest.

PLATE 3 *Podocarpus totara* var. *totara* **tōtara**

Podocarpus: from the Greek, podos, foot; karpos, fruit
totara: from the Māori name

Tree to 30 m, rarely to 40 m tall. Trunk to 1.5 m or occasionally 2.5 m diameter. Trees live at least 900 years.

DISTRIBUTION: North, South and Stewart Islands. Common in lowland forest.

Podocarpus totara var. *waihoensis*

Podocarpus: from the Greek, podos, foot; karpos, fruit
totara: from the Māori name
waihoensis: from the Waiho River

Bushy tree, to 19 m tall and broad.

DISTRIBUTION: South Westland. Lowland forest edges and stream margins.

PLATE 4 *Dacrydium cupressinum* **rimu, red pine**

Dacrydium: tear-like, for the resinous exudations; from the Greek, dakryon, tears
cupressinum: *Cupressus*-like

Tree to 35 m tall, trunk up to 2 m diameter. Trees may live to 900 years.

DISTRIBUTION: North, South and Stewart Islands. Common in lowland and montane forest.

PLATE 5 *Manoao colensoi* **silver pine, manoao**

Manoao: from the Māori name
colensoi: after W Colenso, 19th century explorer and botanist

Tree to 15 m tall, trunk to 60 cm diameter or more.

DISTRIBUTION: North Island, from Te Paki (Radar Bush) southwards to Mt Ruapehu but only common in central North Island; western South Island to 44°S. Lowland and montane forest, a characteristic tree of bog forest in high-rainfall regions.

PHYLLOCLADACEAE

PLATE 6 *Phyllocladus trichomanoides* **tānekaha, celery pine**

Phyllocladus: leaf-like branches; from the Greek, phyllon, leaf; klados, young shoot
trichomanoides: resembling *Trichomanes*

Tree to 25 m tall, trunk to 1 m diameter. Trees may live for 500 years.

DISTRIBUTION: North Island, lowland forest from Te Paki to 40°S; South Island, northern Marlborough and Nelson, to nearly 41°30'S.

PLATE 7 *Phyllocladus toatoa* **toatoa**

Phyllocladus: leaf-like branches; from the Greek, phyllon, leaf; klados, young shoot
toatoa: from the Māori name

Tree to 15 m tall, trunk to 60 cm diameter.

DISTRIBUTION: North Island. Lowland and montane forest from Bay of Islands to Volcanic Plateau.

ARAUCARIACEAE

PLATE 8 *Agathis australis* **kauri**

Agathis: a ball of thread, referring to the appearance of the cone; from the Greek, agathis
australis: southern; from the Latin, australis, southern

Tree to more than 50 m tall, trunk to 7 m diameter. Trees have been estimated at up to 4000 years old.

DISTRIBUTION: North Island. North of 38°S. Lowland forest.

WINTERACEAE

PLATE 9 *Pseudowintera colorata* **mountain horopito, pepper tree**

Pseudowintera: false *Wintera*; from the Greek, pseudo-, false; alluding to the close resemblance to *Wintera*
colorata: coloured, since the leaves are blotched with red

Erect shrub to 2 m tall. Leaves aromatic and peppery to the taste.

DISTRIBUTION: North, South and Stewart Islands. Montane and subalpine forest, scarce north of Auckland, lowland in South Island.

LAURACEAE

PLATE 10 *Beilschmiedia tarairi* **taraire**

Beilschmiedia: after KT Beilschmied, 18th–19th century Polish plant geographer
tarairi: from the Māori name

Tree to 20 m tall or more, trunk to 60 cm diameter, rarely 1 m.

DISTRIBUTION: North Island. Coastal and lowland forest from near North Cape to Kaitaia and East Cape.

PLATE 11 *Litsea calicaris* **mangeao, tangeao**

Litsea: from the Chinese, li, little; tse, plum
calicaris: cup bearing, from cup-shaped calyx in which berries are seated

Tree to 15 m tall or more, trunk to 80 cm diameter. Flowers profuse, sweet scented.

DISTRIBUTION: North Island. North of 38°S. Lowland forest, most abundant near Rotorua and Kawhia.

MONIMIACEAE

PLATE 12 *Hedycarya arborea* **porokaiwhiri, pigeonwood**

Hedycarya: sweet seed; from the Greek, hedys, sweet; karyon, nut
arborea: tree-like; from the Latin, arbor, tree

Tree to 12 m tall, trunk to 50 cm diameter. Male flowers fragrant and short-lived.

DISTRIBUTION: North and South Islands. Lowland and montane forest, reaching Banks Peninsula on east and throughout on west of South Island.

Hedycarya* aff. *arborea (northern offshore islands form)

Plants on the Three Kings and Poor Knights Islands differ from mainland populations by their much larger, broader leaves.

RANUNCULACEAE

PLATE 13 *Clematis paniculata* **puawānanga, white clematis**

Clematis: vine branch; from the Greek, klema
paniculata: bearing panicles; from the Latin, panicula, tuft

Liane with main stems up to 10 cm diameter, or more at base. Flowers fragrant.

DISTRIBUTION: Three Kings, North, South and Stewart Islands. Common throughout lowland and lower montane forest, especially forest margins.

PLATE 14 *Clematis forsteri* (South Island form)

Clematis: vine branch; from the Greek, klema
forsteri: named for JR and G Forster, botanists on Cook's second voyage to New Zealand in 1773

Liane with slender main stems and branches.

DISTRIBUTION: South Island. Widespread. Open shrubland.

PLATE 15 *Clematis quadribracteolata*

Clematis: vine branch; from the Greek, klema

quadribracteolata: having four bracts, referring to the floral bracts; from the Latin, quadr-, four; bractea, thin plate

Trailing or climbing to 2 m tall.

DISTRIBUTION: North Island, 38°20' to 40°S; South Island, throughout except Fiordland. Lowland to montane shrubland and grassland.

PIPERACEAE

PLATE 16 *Macropiper melchior*

Macropiper: large pepper

melchior: after one of the biblical Three Kings after whom the islands are named

Densely branched, compact shrub to *c.* 2 m tall. Leaves lack the peppery taste of other members of this genus.

DISTRIBUTION: Three Kings Islands, South West and West Islands. Stony ground in shaded or semi-exposed sites under *Meryta* forest.

PLATE 17 *Macropiper excelsum* subspecies *excelsum* **kawakawa, pepper tree**

Macropiper: large pepper

excelsum: from the Latin, excelsus, lofty

Shrub or tree to 6 m tall. Leaves aromatic and peppery to taste.

DISTRIBUTION: North, South and Chatham Islands. South of 38°30'S, mainly coastal to Okarito and Banks Peninsula in South Island.

CHLORANTHACEAE

PLATE 18 *Ascarina lucida* var. *lanceolata* **hutu**

Ascarina: because the anthers are worm-shaped; from the Latin, ascaris, intestinal worm

lucida: shining; from the Latin, lucidus, lux, light

lanceolata: variety with lanceolate leaves; from the Latin lancea, light lance, spear

Shrub or small tree to 5 m tall.

DISTRIBUTION: Kermadec Islands, Raoul Island. In 'wet' forest, above 245 m above sea level.

VIOLACEAE

PLATE 19 *Melicytus macrophyllus* **large-leaved māhoe**

Melicytus: honey cavity, named for the cavity below the stamens; from the Greek, meli, honey; kytos, cavity

macrophyllus: large-leaved; from the Greek, macros, large; phyllon, leaf

Tree to 6 m tall, trunk to 40 cm diameter.

DISTRIBUTION: North Island, north of 37°S. Lowland to lower montane forest.

PLATE 20 *Melicytus* aff. *obovatus*

Melicytus: honey cavity, named for the cavity below the stamens; from the Greek, meli, honey; kytos, cavity

obovatus: referring to the inversely egg-shaped leaf; from the Latin, ob-, reversed; ovum, egg

This is an unnamed plant similar to *Melicytus obovatus*, from the Cook Strait coasts. Bushy shrub to 1.5 m tall.

DISTRIBUTION: North and South Islands. Cook Strait islands and coasts.

PLATE 21 *Melicytus* aff. *crassifolius*

Melicytus: honey cavity, named for the cavity below the stamens; from the Greek, meli, honey; kytos, cavity
crassifolius: thick-leaved; from the Latin, crassus, thick, coarse; folium, leaf

This is an unnamed plant similar to *Melicytus crassifolius*, which grows on cliffs. Shrub to 1 m tall.

DISTRIBUTION: South Island. Inland Kaikoura Range, Woodside gorge and Waima River, inland Marlborough and Canterbury. Gorges and steeplands.

Melicytus chathamicus

Melicytus: honey cavity, named for the cavity below the stamens; from the Greek, meli, honey; kytos, cavity
chathamicus: of the Chatham Islands

Dioecious shrub to 3 m tall.

DISTRIBUTION: Chatham Islands.

Melicytus alpinus **porcupine shrub**

Melicytus: honey cavity, named for the cavity below the stamens; from the Greek, meli, honey; kytos, cavity
alpinus: belonging to the alps; from the Latin, alpinus

Shrub to 50 cm, occasionally taller, low growing in shade.

DISTRIBUTION: North and South Islands. Wairarapa and east of the Southern Alps, especially in Marlborough and Canterbury. Also drier parts of Nelson. Hillsides and alluvial terraces.

POLYGONACEAE

PLATE 22 *Muehlenbeckia* species

Muehlenbeckia: after HG Muehlenbeck, 18th–19th century Alsatian physician

This is an unnamed plant occurring in Northland and on the Chatham Islands. Liane with many branchlets twisting round branches of host plant, or self-entangled, climbing to 15 m.

DISTRIBUTION: North Island (Northland), Chatham Islands. Coastal or lowland.

PLATE 23 *Muehlenbeckia axillaris* **pōhuehue, creeping muehlenbeckia**

Muehlenbeckia: after HG Muehlenbeck, 18th–19th century Alsatian physician
axillaris: flowers borne in the leaf axils; from the Latin, axilla, armpit

Prostrate, dioecious plant forming loosely to densely matted patches up to 1 m or more across.

DISTRIBUTION: North and South Islands. Montane and subalpine riverbeds, gravelly and rocky places south of 38°S, also found locally in lowland riverbeds and on the coast.

Muehlenbeckia astonii

Muehlenbeckia: after HG Muehlenbeck, 18th–19th century Alsatian physician
astonii: after BC Aston, 19th–20th century chemist

Rounded, dioecious shrub to 2 m tall.

DISTRIBUTION: North and South Islands. Southern coasts of Wairarapa and Wellington, coastal and inland Marlborough, North Canterbury, to southern Banks Peninsula.

ONAGRACEAE

PLATE 24 *Fuchsia procumbens*

Fuchsia: after L Fuchs, 16th century German academic and botanical illustrator

procumbens: trailing without rooting; from the Latin, procumbere, prostrate

Shrub, prostrate or scrambling over other small plants. Stems very slender, to 50 cm long or more.

DISTRIBUTION: North Island (Northland), Great Barrier Island, Coromandel Peninsula. Coastal forest and scrubland.

PLATE 25 *Fuchsia excorticata* **kōtukutuku, kōnini (berry)**

Fuchsia: after L Fuchs, 16th century German academic and botanical illustrator

excorticata: with peeling bark; from the Latin, ex-, from; corticis, bark

Tree to 12 m, trunk to 30 cm, rarely to 60 cm diameter.

DISTRIBUTION: North, South, Stewart and Auckland Islands. Common in lowland, montane and subalpine forest, especially in gullies.

AIZOACEAE

PLATE 26 *Tetragonia tetragonioides* **kōkihi, rengamutu, rengarenga, tūtae-ika-moana, New Zealand spinach**

Tetragonia: four angles, refers to the shape of the fruit; from the Greek, tetra, four; gonia, angle

Subshrub with trailing stems to 1 m or more long.

DISTRIBUTION: Kermadec, Three Kings, North, South, Stewart and Chatham Islands. Also in Tasmania, Australia, Pacific islands, southern South America. Coastal.

NYCTAGINACEAE

PLATE 27 *Pisonia brunoniana* **parapara, bird-catching tree**

Pisonia: after W Piso, 17th century Dutch physician

brunoniana: after R Brown, 18th–19th century English botanist

Shrub or tree to 6 m tall or more, trunk to 60 cm diameter.

DISTRIBUTION: Kermadec, Three Kings and North Islands. Mainly on offshore islands north of East Cape. Coastal forest.

THYMELAEACEAE

PLATE 28 *Pimelea longifolia* **tāranga**

Pimelea: from the Greek, pimele, fat; referring to the oily seeds

longifolia: long-leaved; from the Latin, longus, long; folium, leaf

Shrub to 2 m tall.

DISTRIBUTION: North and South Islands. From 35°S to the southern end of the Paparoa Range. Lowland to subalpine and open places.

PLATE 29 *Pimelea barbata* **subspecies** *barbata*

Pimelea: from the Greek, pimele, fat; referring to the oily seeds
barbata: shaggy, with tufts of long, weak hairs; from the Latin, barbatus, bearded

DISTRIBUTION: North Island. Te Waka and Maungaharuru Ranges, Hawke's Bay. Limestone cliffs.

Pimelea **species**

Pimelea: from the Greek, pimele, fat; referring to the oily seeds

This is an unnamed plant from the eastern South Island. Low-growing shrub with many upright branchlets to 30 cm tall.

DISTRIBUTION: South Island. Mountains east of Main Divide from southern Marlborough to Waitaki Valley, South Canterbury. Limestone.

PROTEACEAE

PLATE 30 *Toronia toru* **toru**

Toronia: from the Māori name
toru: from the Māori name

Tree to 10 m tall or more, at higher altitudes shrub-like, trunk *c.* 20 cm diameter. Flowers fragrant.

DISTRIBUTION: North Island. Lowland to montane shrubland and forest 34°28'–38°30'S.

PLATE 31 *Knightia excelsa* **rewarewa, New Zealand honeysuckle**

Knightia: after TA Knight, 18th–19th century plant physiologist
excelsa: tall; from the Latin, excelsus, lofty, high

Tree to 30 m tall, trunk to 1 m or more diameter.

DISTRIBUTION: North and South Islands. Lowland to lower montane forest from near North Cape to Marlborough Sounds.

CORIARIACEAE

PLATE 32 *Coriaria arborea* **tutu**

Coriaria: from the Latin, corium = Greek, chorion, leather
arborea: tree-like; from the Latin, arbor, tree

Shrub or rarely a small tree to 6 m tall.

DISTRIBUTION: North, South, Stewart and Chatham Islands. Lowland to montane secondary forest and shrubland, especially on cliffs, debris fans and alluvial ground.

PLATE 33 *Coriaria sarmentosa*

Coriaria: from the Latin, corium = Greek, chorion, leather
sarmentosa: from the Latin, sarmentosus, full of twigs

Shrub 1–2 m tall with stout branching stems.

DISTRIBUTION: North, South and Stewart Islands. South Wellington coast southwards. Sea level to the mountains, on debris slopes and streamsides.

PLATE 34 *Coriaria pottsiana*

Coriaria: from the Latin, corium = Greek, chorion, leather
pottsiana: after N Potts, 19th–20th century plant collector

Shrub with slender branches.

DISTRIBUTION: North Island. Mt Hikurangi, Raukumara Range. Shingle slides *c.* 1200–1400 m above sea level.

PITTOSPORACEAE

PLATE 35 *Pittosporum ralphii* **karo**

Pittosporum: seeds covered in resinous pulp; from the Greek, pitta, pitch; spora, seed
ralphii: after TS Ralph, 19th century medical doctor and author

Tree to 10 m tall.

DISTRIBUTION: North Island. East Cape to north Wairarapa; also Patea and Whanganui rivers. From near sea level up to 900 m in hill country; especially abundant on mudstone cliffs and in gorges.

Pittosporum ellipticum

Pittosporum: seeds covered in resinous pulp; from the Greek, pitta, pitch; spora, seed
ellipticum: elliptical; from the Greek, ellipsis, a defect, falling short

Tree to 8 m tall.

DISTRIBUTION: North Island. Kaitaia to Mt Pirongia in the west and Waioeka gorge in the east. Forest in hill country up to 1000 m above sea level.

PLATE 36 *Pittosporum eugenioides* **tarata, lemonwood**

Pittosporum: seeds covered in resinous pulp; from the Greek, pitta, pitch; spora, seed
eugenioides: resembling the genus *Eugenia*

Tree to 12 m tall, trunk to 60 cm diameter. Flower scented, strong and sweet.

DISTRIBUTION: North and South Islands. Lowland to montane forest.

PLATE 37 *Pittosporum* aff. *crassifolium*

Pittosporum: seeds covered in resinous pulp; from the Greek, pitta, pitch; spora, seed
crassifolium: thick-leaved; from the Latin, crassus, thick, coarse; folium, leaf

This is an unnamed plant similar to *Pittosporum crassifolium*, from the Kermadec Islands. Tree up to 3 m tall.

DISTRIBUTION: Kermadec Islands, Raoul Island. On cliff faces.

PLATE 38 *Pittosporum obcordatum*

Pittosporum: seeds covered in resinous pulp; from the Greek, pitta, pitch; spora, seed
obcordatum: referring to the inversely heart-shaped leaf; from the Latin, ob-, reversed; cordatus, heart

Divaricating shrub or tree to *c.* 10 m tall. Flower fragrant.

DISTRIBUTION: North Island, Kaitaia to the Wairarapa; South Island, Catlins, Fiordland. Alluvial forest.

PASSIFLORACEAE

PLATE 39 *Passiflora tetrandra* **kōhia, pōhue, akakaikū, New Zealand passion flower**

Passiflora: passion flower; from the Latin, passio, suffering; flos, flower

tetrandra: four divisions; from the Greek, tetra, four; andr-, male

Liane to *c.* 10 m tall.

DISTRIBUTION: North and South Islands. Lowland forest to 44°S.

MYRTACEAE

PLATE 40 *Leptospermum scoparium* (i) **kahikātoa, kānuka (Te Aupouri only), mānuka**

Leptospermum: slender seed; from the Greek, leptos, thin; sperma, seed

scoparium: like a broom; from the Latin, scopae, twigs, sweeping bush made of twigs

This is an unnamed plant similar to *Leptospermum scoparium* (ii), from Karikari Peninsula. Shrub or small tree to 5 m tall.

DISTRIBUTION: Three Kings Islands, Great Island only; North Island, northern Northland. Usually coastal.

Leptospermum scoparium (ii) **mānuka, kahikātoa, red tea-tree**

Shrub or tree to 10 m tall. Distinctive gingery peppery smell.

DISTRIBUTION: North, South and Stewart Islands. On mountains, on the seashore, in swamps, on dry land and on rocky cliffs.

PLATE 41 *Kunzea* aff. *ericoides* **mānuka, kānuka, white tea-tree**

Kunzea: after G Kunze, 18th–19th century botanist and academic

ericoides: resembling heath (*Erica*), which has similar leaves

This is an unnamed plant similar to *Kunzea ericoides*. Tree to 30 m tall.

DISTRIBUTION: North and South Islands. Apparently absent from Taranaki/Egmont and surrounding plain. Scarce in South Island. Disturbed forest, especially on clay soils and in hill country. Favouring free-draining situations.

PLATE 42 *Metrosideros excelsa* **pōhutukawa, New Zealand Christmas tree**

Metrosideros: iron-hearted, referring to the hardness of the timber; from Greek, metra, womb, core; sideron, iron

excelsa: from the Latin, excelsum, high, elevated

Tree to 20 m tall or more, trunk to 2 m diameter.

DISTRIBUTION: Three Kings Islands, North Island to *c.* 38°S. Adventive further south. Mainly coastal.

PLATE 43 *Metrosideros parkinsonii*

Metrosideros: iron-hearted, referring to the hardness of the timber; from the Greek, metra, womb, core; sideron, iron

parkinsonii: after S Parkinson, 18th century botanical artist on Cook's first voyage to New Zealand

Semi-lianoid, multi-trunked shrub or small tree to 7 m tall.

DISTRIBUTION: North Island, Great Barrier and Little Barrier Islands only; South Island, north-west Nelson to central Westland. Sea level to 900 m.

PLATE 44 *Metrosideros carminea* **akakura, carmine rātā, crimson rātā**

Metrosideros: iron-hearted, referring to the hardness of the timber; from the Greek, metra, womb, core; sideron, iron

carminea: crimson; from the Latin, carminus

Liane to 15 m tall or more.

DISTRIBUTION: North Island. North Cape to East Cape and northern Taranaki. Lowland forest.

PLATE 45 *Metrosideros fulgens* **akatawhiwhi, scarlet rātā vine**

Metrosideros: iron-hearted, referring to the hardness of the timber; from the Greek, metra, womb, core; sideron, iron

fulgens: shining; from the Latin, fulgere, to shine

Liane or bush to 10 m tall or more.

DISTRIBUTION: North and South Islands. North of 44°30'S. Northern Marlborough, Nelson and west of Main Divide. Lowland forest.

PLATE 46 *Lophomyrtus bullata* **ramarama**

Lophomyrtus: from the Greek, lophos, crested; Latin, myrtus = Greek, murtos, myrtle

bullata: blistered or puckered; from the Latin, bulla, bubble

Shrub or small tree to 8 m tall or more.

DISTRIBUTION: North and South Islands. In South Island only in Northern Marlborough and eastern Nelson. Lowland forest and swamps.

ELAEOCARPACEAE

PLATE 47 *Elaeocarpus dentatus* **hīnau**

Elaeocarpus: olive fruit; from the Greek, elaia, the olive tree; karpos, fruit

dentatus: toothed, referring to the leaves; from the Latin, dentis, tooth

Tree to 20 m tall or more, trunk to 1 m diameter.

DISTRIBUTION: North and South Islands. Lowland to montane forest north of 43°30'S.

PLATE 48 *Aristotelia serrata* **makomako, wineberry**

Aristotelia: after Aristotle, the Greek philosopher

serrata: serrated, referring to the leaves; from the Latin, serratus, shaped like a saw

Tree to *c.* 10 m tall, trunk to *c.* 30 cm diameter.

DISTRIBUTION: North, South and Stewart Islands. Lowland to montane forest throughout.

PLATE 49 *Aristotelia fruticosa* **mountain wineberry**

Aristotelia: after Aristotle, the Greek philosopher

fruticosa: shrub-like, bushy; from the Latin, fruticis, a shrub

Shrub to 2 m tall.

DISTRIBUTION: North, South and Stewart Islands. South of 37°30'S. Montane and subalpine forest and shrubland.

MALVACEAE

PLATE 50 *Entelea arborescens* **whau, cork tree**

Entelea: perfect, from the stamens being all fertile; from the Greek, enteles, complete, perfect
arborescens: becoming tree-like; from the Latin, arborescere, to become a tree

Shrub or small tree to 6 m tall or more, trunk to 25 cm diameter.

DISTRIBUTION: Three Kings, North and South Islands. To 41°30'S. In South Island, only a few sites in Golden Bay and Marlborough Sounds. Coastal and lowland forest.

PLATE 51 *Plagianthus regius* subspecies *chathamicus* **Chatham Island ribbonwood**

Plagianthus: oblique flower, named for the shape of the petals; from the Greek, plagios, oblique, sideways
regius: royal; from the Latin, rex, Regis, king
chathamicus: of the Chatham Islands

Tree to *c.* 15 m tall.

DISTRIBUTION: Chatham Islands. Lowland forest.

PLATE 52 *Hoheria glabrata* **houhere, whauwhau, mountain ribbonwood**

Hoheria: from the Māori name, houhere
glabrata: without hairs; from the Latin, glaber

Tree to 10 m tall or more.

DISTRIBUTION: North Island, Mt Taranaki/Egmont; South Island, lowland to subalpine forest and shrubland, mainly west of Main Divide.

PLATE 53 *Hoheria angustifolia*

Hoheria: from the Māori name, houhere
angustifolia: narrow-leaved; from the Latin, angustus, narrow; folium, leaf

Tree to 10 m tall or more, trunk to 30 cm diameter.

DISTRIBUTION: North and South Islands. Lowland forest and swamp margins, sometimes forming groves, especially on alluvium. Mostly in east in South Island.

PLATE 54 *Hibiscus diversifolius*

Hibiscus: from the Latin for marsh mallow
diversifolius: with leaves of more than one shape; from the Latin, diversus, different; folium, leaf

A stout, rigid perennial *c.* 2 m tall.

DISTRIBUTION: North Island. North Cape to 35°30'S. Coastal.

EUPHORBIACEAE

PLATE 55 *Homalanthus polyandrus*

Homalanthus: equal flowered, refers to the flattened male flowers; from the Greek, homalos, even, alike; anthos, flower
polyandrus: with many stamens; from the Greek, poly, many; andros, man

Tree to 9 m tall, trunk *c.* 3 m high, to 40 cm diameter.

DISTRIBUTION: Kermadec Islands, Raoul Island.

CUNONIACEAE

PLATE 56 *Ackama rosifolia* **makamaka**

Ackama: formed from the Māori name, makamaka

rosifolia: rose-leaved; from the Latin, rosa, rose; folium, leaf

Shrub or tree to 12 m tall, trunk to 60 cm diameter.

DISTRIBUTION: North Island. Lowland forest, 34°28'–36°S.

PLATE 57 *Weinmannia racemosa* **kāmahi, tawhero**

Weinmannia: after JW Weinmann, 17th–18th century Bavarian apothecary

racemosa: raceme bearing, from the Latin, racemes, bunch of grapes or fruit

Tree to 25 m tall or more, trunk to 1.2 m diameter.

DISTRIBUTION: North, South and Stewart Islands. Lowland to montane forest south of 37°S.

QUINTINIACEAE

PLATE 58 *Quintinia serrata* **tāwheowheo**

Quintinia: after J de la Quintinie, 17th century French botanist

serrata: serrated, referring to leaves; from the Latin, serratus, shaped like a saw

Tree to 12 m tall, trunk to 50 cm or more diameter.

DISTRIBUTION: North and South Islands. Lowland and montane forest from 35° to 43°40'S. Absent from Mt Taranaki/Egmont.

IXERBACEAE

PLATE 59 *Ixerba brexioides* **tāwari**

Ixerba: an anagram of *Brexia*, which it closely resembles

brexioides: resembling *Brexia*; from *Brexia* and the Greek, -oides, resembling

Tree to 10 m tall or more, trunk 60 cm to 1 m diameter.

DISTRIBUTION: North Island. Forests, mainly montane, 35°30'–39°S.

ROSACEAE

PLATE 60 *Rubus parvus* **creeping lawyer**

Rubus: bramble bush; from the Latin, rubus, bramble

parvus: from the Latin, parvus, small

Prostrate plant with reddish stems.

DISTRIBUTION: South Island. North-west Nelson, Westland and northern Fiordland. Forest and on river flats to 1000 m above sea level.

PLATE 61 *Rubus australis* **tātarāmoa, taraheke, taramoa, tātaraheke, bush lawyer**

Rubus: bramble bush; from the Latin, rubus, bramble

australis: southern; from the Latin, australis, southern

Liane to 10 m or more. Rope-like stems to 8 cm diameter.

DISTRIBUTION: North, South and Stewart Islands. Lowland forest.

FABACEAE

PLATE 62 *Clianthus puniceus* **kōwhai-ngutu-kākā, kakabeak**

Clianthus: splendid flower; from the Greek, kelos, glory; anthos, flower
puniceus: reddish crimson; from the Latin, puniceus, scarlet

Shrub with spreading branches, to 1–2 m tall.

DISTRIBUTION: Exact distribution uncertain, mainly because the majority of records are based on early literature references, which are not supported by herbarium vouchers. Cultivated by Māori in the past.

Clianthus maximus **kōwhai-ngutu-kākā, kakabeak**

Clianthus: splendid flower; from the Greek, kelos, glory; anthos, flower
maximus: possibly refers to the larger growth habit; from the Latin, maximus, largest

Shrubs 1–4 m tall. Plants multi-stemmed from the base with branches layering, or with 10 cm diameter trunk.

DISTRIBUTION: North Island. East coast.

ROUSSEACEAE

PLATE 63 *Carpodetus serratus* **putaputawētā, marble leaf**

Carpodetus: seeds bound together, referring to fruit being girdled by rim of calyx; from the Greek, karpos, fruit; detos, bound together
serratus: with saw-like, toothed leaves; from the Latin, serratus, shaped like a saw

Tree to 10 m tall, trunk to 20 cm diameter or more.

DISTRIBUTION: North, South and Stewart Islands. Lowland to montane forest.

PLATE 64 *Sophora godleyi*

Sophora: from the Arabic, sufayra, the name for a leguminous tree
godleyi: after EJ Godley, 20th–21st century New Zealand botanist, specialist of *Sophora*

Tree to 20 m tall, trunk to 86 cm diameter.

DISTRIBUTION: North Island. Northern, inland and south-eastern Taranaki, upper Wanganui River and Taihape, Rangitikei, northern Manawatu. Forest on Tertiary siltstones, sandstones and mudstones (papa).

PLATE 65 *Sophora microphylla* **kōwhai**

Sophora: from the Arabic, sufayra, the name for a leguminous tree
microphylla: small-leaved, referring to the leaf size compared with those of *Sophora tetraptera*; from the Greek, mikros, small; phyllon, leaf

Tree to 25 m tall, trunk to 60 cm diameter.

DISTRIBUTION: North and South Islands. River valleys, coastal areas, lake margins, forest margins and on hill slopes among loose and rubbly rock. Absent from East Cape and northern Hawke's Bay.

PLATE 66 *Sophora tetraptera* **kōwhai**

Sophora: from the Arabic, sufayra, the name for a leguminous tree
tetraptera: four-winged, referring to seed pods; from the Latin tetra-, four; pteron, wing

Tree to 12 m tall, trunk to 60 cm.

DISTRIBUTION: North Island. East Cape to Cape Palliser. Lowland.

PLATE 67 ***Carmichaelia muritai***

Carmichaelia: after D Carmichael, 18th–19th century Scottish army officer and botanical author
muritai: from the Māori, muritai, sea breeze

Tree to 6 m tall, trunk to 25 cm diameter.

DISTRIBUTION: South Island. Marlborough, north of Clarence River. Coastal and lowland.

PLATE 68 ***Carmichaelia appressa* (i)**

Carmichaelia: after D Carmichael, 18th–19th century Scottish army officer and botanical author
appressa: lying flat; from the Latin, appressus, lying flat, pressed together

Many-branched, mainly prostrate shrub to 1.5 m tall, branches to 1 m long.

DISTRIBUTION: South Island. Kaitorete Spit, Canterbury. Very localised.

***Carmichaelia appressa* (ii)** (from inland Canterbury)

Sprawling, prostrate shrub covering an area of 1 m or more diameter.

DISTRIBUTION: South Island. Canterbury, upper Rangitata River and its tributaries. Gravels on braided river flats. Very localised.

PLATE 69 ***Carmichaelia carmichaeliae*** **pink broom**

Carmichaelia: after D Carmichael, 18th–19th century Scottish army officer and botanical author
carmichaeliae: resembling *Carmichaelia*, named when it was known as *Notospartium carmichaeliae*

Shrub to 4 m tall.

DISTRIBUTION: South Island. Marlborough, valleys of the Awatere River, lower Wairau River. On cliffs and, less commonly, on river terraces.

NOTHOFAGACEAE

PLATE 70 ***Nothofagus fusca*** **tawhai raunui, red beech**

Nothofagus: false beech; from the Greek, nothos, spurious; Latin, fagus, beech
fusca: dark brown, referring to the bark; from the Latin, fuscus, dark, dusky, swarthy

Tree to 30 m tall, trunk to 2 m or more diameter, often strongly buttressed.

DISTRIBUTION: North and South Islands. Lowland to montane forest. Absent from western Waikato, Taranaki, central Westland, eastern Otago and eastern Southland.

PLATE 71 ***Nothofagus solandri* var. *solandri*** **tawhai rauriki, black beech**

Nothofagus: false beech; from the Greek, nothos, spurious; Latin, fagus, beech
solandri: after D Solander, 18th century Swedish botanist on Cook's first voyage to New Zealand

Tree to 25 m tall, trunk to 1 m or more diameter.

DISTRIBUTION: North and South Islands. Lowland and lower montane forest. Scarce north of Volcanic Plateau and East Cape.

MORACEAE

PLATE 72 ***Streblus smithii***

Streblus: from the Greek, streblos, twisted; referring to the contorted branches of some of the species
smithii: after SP Smith, 19th–20th century surveyor and ethnologist

Shrub or tree to 5 m tall.

DISTRIBUTION: Three Kings Islands.

CORYNOCARPACEAE

PLATE 73 ***Corynocarpus laevigatus*** **karaka**

Corynocarpus: club-shaped fruit; from the Greek, koryne, club; karpos, fruit
laevigatus: smooth; from the Latin, laevis, smooth; referring to the leaves

Tree to 16 m tall.

DISTRIBUTION: Kermadec, North, South and Chatham Islands. North of 44°S. Mainly coastal. Cultivated by Māori in the past.

PENNANTIACEAE

PLATE 74 ***Pennantia baylisiana***

Pennantia: after T Pennant, 18th century British zoologist, antiquarian and author
baylisiana: after GTS Baylis, 20th century mycologist, botanist and academic

Multi-trunked tree to 5 m in wild (8 m in cultivation).

DISTRIBUTION: Three Kings Islands. Only one tree known in the wild on Great Island.

SANTALACEAE

PLATE 75 ***Mida salicifolia*** **maire taiki, New Zealand sandalwood**

Mida: an imperfect rendering of the Māori name, maire
salicifolia: willow-leaved; from the Latin, salicis, willow; folium, leaf

Tree to 6 m tall or more.

DISTRIBUTION: North Island. Lowland forest.

LORANTHACEAE

PLATE 76 ***Ileostylus micranthus*** **pirita**

Ileostylus: from the Greek, eileos, twisted; stylos, style; referring to the contorted style
micranthus: small-flowered; from the Greek, micros, small; anthos, flower

Hemiparasitic, bushy shrub to 1 m diameter or more.

DISTRIBUTION: North, South and Stewart Islands. On a wide range of native hosts and many species of introduced trees.

PLATE 77 *Peraxilla colensoi* **pirita, korukoru (when in flower), scarlet-flowered beech mistletoe**

Peraxilla: completely axillary; from the Latin, per-, through; axilla, armpit; referring to position of flowers
colensoi: after W Colenso, 19th century New Zealand explorer and botanist

Hemiparasitic, bushy shrub to 1 m or more diameter.

DISTRIBUTION: North and South Islands. Now very scarce in North Island. Mainly on *Nothofagus*, especially *Nothofagus menziesii*, also *Metrosideros* and *Pittosporum* species, and some introduced trees (*Pyrus*, *Robinia*, *Rosa*).

PLATE 78 *Trilepidea adamsii* **Adams' mistletoe**

Trilepidea: three scales; from the Greek, tri-, three; lepis, scale; probably referring to the three scales at the base of the flower
adamsii: after J Adams, 19th–20th century school teacher and plant collector

Hemiparasitic shrub with a spread of 1 m.

DISTRIBUTION: North Island. On *Coprosma arborea*, *Melicope ternata* and *Myrsine australis*. Now believed to be extinct.

PLATE 79 *Tupeia antarctica* **tāpia, pirita, white mistletoe**

Tupeia: derived from the name of the Tahitian priest Tupaia, who accompanied Cook on his first voyage to New Zealand
antarctica: southern polar regions; from the Latin, antarcticus = Greek, anti-, opposed to; arktikos, North Pole

Hemiparasitic shrub to 1 m diameter.

DISTRIBUTION: North and South Islands. On *Pittosporum eugenioides*, *Nestegis cunninghamii* and *Pseudopanax arboreus*.

RHAMNACEAE

PLATE 80 *Pomaderris kumeraho* **kūmarahou**

Pomaderris: from the Greek, poma, lid; derris, skin; for the membranous covering of the seed vessel
kumeraho: from the Māori name

Shrub 2–3 m tall.

DISTRIBUTION: North Island. North Cape to Bay of Plenty and Kawhia Harbour. Hills, on clay soils.

PLATE 81 *Pomaderris hamiltonii*

Pomaderris: from the Greek, poma, lid; derris, skin; for the membranous covering of the seed vessel
hamiltonii: after the family of WM Hamilton, 20th century Director-General of DSIR

Shrub or tree to 4 m tall.

DISTRIBUTION: North Island. Northland, near Leigh and Warkworth, Great Barrier Island, Waikato, inland between Kaiaua and Miranda. Poor clay soil in shrubland.

PLATE 82 *Discaria toumatou* **tūmatakuru, wild Irishman, matagouri**

Discaria: from the Latin, discus = Greek, diskos, a quoit; referring to the disc in the flowers
toumatou: from the Māori name tou matou

Much-branched, more or less deciduous, spiny shrub or tree, to 5 m tall or more.

DISTRIBUTION: North and South Islands. South of 37°S. Coastal to montane scrub.

MELIACEAE

PLATE 83 *Dysoxylum spectabile* **kohekohe**

Dysoxylum: bad wood; from the Greek, dys-, bad; xylon, wood; referring to the bitterness of the wood
spectabile: beautiful; from the Latin, spectabilis, notable, remarkable, showy, beautiful

Tree to 15 m tall.

DISTRIBUTION: North and South Islands. North Cape to Nelson and Marlborough Sounds. Coastal and lowland forest. Absent from Napier south in eastern North Island.

SAPINDACEAE

PLATE 84 *Alectryon excelsus* subspecies *excelsus* **tītoki**

Alectryon: a rooster; from the Greek, alektryon, a cock; from the resemblance of the scarlet aril to a cock's comb
excelsus: from the Latin, excelsus, tall; big and showy

Tree to 10 m tall or more, trunk to 60 cm or more diameter.

DISTRIBUTION: North and South Islands. North Cape to Banks Peninsula and just north of Karamea. Lowland and coastal forest.

PLATE 85 *Alectryon excelsus* subspecies *grandis* **Three Kings tītoki**

Alectryon: a rooster; from the Greek, alektryon, a cock; from the resemblance of the scarlet aril to a cock's comb
excelsus: from the Latin, excelsus, tall; big and showy
grandis: fully grown

Tree to 5 m or more tall.

DISTRIBUTION: Three Kings Islands. Coastal forest.

ARALIACEAE

PLATE 86 *Schefflera digitata* **patē, seven-finger**

Schefflera: after JC Scheffler, 18th century botanical author
digitata: finger-like parts; from the Latin, digitus, finger

Tree to 8 m tall.

DISTRIBUTION: North, South and Stewart Islands. Lowland to lower montane forest.

PLATE 87 *Pseudopanax macintyrei* **limestone five-finger**

Pseudopanax: false *Panax*; from the Greek, pseudo-, false; panax, cure-all
macintyrei: after JW McIntyre, 19th–20th century garden foreman

Tree to 5 m tall.

DISTRIBUTION: South Island. Eastern Nelson, north-west Nelson and central Westland. On limestone and marble, to 750 m above sea level.

PLATE 88 *Pseudopanax chathamicus* **hoho, Chatham Island lancewood**

Pseudopanax: false *Panax*; from the Greek, pseudo-, false; panax, cure-all
chathamicus: of the Chatham Islands

Tree to 7 m tall.

DISTRIBUTION: Chatham Islands. Lowland and coastal forest.

PLATE 89 ***Pseudopanax ferox***

Pseudopanax: false *Panax*; from the Greek, pseudo-, false; panax, cure-all

ferox: fierce, dangerous; from the Latin, ferox, wild, savage; referring to the tough pointed teeth on juvenile leaf margins

Shrub or tree to 5 m tall.

DISTRIBUTION: North and South Islands. Lowland forest and scrub.

PLATE 90 ***Raukaua simplex* var. *simplex*** **haumakōroa**

Raukaua: derived from the Māori name, Raukawa

simplex: simple; from the Latin, simplex, one-fold, unmixed; referring to the simple leaf

Shrub or tree to 8 m tall.

DISTRIBUTION: North, South, Stewart and Auckland Islands. Montane and subalpine forest south of 39°S.

ARGOPHYLLACEAE

PLATE 91 ***Corokia buddleioides* var. *linearis*** **korokio**

Corokia: adaptation of the Māori name, korokio

buddleioides: resembling *Buddleja*

linearis: from the Latin, linea, string line

Shrub to 3 m tall.

DISTRIBUTION: North Island. Inland Bay of Plenty. Shrubland and open forest.

PLATE 92 ***Corokia cotoneaster* (i)** **korokio**

Corokia: adaptation of the Māori name, korokio

cotoneaster: resembling *Cotoneaster*

Bush to 3 m tall.

DISTRIBUTION: Three Kings, North and South Islands. Very adaptable.

***Corokia cotoneaster* (ii)** (from Paritutu)

Divaricating shrub with leaves to 3 cm long.

DISTRIBUTION: North Island. Northern Aotea Harbour, Waikato, to Paritutu Rock, New Plymouth.

APIACEAE

PLATE 93 ***Scandia rosifolia* (i)** **koheriki**

Scandia: from the Latin, scandere, to climb

rosifolia: rose-leaved; from the Latin, rosa, rose; folium, leaf

Prostrate or scrambling, glabrous perennial herb or subshrub with stems to 1 m long.

DISTRIBUTION: Three Kings and North Islands. North of Taranaki and Napier. Usually coastal.

***Scandia rosifolia* (ii)** (from Raglan Harbour to Ruahine Range)

With narrow leaflets and smaller flowers than those of *Scandia rosifolia* (i).

DISTRIBUTION: North Island. From Raglan, Kawhia Harbour and East Cape south to the Ruahine Range. Primarily on limestone. Coastal to montane areas.

PLATE 94 *Scandia geniculata*

Scandia: from the Latin, scandere, to climb

geniculata: jointed; from the Latin, geniculum, joint

Slender liane becoming woody on older branches, or tangled mass to 3 m in height.

DISTRIBUTION: North and South Islands. Mostly coastal.

ERICACEAE

PLATE 95 ***Gaultheria depressa* var. *novae-zealandiae*** **mountain snowberry**

Gaultheria: after JF Gaulthier, 18th century French physician of Quebec

depressa: from the Latin, depressus, to keep down

novae-zealandiae: of New Zealand

Prostrate plant seldom more than 10 cm high.

DISTRIBUTION: North, South and Stewart Islands. Montane and subalpine grassland, herbfield and open scrub south of 37°30'S.

Gaultheria macrostigma

Gaultheria: after JF Gaulthier, 18th century French physician of Quebec

macrostigma: from the Greek, macros, large; stigma, point

Prostrate or straggling shrub to 20 cm tall.

DISTRIBUTION: North, South and Stewart Islands. Montane to lower subalpine open scrub, grassland and rocky places.

PLATE 96 ***Cyathodes pumila***

Cyathodes: cup-like; from the Greek, kyathos, ladle; -oides, resembling; referring to flower's cup-shaped disc

pumila: very small; from the Latin, pumilus, dwarfish

Shrub forming patches to 40 cm diameter.

DISTRIBUTION: North, South and Stewart Islands. Boggy ground in mountain areas. Uncommon on Tararua Range.

Leptecophylla robusta **poteriteri, Chatham Island mingimingi**

Leptecophylla: fine-pointed leaf; from the Greek, lepteces, fine-pointed; phyllon, leaf

robusta: stout, strong; from the Latin, robustus, made of oak wood, solid, firm

Shrub or tree to 5 m tall.

DISTRIBUTION: Chatham Islands. Dry rocky places and shrubland.

PLATE 97 ***Archeria racemosa***

Archeria: after W Archer, 19th century Tasmanian artist and botanist

racemosa: bearing racemes; from the Latin, racemes, bunch of grapes or fruit

Shrub to 5 m tall.

DISTRIBUTION: North Island. Great and Little Barrier Islands to near East Cape and a little south of 38°S. Lowland to montane forest.

Archeria traversii

Archeria: after W Archer, 19th century Tasmanian artist and botanist
traversii: after WTL Travers, 18th–19th century New Zealand politician and naturalist

Shrub to 5 m tall.

DISTRIBUTION: South and Stewart Islands. Montane. Absent from Marlborough and much of the east.

PLATE 98 ***Dracophyllum latifolium*** **neinei, taritari āwhā, spiderwood**

Dracophyllum: resembling *Dracaena*, dragon tree; from the Greek, drakon, dragon; phyllon, leaf
latifolium: broad-leaved; from the Latin, latus, broad; folium, leaf

Tree to 10 m tall.

DISTRIBUTION: North Island, Kaitaia to north Taranaki, East Cape to Lake Waikaremoana; South Island, Puponga to Nile River. Hill and mountain forest.

PLATE 99 ***Dracophyllum elegantissimum***

Dracophyllum: resembling *Dracaena*, dragon tree; from the Greek, drakon, dragon; phyllon, leaf
elegantissimum: very elegant; from the Latin, elegans, choice, fine, tasteful

Tree to 14 m tall.

DISTRIBUTION: South Island. North-west Nelson. Lowland to lower montane beech forest up to *c.* 700 m above sea level.

PLATE 100 ***Dracophyllum longifolium* var. *cockayneanum***

Dracophyllum: resembling *Dracaena*, dragon tree; from the Greek, drakon, dragon; phyllon, leaf
longifolium: long-leaved; from the Latin, longus, long; folium, leaf
cockayneanum: after L Cockayne, 19th–20th century New Zealand botanist and plant geographer

Shrub or tree to 12 m tall.

DISTRIBUTION: Campbell and Auckland Islands. Open shrubland on exposed coastal areas and on hillsides.

PLATE 101 ***Dracophyllum sinclairii*** (from Te Paki Trig bush)

Dracophyllum: resembling *Dracaena*, dragon tree; from the Greek, drakon, dragon; phyllon, leaf
sinclairii: after A Sinclair, 18th–19th century naval surgeon and naturalist

Shrub or tree to 6 m tall.

DISTRIBUTION: North Spirits Bay to Auckland, Coromandel Peninsula, Kawhia, Opotiki to Tokomaru Bay.

PLATE 102 ***Dracophyllum menziesii***

Dracophyllum: resembling *Dracaena*, dragon tree; from the Greek, drakon, dragon; phyllon, leaf
menziesii: after A Menzies, 18th–19th century Scottish surgeon

Much branched shrub to 1 m tall.

DISTRIBUTION: South and Stewart Islands. High rainfall mountainous regions.

MYRSINACEAE

PLATE 103 ***Myrsine australis*** **māpou, māpau, red matipo**

Myrsine: from the Greek, myrsine, myrtle bush
australis: southern; from the Latin, australis, southern

Shrub or small tree to 7 m tall.

DISTRIBUTION: North, South and Stewart Islands. Lowland to montane forest and shrubland.

PLATE 104 ***Myrsine argentea***

Myrsine: from the Greek, myrsine, myrtle bush
argentea: silvery; from the Latin, argentum, silver; referring to foliage and fruit

Shrub or tree to 4 m tall.

DISTRIBUTION: South Island. Mt Burnett, north-west Nelson. In forest and scrub, 500–600 m above sea level.

Myrsine aquilonia

Myrsine: from the Greek, myrsine, myrtle bush
aquilonia: northern; from the Latin, aquilonius, northerly

Shrub or tree to 4 m tall.

DISTRIBUTION: North Island. Northland, near Te Kao, Kaitaia, and Tukukaka and Poor Knights Islands.

PLATE 105 ***Myrsine coxii***

Myrsine: from the Greek, myrsine, myrtle bush
coxii: after FAD Cox, 19th–20th century New Zealand farmer

Shrub or tree to 6 m or more tall.

DISTRIBUTION: Chatham Islands. Swamps.

Myrsine oliveri

Myrsine: from the Greek, myrsine, myrtle bush
oliveri: after WRB Oliver, 19th–20th century botanist, Director of Dominion Museum, Wellington

Tree to 6 m tall.

DISTRIBUTION: Three Kings Islands.

OLEACEAE

PLATE 106 ***Nestegis apetala*** **coastal maire**

Nestegis: lacking shelter; from the Greek, ne-, not, without; stege, cover, roof; referring to the absence of a corolla
apetala: without petals; from the Greek, a-, without, not; petalon, thin plate

Tree 3–6 m tall.

DISTRIBUTION: Three Kings and North Islands. Coastal cliffs, mainly on offshore islands from Bay of Islands to Whangamata in North Island.

PLATE 107 *Nestegis montana* **rōroro, maire roro, narrow-leaved maire**

Nestegis: lacking shelter; from the Greek, ne-, not, without; stege, cover, roof; referring to the absence of a corolla

montana: montane; from the Latin, montanus, highland

Tree to 10 m tall or more.

DISTRIBUTION: North and South Islands. North of 41°30'S. Lowland and lower montane forest. Rare in South Island where it is confined to a few small stands in the Waimea Basin and near the coast north of Kaikoura.

APOCYNACEAE

PLATE 108 *Parsonsia heterophylla* **kaihua, akakiore, New Zealand jasmine**

Parsonsia: after J Parsons, 18th century Scottish medical doctor and natural historian

heterophylla: differing leaves; from the Greek, heteros, other, differing; phyllon, leaves

Liane to 10 m tall, stems to 10 cm diameter. Flowers fragrant.

DISTRIBUTION: Three Kings, North, South and Stewart Islands. Coastal, lowland and lower montane forest.

PLATE 109 ***Parsonsia capsularis* var. *tenuis***

Parsonsia: after J Parsons, 18th century Scottish medical doctor and natural historian

capsularis: capsule-bearing; from the Latin, capsa, box

tenuis: slender, slight

Liane to *c.* 5 m tall.

DISTRIBUTION: South Island. 44° to 46°S. Lowland to montane shrubland.

Parsonsia praeruptis

Parsonsia: after J Parsons, 18th century Scottish medical doctor and natural historian

praeruptis: referring to plant's habitat on cliffs

Trailing or scrambling plant with no twining or climbing tendencies, stems to 2 m long.

DISTRIBUTION: North Island. Surville Cliffs, North Cape. Ultramafic scree, cliffs.

Parsonsia capsularis

Parsonsia: after J Parsons, 18th century Scottish medical doctor and natural historian

capsularis: capsule-bearing; from the Latin, capsa, box

LOGANIACEAE

PLATE 110 ***Geniostoma rupestre* var. *ligustrifolium*** **hangehange, New Zealand privet**

Geniostoma: hairy throat; from the Greek, geneion, chin or beard; stoma, mouth; alluding to the hairs in the corolla

rupestre: growing in rocks; from the Latin, rupes, steep rocky cliff

ligustrifolium: privet-leaved; from the Latin, ligustrum, privet; folium, leaf

Shrub or tree to 3 m tall or more.

DISTRIBUTION: Three Kings, North and South Islands. North of 41°30'S. Coastal and lowland forest.

ALSEUOSMIACEAE

PLATE 111 ***Alseuosmia macrophylla*** **toropapa**

Alseuosmia: from the Greek, alsos, grove; eu, well; osme, perfume
macrophylla: large-leaved; from the Greek, macros, large; phyllon, leaf

Shrub or tree to 5 m tall.

DISTRIBUTION: North Island, North Cape to 39°30'S, west of the main ranges; South Island, north-west Nelson and northern Westland to Charleston. Lowland and montane forest.

PLATE 112 ***Alseuosmia***

Alseuosmia: from the Greek, alsos, grove; eu, well; osme, perfume

Leaf variations from Northland localities and sources of material painted (*a*) Near Brynderwyn; (*b*) Manginangina Kauri Reserve, Puketi Forest; (*c*) Near Whangaruru; (*d*) Parahaki Hill, Whangarei, at 200 m; (*e*) and (*f*) Mangamuka Hill; (*g*) Omahuta Forest at 300 m; (*h*) Monument Hill, Kaitaia; (*i*) Warawara Forest at 300 m; (*j*) near Kaitaia.

PLATE 113 ***Alseuosmia***

Alseuosmia: from the Greek, alsos, grove; eu, well; osme, perfume

Leaf variations and sources of material painted continued (*k*) Taumatamahoe, near summit, 500 m; (*l*) Te Paki Trig Bush; (*m–r*) Herekino Gorge, near north side.

PLATE 114 ***Alseuosmia pusilla***

Alseuosmia: from the Greek, alsos, grove; eu, well; osme, perfume
pusilla: small; from the Latin, pusillus, insignificant, petty

Shrub to 1 m tall.

DISTRIBUTION: North and South Islands. Te Moehau and Mt Pirongia to southern Westland. Wet lowland and montane forests.

PLATE 115 ***Alseuosmia banksii***

Alseuosmia: from the Greek, alsos, grove; eu, well; osme, perfume
banksii: after J Banks, 18th century English botanist on Cook's first voyage to New Zealand

Spreading or trailing shrub to 1.5 m tall.

DISTRIBUTION: North Island. Forest.

RUBIACEAE

PLATE 116 ***Coprosma waima***

Coprosma: manure smell; from the Greek, kopros, dung; osme, odour
waima: after its locality, the Waima Range in Northland

Shrub to 3 m tall.

DISTRIBUTION: North Island. Waima Forest, Northland.

PLATE 117 ***Coprosma chathamica***

Coprosma: manure smell; from the Greek, kopros, dung; osme, odour
chathamica: of the Chatham Islands

Tree to 15 m tall.

DISTRIBUTION: Chatham Islands.

PLATE 118 *Coprosma acutifolia*

Coprosma: manure smell; from the Greek, kopros, dung; osme, odour
acutifolia: pointed-leaved; from the Latin, acutus, sharp; folium, leaf

Tree to 10 m tall.

DISTRIBUTION: Kermadec Islands. Raoul Island only.

PLATE 119 *Coprosma acerosa* (from Buller Gorge)

Coprosma: manure smell; from the Greek, kopros, dung; osme, odour
acerosa: needle-like; from the Latin, acer, pointed; -osus, full of; referring to the leaves

Prostrate or sprawling plant with slender, interlacing branches that form rather open mats.

DISTRIBUTION: South Island. Lowland and montane.

PLATE 120 *Coprosma spathulata* subspecies *hikuruana*

Coprosma: manure smell; from the Greek, kopros, dung; osme, odour
spathulata: from the Latin, spatula, from the Greek, spathe, broad blade, paddle
hikuruana: from a Māori name for the far north, Hikurua

Scrambling or trailing shrub with slender stems to 10 m long.

DISTRIBUTION: North Island. Surville Cliffs, North Cape. On serpentinite rock.

PLATE 121 *Coprosma arborea* **māmāngi, tree coprosma**

Coprosma: manure smell; from the Greek, kopros, dung; osme, odour
arborea: tree-like; from the Latin, arbor, tree

Tree to 10 m tall.

DISTRIBUTION: North Island. North Cape to Waitomo. Lowland forest from sea level to *c.* 450 m.

PLATE 122 *Coprosma microcarpa*

Coprosma: manure smell; from the Greek, kopros, dung; osme, odour
microcarpa: bearing small fruit; from the Greek, mikros, small; karpos, fruit

Shrub or tree to 4 m tall.

DISTRIBUTION: North and South Islands. Central volcanoes and Huiarau Range to Ashburton River catchment and Karamea. Lowland and montane.

Coprosma cheesemanii

Coprosma: manure smell; from the Greek, kopros, dung; osme, odour
cheesemanii: after TF Cheeseman, 19th–20th century New Zealand botanist

Prostrate or sprawling shrub to 1 m tall.

DISTRIBUTION: North, South and Stewart Islands. Montane, subalpine and alpine shrubland, tussockland and bog. In North Island, restricted to Volcanic Plateau and northern Ruahine Range.

PLATE 141 *Calystegia sepium* subspecies *roseata* **pōhue, akapōhue, nahinahi, rauparaha, New Zealand bindweed**

Calystegia: covered calyx; from the Greek, kamara, arch, vault; phyllon, leaf; referring to the bracts which enclose the calyx
sepium: from the Latin, sepium, of hedges
roseata: rose-coloured

Long, slender rhizomes and twining stems to 3 m or more long, smothering plants or forming tangled masses in the absence of support.

DISTRIBUTION: Three Kings and North Islands; South Island, to Otago but local; Chatham Islands. Coastal, also along stream edges and forest margins.

PLATE 142 *Ipomoea pes-caprae* subspecies *brasiliensis*

Ipomoea: resembling *Convolvulus*; from the Greek, ips, worm
pes-caprae: goat's foot; from the Latin, pes, foot, caprae, she-goat
brasiliensis: of Brazil; named for shape of leaf

Prostrate plant with trailing stems, sometimes to 10 m.

DISTRIBUTION: Kermadec and North Islands. Only on Ninety Mile Beach in North Island. Mainly coastal.

PLATE 143 *Ipomoea cairica* **panahi, pōwhiwhi**

Ipomoea: resembling convolvulus; from the Greek, ips, worm
cairica: of Cairo, Egypt

Climber or scrambler on the ground. Stems *c.* 5 mm diameter, to 3 m long.

DISTRIBUTION: Kermadec, Three Kings and North Islands. Coastal cliffs, banks and old fore-dunes north of 35°30'S.

PLANTAGINACEAE

PLATE 144 *Parahebe catarractae*

Parahebe: from the Greek, para, beside; indicates the close relationship with genus *Hebe*
catarractae: of the waterfall; from the Greek, katarrhaktes, striking down, rapids

Prostrate or erect shrub to 60 cm tall.

DISTRIBUTION: South Island. Fiordland. Lowland to subalpine, to 1400 m, beside streams.

PLATE 145 *Heliohebe hulkeana* subspecies *evestita*

Heliohebe: from the Greek, helio, the sun; referring to the sunny habitats where the plants are found
hulkeana: after WK Hulke, 19th–20th century nurseryman in New Zealand
evestita: lack of a dense covering of short, eglandular hairs; from the Latin, e-, lacking; vestitus, clothed

Shrub in hanging clumps to 1 m long × 0.5 m wide.

DISTRIBUTION: South Island. Southern Marlborough, Waima (Ure) River to Boundary Stream (Clarence River), also Whernside Ridge and Mt Ben More. Rocky gorges, cliffs and outcrops, on limestone.

PLATE 146 ***Hebe* aff. *brevifolia***

Hebe: Greek goddess of youth and spring

brevifolia: short-leaved; from the Latin, brevis, short; folium, leaf

This is an unnamed plant similar to *Hebe brevifolia*.

Shrub to 80 cm tall.

DISTRIBUTION: North Island. Surville Cliffs, North Cape. Shrubland on serpentine rock.

Heliohebe acuta

Heliohebe: from the Greek, helio, the sun; referring to the sunny habitats where the plants are found

acuta: sharp, tapering to a point; from the Latin, acutus, sharp, pointed; referring to the narrowly pointed leaf tip of the keeled leaf

Procumbent shrub to 20 cm tall.

DISTRIBUTION: South Island. Eastern Marlborough, Seaward and Inland Kaikoura ranges, Clarence River from George Stream to Dillon River. Montane. 700–*c.* 1000 m above sea level.

PLATE 147 ***Hebe speciosa*** **tītīrangi, napuka**

Hebe: Greek goddess of youth and spring

speciosa: beautiful; from the Latin, speciosus, showy, beautiful

Shrub 1–2 m tall.

DISTRIBUTION: North and South Islands. Coastal cliff tops at Scotts Point, South Hokianga Head, Maunganui Bluff, Muriwai, Ngatutura Point, Taranaki bluffs near Aotea Harbour, Mokau, Tongaporutu, Urenui, Titirangi Bay (Marlborough Sounds), though thought to be extinct in many of these places.

PLATE 148 ***Hebe salicifolia*** **koromiko, koromuka, kōkoromiko, korokio, kōkōmuka, kōkoromuka**

Hebe: Greek goddess of youth and spring

salicifolia: willow-leaved; from the Latin, salicis, willow; folium, leaf

Shrub or tree to 5 m tall.

DISTRIBUTION: South and Stewart Islands. Lowland and montane scrub. Absent from Marlborough Sounds. Adventive on Auckland Island.

PLATE 149 ***Hebe* aff. *stricta***

Hebe: Greek goddess of youth and spring

stricta: from the Latin, strictus, from stringere, to compress, draw tight

This is an unnamed plant similar to *Hebe stricta*.

Shrub to 4 m tall.

DISTRIBUTION: North Island. Local in Northland (Pipiwai) and Waikato (Wairongomai Valley and Mt Pirongia). Swampy ground.

Hebe* aff. *bishopiana

Hebe: Greek goddess of youth and spring

bishopiana: after JJ Bishop, 19th–20th century New Zealand botanist who first recognised the species as distinct

This is an unnamed plant similar to *Hebe bishopiana*.

Shrub to 6 m tall.

DISTRIBUTION: North Island. Wairua Management Reserve and Hikurangi Swamp, Northland. Swampy ground.

PLATE 150 ***Hebe* aff. *pinguifolia* (i)**

Hebe: Greek goddess of youth and spring
pinguifolia: thick-leaved; from the Latin, pinguis, fat; folium, leaf

This is an unnamed plant similar to *Hebe pinguifolia*, from Mt Peel, Canterbury.

Shrub to *c.* 30 cm.

DISTRIBUTION: South Island. Eastern Canterbury mountains.

***Hebe* aff. *pinguifolia* (ii)**

This is an unnamed plant similar to *Hebe pinguifolia*, from Mt Nimrod, Canterbury.

Prostrate shrub to *c.* 8 cm tall.

DISTRIBUTION: South Island. Mid and South Canterbury. Ridge tops and rocky slopes.

Hebe* aff. *ligustrifolia

Hebe: Greek goddess of youth and spring
ligustrifolia: privet-leaved; from the Latin, ligustrum, privet; folium, leaf

This is an unnamed plant similar to *Hebe ligustrifolia*, from Surville Cliffs, North Cape.

Compact shrub 10–20 × 10–20 cm.

DISTRIBUTION: North Island. Northland, Cape Reinga to Surville Cliffs, North Cape and coastal sites to Tutukaka. Ultramafic rocks and soils.

Hebe brockiei

Hebe: Greek goddess of youth and spring
brockiei: after WB Brockie, 19th–20th century botanist and Curator of Otari Open Air Native Plant Museum, Wellington

Prostrate shrub to *c.* 20 cm tall.

DISTRIBUTION: South Island. North-western Canterbury, Amuri Pass and Lake Man at head of Doubtful River. Grassland, 1200–1500 m above sea level.

PLATE 151 ***Hebe salicornioides***

Hebe: Greek goddess of youth and spring
salicornioides: resembling Salicornia

Shrub to 1 m tall.

DISTRIBUTION: South Island. Southern Marlborough and North Canterbury, upper Wairau, Clarence and Waiau catchments. Mountains, usually in very wet ground, 750–1500 m above sea level.

Hebe armstrongii

Hebe: Greek goddess of youth and spring
armstrongii: after JF Armstrong, 19th–20th century gardener, Curator of Christchurch Botanic Gardens

Shrub to 1 m tall.

DISTRIBUTION: South Island. Mid Canterbury. Mountains, 750–1500 m above sea level.

Hebe tetragona

Hebe: Greek goddess of youth and spring

tetragona: four-angled; from the Greek, tetra-, four; gonia, angle

Shrub to 1 m tall.

DISTRIBUTION: North Island. Raukumara Range to northern Ruahine Range and Volcanic Plateau. In tussockland and scrub, 900–1500 m above sea level.

Hebe ochracea

Hebe: Greek goddess of youth and spring

ochracea: yellowish brown; from the Latin, ochrus = Greek ochros, yellow, sallow, pale

Shrub to 70 cm tall.

DISTRIBUTION: South Island. North-west Nelson, Mt Arthur, Mt Cobb, Kakapo Peak, Snowden Range, Mt Mytton, Mt Peel, Mt Owen and Buckland Peaks, Paparoa Range. In tussockland from 1000–1500 m above sea level.

BIGNONIACEAE

PLATE 152 ***Tecomanthe speciosa***

Tecomanthe: from the Aztec, tecomatl, clay pot; Greek, anthos, flower

speciosa: beautiful; from the Latin, speciosus, showy, beautiful

Liane to 10 m tall.

DISTRIBUTION: Three Kings Islands. Great Island, Tasman Stream.

GESNERIACEAE

PLATE 153 ***Rhabdothamnus solandri*** **taurepo, waiūatua, New Zealand gloxinia**

Rhabdothamnus: twiggy shrub; from the Greek, rhamnos, thorny shrub

solandri: after D Solander, 18th century Swedish botanist on Cook's first voyage to New Zealand

Shrub to 2 m tall.

DISTRIBUTION: North Island. Lowland forest.

SCROPHULARIACEAE

PLATE 154 ***Myoporum laetum* var. *laetum*** **ngaio**

Myoporum: closed pores; from the Greek, myeo, closed, shut; poros, pore; referring to the glands in the leaves

laetum: pleasing, fruitful; from the Latin, laetus, joyful

Tree to 10 m tall.

DISTRIBUTION: Three Kings, North, South and Chatham Islands. Coastal and lowland forest.

LAMIACEAE

PLATE 155 *Vitex lucens* **pūriri**

Vitex: origin obscure
lucens: shining; from the Latin, lucere, to shine

Tree to 20 m tall.

DISTRIBUTION: North Island. From near North Cape to 39°30'S, south of New Plymouth in the west and south of Gisborne in the east. Coastal and lowland forest.

ACANTHACEAE

PLATE 156 *Avicennia marina* **subspecies** *australasica* **mānawa, mangrove**

Avicennia: after Avicenna, Latin name for the 10th–11th century Arabian philosopher and physician Abu Ali Hussain ibn-Abdallah ibn-Sina
marina: by the sea; from the Latin, mare, sea
australasica: of the Australasian region

Tree to 8 m tall.

DISTRIBUTION: North Island. From near North Cape to Kawhia and Ohiwa harbours. Tidal creeks and estuaries.

SMILACACEAE

PLATE 157 *Ripogonum scandens* **kareao, supplejack**

Ripogonum: flexible and jointed; from the Greek, rhipos, wickerwork, pliant; gonu, knee, joint, node; referring to the stems
scandens: climbing; from the Latin, scandere, to climb

Woody, twining liane growing from a stout rhizome.

DISTRIBUTION: North, South, Stewart and Chatham Islands. Lowland forest.

ASPARAGACEAE

PLATE 158 *Cordyline banksii* **tī ngahere, forest cabbage tree**

Cordyline: club palm; from the Greek, kordyle, a club; referring to the fleshy roots
banksii: after J Banks, 18th century English botanist on Cook's first voyage to New Zealand

Tree to 4 m tall, stems 10–15 cm diameter.

DISTRIBUTION: North and South Islands. North Cape to Moeraki and southern Westland, mainly west of Main Divide. Lowland forest and scrub.

PLATE 159 *Cordyline obtecta*

Cordyline: club palm; from the Greek, kordyle, a club; referring to the fleshy roots
obtecta: covered, concealed; from the Latin, obtegere, to cover over, conceal; referring to the inflorescence scarcely protruding beyond the leaves in the plant first described

Tree to 4 m tall, single- or multi-stemmed.

DISTRIBUTION: Three Kings Islands; North Island, North Cape, Murimotu Island, Poor Knights Islands. Coastal forest.

XANTHORRHOEACEAE

PLATE 160 ***Phormium tenax*** **harakeke, kōrari (the flowering stem), New Zealand flax**

Phormium: basket; from the Greek, phormos, basket, wickerwork; referring to the use made of the leaves

tenax: tenacious; from the Latin, tenax, holding fast; referring to the strong leaf fibres

Leaves 1–3 m long, 5–12 cm broad.

DISTRIBUTION: North, South, Stewart, Chatham and Auckland Islands. Mainly in damp places.

PLATE 161 ***Phormium cookianum*** **wharariki, mountain flax, coastal flax**

Phormium: basket; from the Greek, phormos, basket, wickerwork; referring to the use made of the leaves

cookianum: after Captain J Cook, 18th century explorer and navigator of New Zealand

Leaves mostly less than 2 m long

DISTRIBUTION: North, South and Stewart Islands. Coastal to subalpine cliffs, tussockland and scrub.

ARECACEAE

PLATE 162 ***Rhopalostylis baueri*** **Kermadec nīkau**

Rhopalostylis: club-shaped style; from the Greek, rhopalon, tapering club; stylos, pillar

baueri: after FL Bauer, 18th–19th century Austrian botanical artist

Palm tree to 10 m tall, trunk to 30 cm diameter.

DISTRIBUTION: Kermadec Islands. Raoul Island. Wet forest and on the higher ridges.

PANDANACEAE

PLATE 163 ***Freycinetia banksii*** **kiekie, tāwhara (edible bracts), ureure (fruit)**

Freycinetia: after HLC Freycinet, 18th–19th century French explorer

banksii: after J Banks, 18th century English botanist on Cook's first voyage to New Zealand

Scrambling liane, reaching tree tops or forming large masses on the ground.

DISTRIBUTION: North and South Islands. To Dusky Sound in the west and Conway River in the east. Coastal and lowland forest and scrub.

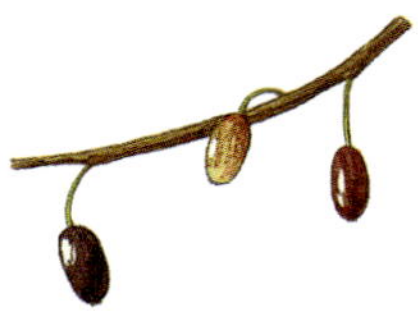

Endnotes

1 JC Beaglehole (ed). 1962. *The* Endeavour *Journal of Joseph Banks, 1768–1771*, vol. 1. Sydney, Angus & Robertson. p. 396.

2 William Curtis. 1791. '*Sophora tetraptera.* Winged-podded *Sophora*'. *Botanical Magazine*, vol. 3. p. 167.

3 Joseph Dalton Hooker. 1852. *The Botany of the Antarctic Voyage*, vol. 2. *Flora Novae-Zelandiae*, part 1, *Flowering Plants*. London, Reeve Brothers. p. xxvii.

4 JC Ross. 1847. *A Voyage of Discovery and Research in the Southern and Antarctic Regions, During the Years 1839–43*, vol. 1. London, John Murray. p. 147.

5 Wilfrid Blunt. 1950. *The Art of Botanical Illustration*. London, Collins. p. 226.

6 Moira M Long. 'Martha King'. *Dictionary of New Zealand Biography, Te Ara – the Encyclopedia of New Zealand*. www.teara.govt.nz/en/biographies/1k12/king-martha (accessed 1 February 2013).

7 James Hector. 1880. 'Preface' in John Buchanan. *The Indigenous Grasses of New Zealand*. Wellington, Government Printer. n.p.

8 June Starke. 'Georgina Burne Hetley'. *Dictionary of New Zealand Biography, Te Ara – the Encyclopedia of New Zealand*. www.teara.govt.nz/en/biographies/2h31/hetley-georgina-burne (accessed 1 February 2013).

9 Mrs Charles Hetley. 1888. *The Native Flowers of New Zealand Illustrated in Colours*. London, Sampson, Low, Marston, Searle & Rivington. p. 3.

10 *Ibid*., p. 6.

11 June Starke. 'Georgina Burne Hetley'.

12 Mr and Mrs EH Featon. 1889. *The Art Album of New Zealand Flora*. Wellington, Bock & Cousins. p. viii.

13 TF Cheeseman. 1914. *Illustrations of the New Zealand Flora*, vol. 1. Wellington, Government Printer. p. 4.

14 PJ Brownsey. 2007. 'Obituary. Nancy Adams, CBE, QSO, 1926–2007'. *New Zealand Botanical Society Newsletter*, vol. 88. p. 11.

15 *Ibid*., p. 12.

16 Audrey Eagle. 2006. *Eagle's Complete Trees and Shrubs of New Zealand*, vol. 1. Wellington, Te Papa Press. p. xii.

17 S Natusch. 1976. 'Book review. Eagle's Trees and Shrubs of New Zealand in Colour: Two Hundred and Twenty-eight Botanical Paintings'. *Wellington Botanical Society Bulletin*, vol. 39. p. 53.

18 S Natusch. 1985. 'Eagle's Trees and Shrubs of New Zealand: Second Series. Audrey's Eagle Eye'. *Wellington Botanical Society Bulletin*, vol. 42. p. 69.

19 Montana New Zealand Book Awards 2007 Judges' Report. ir.canterbury.ac.nz/bitstream/10092/6118/1/12616830_MNZBA07JudgesReport.pdf. p. 5 (accessed 1 February 2013).

Selected Bibliography

Adams, NM. 1983. *New Zealand Native Trees 2. Mobil New Zealand Nature Series.* Wellington, Reed.

———,1985. 'The Banks Lecture 1985. Botanical illustration in New Zealand – the Kew connection'. *Annual Journal of the Royal New Zealand Institute of Horticulture*, vol. 13: 4–12.

———, 1994. *Seaweeds of New Zealand.* Christchurch, Canterbury University Press.

Blunt, W and WT Stearn. 1994. *The Art of Botanical Illustration.* London, Antique Collectors' Club and Royal Botanic Gardens, Kew.

Breitwieser, I and P Brownsey, K Ford, D Glenny, PB Heenan, WA Nelson, A Wilton (eds). 2010. *Flora of New Zealand.* www.nzflora.info (accessed 1 February 2013).

Brownsey, PJ. 2007. 'Obituary. Nancy Adams, CBE, QSO, 1926–2007'. *New Zealand Botanical Society Newsletter*, vol. 88: 11–13.

———, 2012. 'The Banks and Solander collections – a benchmark for understanding the New Zealand flora'. *Journal of the Royal Society of New Zealand*, vol. 42: 131–37.

Buchanan, J. 1880. *The Indigenous Grasses of New Zealand.* Wellington, Government Printer.

Carr, DJ (ed). 1983. *Sydney Parkinson, Artist of Cook's* Endeavour *Voyage.* London, British Museum (Natural History) with Australian National University Press.

Cheeseman, TF (with plates drawn by Matilda Smith). 1914. *Illustrations of the New Zealand Flora.* Wellington, Government Printer.

Curtis, W. 1791. '*Sophora tetraptera.* Winged-podded *Sophora*'. *Botanical Magazine*, vol. 3: 167.

Davies, WC. 1956. *New Zealand Native Plant Studies.* Wellington, Reed.

Dawson, JW and R Lucas. 2011. *New Zealand's Native Trees.* Nelson, Craig Potton Publishing.

de Bray, L. 1989. *The Art of Botanical Illustration.* Secaucus, New Jersey, Wellfleet Press.

Diment, JA and CJ Humphries (eds). 1980–90. *Banks' Florilegium.* London, Alecto Historical Editions with British Museum (Natural History).

Dumont d'Urville, JSC (ed). 1832–34. 'Voyage de découvertes de l'*Astrolabe*'. *Botanique* (parts 1–2 and atlas). Paris, Tastu.

———, (ed). 1853. *Voyage au Pole Sud et dans l'Océanie sur les Corvettes l'*Astrolabe *et la Zelée* (vols 1–2 and atlas). Paris, Gide.

Eagle, A. 1975. *Eagle's Trees and Shrubs of New Zealand in Colour.* Auckland, Collins.

———, 1982. *Eagle's Trees and Shrubs of New Zealand*, 2nd series. Auckland, Collins.

———, 1986. *Eagle's Trees and Shrubs of New Zealand*, vols 1 and 2 revised. Auckland, Collins.

———, 2006. *Eagle's Complete Trees and Shrubs of New Zealand.* Wellington, Te Papa Press.

Ebes, H. 1988. *The Florilegium of Captain Cook's First Voyage to Australia 1768–1771.* Melbourne, Ebes Douwma Antique Prints and Maps.

Featon, Mr and Mrs EH. 1889. *The Art Album of New Zealand Flora.* Wellington, Bock & Cousins.

Hetley, Mrs Charles (Georgina B). 1888. *The Native Flowers of New Zealand Illustrated in Colours.* London, Sampson, Low, Marston, Searle & Rivington.

Hooker, JD. 1844. *The Botany of the Antarctic Voyage*, vol. 1. *Flora Antarctica* (parts 1 and 2). London, Reeve Brothers.

———, 1855. *The Botany of the Antarctic Voyage*, vol. 2. *Flora Novae-Zelandiae* (parts 1 and 2). London, Reeve Brothers.

Kirk, T. 1889. *Forest Flora of New Zealand.* Wellington, Government Printer.

Long, MM. 2012. 'Martha King'. *Dictionary of New Zealand Biography, Te Ara – the Encyclopedia of New Zealand.* www.teara.govt.nz/en/biographies/1k12/king-martha (accessed 1 Febuary 2013).

Mark, AF and NM Adams. 1973. *New Zealand Alpine Plants.* Wellington, Reed.

Matthews, BA. 1977. 'Illustrations of indigenous New Zealand plants in *Curtis's Botanical Magazine*'. *New Zealand Journal of Botany*, vol. 15: 767–73.

Norman, G. 2012. *Buller's Birds of New Zealand: The Complete Works of JG Keulemans*. Wellington, Te Papa Press.

Poole, AL and NM Adams. 1986. *Trees and Shrubs of New Zealand*. Wellington, Government Printer.

———, 1994. *Trees and Shrubs of New Zealand*. Christchurch, Manaaki Whenua Press.

Salmon, JT. 1980. *The Native Trees of New Zealand*. Wellington, Reed.

Sampson, FB. 1985. *Early New Zealand Botanical Art*. Auckland, Reed Methuen.

Starke, J. 2012. 'Georgina Burne Hetley'. *Dictionary of New Zealand Biography, Te Ara – the Encyclopedia of New Zealand*. www.teara.govt.nz/en/biographies/2h31/hetley-georgina-burne (accessed 1 February 2013).

Studholme, JG. 1981. '*Banks' Florilegium*'. *The Australian Antique Collector*, January–June: 44–51.

Tyler, L. 2012. *Picturing the Plants: John Buchanan's Illustrations of New Zealand Flora*. Auckland Museum blog. blog.aucklandmuseum.com/tag/illustrations (accessed 1 February 2013).

Online Access to Botanical Artwork

Adams, Nancy. Collections Online. Museum of New Zealand Te Papa Tongarewa. collections.tepapa.govt.nz/search.aspx?term=Nancy+Adams (accessed 1 February 2013).

Edwards, Sydenham. Collections Online. Museum of New Zealand Te Papa Tongarewa. collections.tepapa.govt.nz/search.aspx?term=Sydenham+Edwards (accessed 1 February 2013).

Featon, Sarah. Collections Online. Museum of New Zealand Te Papa Tongarewa. collections.tepapa.govt.nz/search.aspx?term=Sarah+Featon (accessed 1 February 2013).

Fitch, Walter. Collections Online. Museum of New Zealand Te Papa Tongarewa. collections.tepapa.govt.nz/search.aspx?term=Walter+Fitch (accessed 1 February 2013).

Hetley, Georgina. National Library of New Zealand collections. natlib.govt.nz/items?utf8=%E2%9C%93&text=Georgina+Hetley (accessed 1 February 2013).

Parkinson, Sydney. The *Endeavour* botanical illustrations. Natural History Museum, London. www.nhm.ac.uk/nature-online/art-nature-imaging/collections/endeavour-botanical/ (accessed 1 February 2013).

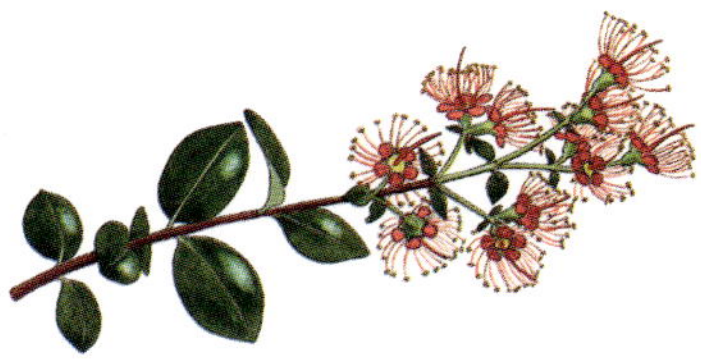

Index

Page numbers in bold indicate illustrations